Inorganic Chemistry: Principles and Properties

Inorganic Chemistry: Principles and Properties

Rabindranath Mukherjee

Indian Institute of Technology Kanpur, India

World Scientific

NEW JERSEY · LONDON · SINGAPORE · BEIJING · SHANGHAI · HONG KONG · TAIPEI · CHENNAI · TOKYO

Published by

World Scientific Publishing Co. Pte. Ltd.
5 Toh Tuck Link, Singapore 596224
USA office: 27 Warren Street, Suite 401-402, Hackensack, NJ 07601
UK office: 57 Shelton Street, Covent Garden, London WC2H 9HE

British Library Cataloguing-in-Publication Data
A catalogue record for this book is available from the British Library.

INORGANIC CHEMISTRY: PRINCIPLES AND PROPERTIES

ISBN 978-981-12-8176-1 (hardcover)
ISBN 978-981-12-8177-8 (ebook for institutions)
ISBN 978-981-12-8178-5 (ebook for individuals)

For any available supplementary material, please visit
https://www.worldscientific.com/worldscibooks/10.1142/13554#t=suppl

Typeset by Stallion Press
Email: enquiries@stallionpress.com

Dedicated to my late parents

Preface

Since my student days and later on as a teacher, given the textbooks available, I realized that for a systematic and comprehensive knowledge of the present-day inorganic chemistry, one has no option but to consult various books on different specialized topics and to collect the reading materials for teaching. Given this situation the students of inorganic chemistry lose interest in the subject. Reacting to this situation, I have attempted to present a book on the principles and properties of inorganic chemistry. My effort to write about inorganic chemistry at the senior/graduate interface arose from the frustrations of our students in dealing with such organizational schemes.

This book is not at all about the chemistry of the elements, which is usually regarded as the subject matter of inorganic chemistry textbooks. There are already several excellent textbooks of inorganic chemistry that treat the subject in considerable detail. What has been attempted to do is to write a book that allows students and the instructors the flexibility to pursue his/her convictions on the topics to be stressed. In this book the focus is on structure, symmetry, bonding, redox properties and reactivity aspects of simple inorganic molecules to transition metal complexes to metal-metal bonded compounds, to organometallic compounds, and finally to bioinorganic proteins and enzymes. It is intended that this book aids the students in integrating these topics. The book is written in line with the historical philosophy of a topical approach and flexible course content.

This book is written from the standpoint of core physical-inorganic principles and is intended to make it student- and teacher-friendly. It is designed to target Master students to graduate research-level university students. For a clear knowledge and appreciation of inorganic compounds one must understand their electronic structure, which is related to their electronic configuration, valence electrons, and bonding concepts.

The author has taught first-year Bachelor of Technology students, five-year integrated Bachelor of Science-Master of Science students, two-year Master of Science students, and PhD (graduate) students during their coursework on topics such as molecular shapes, symmetry, molecular orbital treatments, redox properties, structure, bonding, and properties of transition metal complexes, fundamentals of organometallic chemistry, and a glimpse of bioinorganic chemistry. This book is the result of a long-perceived need for the development of extensive supplements for the teaching of such courses. For researchers looking for reading material to get a clear picture of basic concepts necessary for pursuing modern inorganic chemistry research, this book is expected to provide a solid platform.

The brief introductions to the chapters set the stage ready for the material that follows. To read the book in a more meaningful way of learning experience, brief summaries highlighting the 'take-home messages' are provided at the end of most subsections throughout the book. These summaries are designed to engage the readers in critical thinking about the concepts and to facilitate review of the topical discussion. Solved *Examples* in the text after most subsections illustrate theory and applications or introduce special points. End-of-chapter *Exercises* will offer students excellent revision aids and overview problems. These *Exercises* are intended to consolidate the understanding of the subject matter. Hints/answers are provided at the end of the book. At the end of each chapter there is a study guide. A supplementary reading list is included after each chapter under the heading *Further Reading* to encourage students and teachers to expand their knowledge base to explore topics in more depth.

The Fundamentals of any topic have been written to provide better flow of thought and improved clarity. Chapters 1 and 2 provide the *Fundamentals* of molecular structure and symmetry. Such concepts are carried forward in discussing molecular orbital approaches in Chapter 3, Chapter 6, and Chapter 8, to the understanding of electronic structure of inorganic compounds. As chemistry deals with bond-breaking and bond-making or in other words, reorganization of electronic charge density, redox reactions are of paramount importance. Chapter 4 discusses the basics and application aspects of redox properties. Chapter 5 deals with spectroscopic term symbols, spin-orbit coupling (*L-S* coupling and *j-j* couplig schemes). Chapter 6 provides a comprehensive focus on various aspects (structural and bonding theories, magnetic and spectroscopic properties) of *Transition Metal Chemistry*. Rates and mechanisms of reactions of transition metal complexes are discussed in Chapter 7. Chapter 8 and Chapter 9 present a brief *albeit* concise view of organometallic chemistry and catalysis of industrial implications. Chapter 10 on bioinorganic chemistry discusses a very general overview of the role of metal ions in proteins and enzymes – the topics of current bioinorganic research.

Tables and figures have been used literally throughout this book. These contain information essential to the portions of the text where they appear and are appropriately indexed. The references included in each chapter are intended to extend the usefulness of the book beyond the classroom. It is hoped that the student will find it a useful reference source, which will continue to be of value after he/she has finished with it as a text.

It is inevitable in a book of this size and complexity that there will be occasional errors. These are mine alone, and I will endeavor to correct them in future editions. I shall welcome constructive criticism and suggestion from the readers.

Finally, I pay my tribute to my late parents (Amarendra Nath Mukherjee and Gita Mukherjee) for inspiring me to write a book on inorganic chemistry. I must thank my wife Mrs. Mita Mukherjee for her constant support in finishing the writing of this book. I offer my respectful regards to my university teachers (late) Prof. S. K. Siddhanta and (late) Prof. R. L. Dutta, and Prof. S. C. Rakshit of the Department of Chemistry, The University of Burdwan, West Bengal, India who influenced me to quality teaching. I also would like to put on record my deep sense of gratitude to my PhD mentor Prof. Animesh Chakravorty of the then Department of Inorganic Chemistry, Indian Association for the Cultivation of Science, Kolkata, India and my postdoctoral mentor (late) Prof. Richard H. Holm of the then

Department of Chemistry, Harvard University, USA who inculcated in me the importance of quality research on fundamental principles and inspiring me to spread knowledge to next generation students through teaching and guiding their research.

In conclusion, it is my pleasant duty to thank my students over the years at Indian Institute of Technology Kanpur (IITK), Indian Institute of Science Education and Research Kolkata (IISERK), and IIT (Indian School of Mines) Dhanbad. Many have been those whose questions have stimulated me to seek better ways to present the subject. I am thankful specially to my IITK students for encouraging me to write this book. I would like to thank my colleagues at IITK for many discussions related to teaching. I am very grateful to my PhD student from IISERK Narottam Mukhopadhyay for his untiring help in the structural drawings and figures. I also thank Narottam and my former PhD student from IITK Dr. Arunava Sengupta for their sincere efforts in executing the book cover drawings.

I am grateful to Sandhya Devi M.G., World Scientific Publishing Company (Singapore), for being helpful at all stages of the project for her understanding attitude and support.

Rabindranath Mukherjee

Dhanbad

May 2, 2023

Contents

Chapter 1

Lewis Structure and Valence Shell Electron Pair Repulsion

The purpose of studying chemistry is to understand the properties and transformations of compounds in terms of molecular structure and bonding. The interpretation of structures and properties rely on phenomenological/semiquantitative theoretical models/theories. To explain the electronic structures and shapes of known inorganic molecules and attempt to predict the shape of molecules whose structures are so far unknown, we need such theories. Each theory has its own strengths and shortcomings. The value of a theory lies more in its usefulness than in its rigorousness.

1.1 Lewis structure

Lewis (American physical chemist G. N. Lewis: 1875–1946) in 1916 presented a simple, but useful, method of describing the arrangement of valence electrons in molecules. The most elementary discussion of covalent bonding is in terms of shared pairs of electrons. Lewis proposed that a covalent bond is formed when two participating atoms share an electron pair. Unshared pairs of valence electrons on atoms are called *lone pairs*. Lone pairs do not contribute directly to the bonding; however, their influence is reflected in the shape of the molecule and its chemical properties. In other words, Lewis proposed that electrons in a molecule should be paired. The presence of a single (odd) electron indicates that the species is a radical (paramagnetic; see Chapter 6).

Lewis structures are widely used and are very helpful for correlating structure and bonding. Lewis could account for the existence of a wide range of molecules by proposing the *octet rule*, which states that *each atom shares electrons with neighboring atoms to achieve a total of eight valence electrons*. Lewis structures help to explain the electronic structure (electronic arrangement of a molecule).

Electron pair approach

Chemically sensible Lewis structures are easily drawn by chemists with experience. Attempts to simplify the process generally rely too much on the *octet rule*, allowing *expanded octets*, only when required.

General considerations[1]:

1) Hydrogen atoms usually form one covalent bond (exception: B_2H_6; see Chapter 3); therefore, they occupy a terminal or peripheral position.
2) Carbon atoms nearly always form four bonds in stable compounds; hence, they almost never have lone pair (exceptions: CO, CN^-, carbanions, and compounds containing the –NC (isocyanide) group).
3) In stable structures, N usually forms three bonds and O forms two bonds (exceptions: NH_4^+, H_3O^+, O_2^{2-}, HNO_3, HN_3, etc.).
4) Oxygen-to-oxygen bonds (peroxide linkages) are rather unstable; therefore, O–O bonds to be avoided, if possible.
5) Symmetrical arrangements are most common.
6) If the molecule or ion is of the form AB_n ('A' is the central atom and 'B' is a terminal/peripheral atom (termed as 'ligand' in coordination chemistry; see Chapter 6) and '*n*' is the number of B atoms), it is quite common for 'A' to occupy central position bonding to all of the 'B' atoms. However, 'B' will be central atom if its covalency or electrovalence is higher than that of 'A'. For example, in N_2O it is N–N–O rather than N–O–N.

The following guidelines for writing Lewis structures are suggested[2]:

i) Only valence electrons are to be considered.
ii) All electrons are assigned as pairs, either to electron pair bonds or lone pairs at an atom, as far as possible.

[1] M. E. Zandler and E. R. Talaty, *J. Chem. Educ.* **1984**, *61*, 124.
[2] (a) J. A. Carroll, *J. Chem. Educ.* **1986**, *63*, 28. (b) W.-Y. Ahmad and M. B. Zakaria, *J. Chem. Educ.* **2000**, *77*, 329.

iii) Lewis structures with more bonds tend to be more important, except that the elements B, C, N, O, and F can accommodate a maximum of four electron pairs (octet rule), and a triple bond must have one of these elements at a terminus to be reasonable.

iv) Lewis structures with less formal charge are more important than with more formal charge.

v) Formal charges allow electronegativities, with any negative formal charge on a more electronegative atom and positive formal charge on a more electropositive atom, in more important structures.

vi) When two or more optimum resonance forms are equally likely, the molecular structure can be considered as an average of these.

Thus, it is worth noting that the above guidelines lead to Lewis structures which indicate the same σ-bonding hybrids, as derived from octet structures. Lewis structures drawn according to the guidelines given cannot be used to suggest π-bonding hybrids.

> Lewis structures are the starting point for both VSEPR model and hybridization theory.
> It is the σ-bonding hybrids of central atoms to be predictable to identify its geometry.

Example 1.1 Draw Lewis structures of SO_2, SOF_4, and NO_2^-.

Answer

SO_2: Number of valence electron $= 6 + 2 \times 2 = 10$ (S belongs to Gr. VI, each neutral O donates 2 electrons)

Number of pairs $= 5$. We are left with a lone pair (remembering, we have considered donation of two electrons by each neutral O; i.e., two $S = O$ bonds).

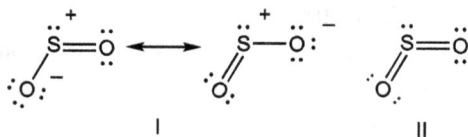

Given the S–O bond length of SO_2 as 143 pm, the more reasonable structure for SO_2 is II than I.

Formal charge and total bond order seem to be good predictive tools.

Both I and II allow explanation of the properties of SO_2, its dual Lewis acid reactivity. Each has a lone pair on S for Lewis base character, and each can be shown to accept another pair as a Lewis acid. As for example, the formation of HSO_3^- through attack of OH^- on S of SO_2.

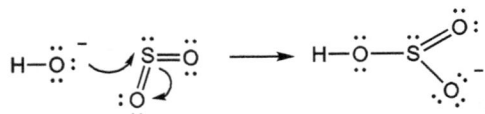

SOF_4: Number of valence electron $= 6 + 1 \times 2 + 4 \times 1 = 12$ (each F donates 1 electron)

Number of pairs $= 6$ (we have considered donation of two electrons by a neutral O; i.e., one S=O bond). Short S–O bond (141 pm) is found in SOF_4. The structure for SOF_4 is III.

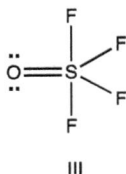

NO_2^-: Number of valence electron $= 5 + 1 \times 2 + 1 \times 1 = 8$ (N belongs to Gr. V, neutral O donates 2 electrons, and uninegative O donates 1 electron)

Number of pairs $= 4$. We are left with a lone pair (we have considered donation of two electrons by a neutral O; i.e., one N=O bond)

The Lewis structure for NO_2^- is IV.

Example 1.2 Draw Lewis structures of O_3, NO_2, and NO_3^-.

Answer

O_3: Number of valence electron $= 6 + 2 \times 2 = 10$ (O belongs to Gr. VI, each neutral O donates 2 electrons).

Number of pairs $= 5$. Thus, we are left with a lone pair (we have considered donation of two electrons by each O; i.e., two O=O bonds).

SO_2-type Lewis structure II is not feasible for O_3 (see guidelines iii). Resonance hybrid V agrees well with the reactivity and structure of O_3. Ozone is appreciably unstable, consistent with residual formal charges.

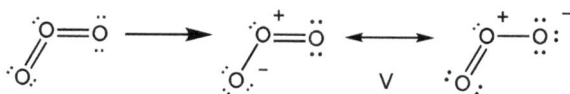

The Lewis structure for O_3 is V.

NO_2: Number of valence electron $= 5 + 2 \times 2 = 9$

Number of pairs $= 4$, with an odd electron (we have considered donation of two electrons by both O's; i.e., two N=O bonds). By this approach, for N the octet rule is violated (see below).

Therefore, the structure for NO_2 is VI.

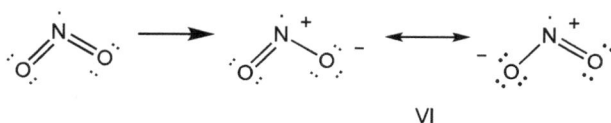

VI

NO_3^-: Number of valence electron $= 5 + 2 \times 2 + 1 \times 1 = 10$

Number of pairs $= 5$ (we have considered donation of two electrons by two neutral O; i.e., two N=O bonds). However, SO_2-type Lewis structure II is not feasible for NO_3^- (see guidelines iii). So, there cannot be two N=O bonds; only one N=O bond is possible. Hence, N donates two electrons to form a N=O bond (2 pairs), one electron each to two mononegative O forming two single-bonded N–O (2 pairs). This arrangement imparts a positive charge on N (4 pairs of electrons, i.e., 8 valence electrons) and, hence an overall negative charge on the molecule.

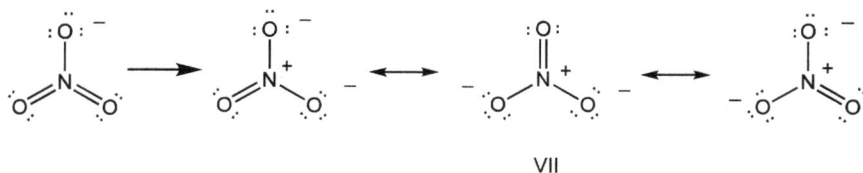

VII

Resonance hybrid VII agrees well with the reactivity and structure of NO_3^-.

1.2 Valence shell electron pair repulsion

In considering the vast number and variety of inorganic compounds, a basic concept is that molecular shapes are far more diverse than in organic chemistry. The idea of correlating molecular geometry (shape) and number of valence electron pairs (both shared and unshared) is extremely important. Several approaches have been offered to account for the experimentally determined molecular geometry of compounds. One method, the Valence Shell Electron Pair Repulsion (VSEPR), was offered by English theoretical chemist N. V. Sidgwick (1873–1952) and H. M. Powell (1906–1991) in 1940, for predicting molecular geometry. The extensions were developed by Australian inorganic chemist R. S. Nyholm (1917–1971) and British-Canadian inorganic chemist R. J. Gillespie (1924–2021) in 1957. Subsequently, this model was refined, further extended, and popularized by Gillespie in 1963.[3,4] The VSEPR model is splendidly successful in predicting the geometry about a central atom that is covalently bonded to two or more atoms. The VSEPR model helps in determining the shape of molecules. The VSEPR requires one to start with a reasonable Lewis (electron-dot) structure (see Section 1.1). The VSEPR is an empirical model, without strong theoretical justification, whose utility lies in the correctness of its structural prediction.

Since it is assumed in VSEPR model that the valence shell electron pairs are equidistant from the central atom, it is useful to envision placing the electron pairs in a Lewis structure as far apart as possible on the surface of a sphere whose midpoint is the central atom. The electron pairs can be regarded as localizing on a sphere of slightly smaller radius at a maximum distance. These arrangements depend on the number of electron pairs around central atom and are the most stable based on charge repulsion, calculated by simple electrostatics. It means that the distance between adjacent electron pairs is maximized, due to the electrostatic repulsions of the electrons. This assumption leads to the well-known shapes for AB_nE_m molecules, where A is the central atom, B a peripheral atom (ligand), E a lone pair, and n and m are their numbers, respectively.

The model is founded on the Pauli Exclusion Principle (Austrian theoretical physicist W. E. Pauli (1900–1958); Nobel Prize in Physics in 1945), which states that *no two electrons in an atom can have the same set of four quantum numbers*. The important physical consequence of the Pauli

---

[3]R. J. Gillespie, *J. Chem. Educ.* **1963**, *40*, 295; *J. Chem. Educ.* **1970**, *47*, 18; *J. Chem. Educ.* **1992**, *69*, 116; *J. Chem. Educ.* **2004**, *81*, 298.
[4]R. J. Gillespie, *Chem. Soc. Rev.* **1992**, 59.

Exclusion Principle, in terms of molecular geometry, is that electrons having parallel spins tend to maximize their angle of separation. Hence, *for the central atom in a molecule the valence shell electrons are viewed as localized pairs that tend to maximize their distance apart.*[5]

> The approximation of molecular shape/geometry, using the VSEPR model, considers that there may be bond pairs and unshared/nonbonded, i.e., lone pairs of electrons in the valence shell of the central atom. It means the electron pairs in a valence shell are not all equivalent.

For molecules with two electron pairs, VSEPR model predicts a linear arrangement. For three electron pairs, an equilateral triangle/a trigonal planar structure is predicted. For four electron pairs, two possible structures must be considered: a square planar and a tetrahedral arrangement. The electron pairs are arranged at the vertices of a regular tetrahedron to minimize repulsion between four bond pairs. For coordination number six, the electron pairs are arranged at the vertices of an octahedron to maximize the distance between the electron pairs. For seven electron pairs, a pentagonal bipyramidal geometry is predicted (Table 1.1). In all of the above structures, the positions of the electron pairs are equivalent and the bond angles are equal. The case of coordination number five — trigonal bipyramidal geometry — in which the electron pairs are nonequivalent — is separately discussed below.

A postulate of VSEPR model states that *nonbonding electron pairs (lone pairs) repel adjacent electron pairs more than bonding electron pairs.* As nonbonding electron pairs are under the influence of only one nucleus, their orbitals are more diffuse and occupy a larger region of space than the orbitals of bonding electron pairs, which are more confined in space because of the influence of two nuclei. A one-electron orbital takes up less room on the surface of an atom than a two-electron orbital, so the repulsive force is less for the one-electron orbital. Thus, the electron pairs will occupy positions which minimize the repulsions in the order given above (Fig. 1.1, Table 1.1).

To apply the VSEPR method to real molecules one needs to determine the number of bond pair and lone pair electrons around the central atom

---

[5]C. D. Mickey, *J. Chem. Educ.* **1980**, *57*, 210

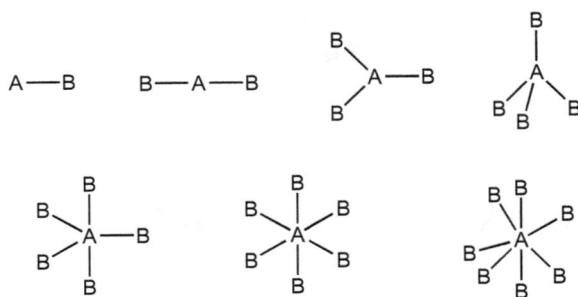

Fig. 1.1   The ideal arrangement of AB_n ($n \leq 7$) molecules.

Table 1.1   The ideal molecular geometry

Type	Ideal geometry	Angle(s)
AB	Linear	180°
AB_2	Linear	180°
AB_3	Equilateral triangular planar/Triangular planar	120°
AB_4	Tetrahedral	109.5°
AB_5	Trigonal bipyramidal	180°, 120°, 90°
AB_6	Octahedral	180°, 90°
AB_7	Pentagonal bipyramidal	180°, 90°, 72°

in the covalent molecule or polyatomic ion. To do so, let us first draw a Lewis structure for the molecule, characterizing each atom or lone pair of electrons associated with the central atom as a group. The groups are subsequently arranged around the central atom in a way that will maximize their separation and concomitantly minimize their repulsions.

To arrive at the correct shape of a neutral molecule or an ion, the following steps are suggested:

1. Draw a chemically meaningful Lewis structure
2. Count the number of valence electron on the central atom
3. Count the number of electrons donated to the central atom by the peripheral atom(s)
4. Count total number of valence electron on the central atom
5. Count the number of pairs
6. Assign the type

7. Assign predicted geometry
8. Assign actual molecular geometry/shape

(a) Application of VSEPR model to some real molecules

A simple illustration of the greater size of a lone pair is given by the decrease in the bond angle (H–A–H) in the series: CH_4 (109.5°) (0 lone pair) $< NH_3$ (107.8°) (1 lone pair) $< H_2O$ (104.5°) (2 lone pairs). As a bonding pair is replaced successively by one and then two lone pairs the tendency of lone pairs to take up more space than bonding pairs squeezes the bonding pairs together and the angle between them decreases.

A lone pair is not seen but the effect of a lone pair is observed.

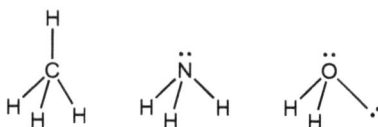

No. of electron pair	4	4	4
No. of bond pair	4	3	2
No. of lone pair	0	1	2
Type	AB_4	AB_3E_1	AB_2E_2
Predicted geometry	Tetrahedral	Tetrahedral	Tetrahedral
Actual geometry	Tetrahedral	Pyramidal	V-shape

Similarly,

H_2O (104.5°) $\gg H_2S$ (92.2°) $> H_2Se$ (91.0°) $> H_2Te$ (89.5°) (type: AB_2E_2)
NH_3 (107.8°) $\gg PH_3$ (93.3°) $> AsH_3$ (91.8°) $> SbH_3$ (91.3°) (type: AB_3E_1)
The angles are H–A–H (A=O, S, Se, Te; N, P, As, Sb).

The above sequences are explained by the fact that repulsions between electron pairs in filled shells are larger than those between electron pairs in incompletely filled shells and the decreasing electronegativity of the central atoms.[6]

---

[6]H. O. Desseyn, M. A. Herman, and J. Mullens, *J. Chem. Educ.* **1985**, *62*, 220.

Again,

$$NH_3 \ (107.8°) < NF_3 \ (102.3°) \ (\text{type: } AB_3E_1)$$

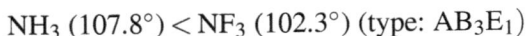

The greater electronegativity of F causes the N–F bond orbital to be smaller than the N–H bond orbital and it takes up less room on the surface of the nitrogen atom. Thus, under the repulsion exerted by the lone pair the angle between the smaller N–F bond orbitals becomes smaller than the angle between the larger N–H bond orbitals.

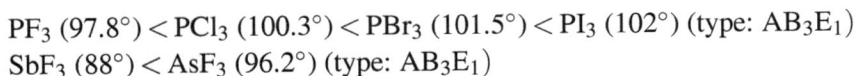

$$PF_3 \ (97.8°) < PCl_3 \ (100.3°) < PBr_3 \ (101.5°) < PI_3 \ (102°) \ (\text{type: } AB_3E_1)$$
$$SbF_3 \ (88°) < AsF_3 \ (96.2°) \ (\text{type: } AB_3E_1)$$

Similarly,

OPF₃ (102.5°) < OPCl₃ (103.6°) < OPBr₃ (106°) (type AB₃B'; B–P–B' angles)

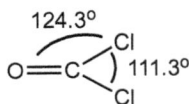

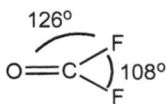

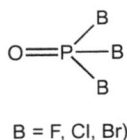

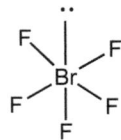

The shape of BrF_5 is distorted square pyramidal (total number of valence electron pair: 6, type: AB_5E_1) with a lone pair in the sixth position of an octahedron. The F–Br–F angle is 84.5°, which is less than the octahedral angle of 90°, because of the greater space taken up by the lone pair.

Example 1.3 The trend of O–N–O angles in the series NO_2^+, NO_2, and NO_2^-.

Answer

| NO_2^+ | NO_2 | NO_2^- |
| Type: AB₂ | AB₂'E₁' | AB₂E₁ |

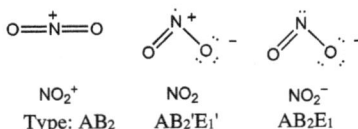

O–N–O angle decreases in the order: $NO_2^+ > NO_2 > NO_2^-$

1) A nonbonding or lone pair is larger and takes up more room on the surface of an atom than a bonding pair.
2) The size of a bonding electron pair, i.e., the space that it takes up on the surface of an atom, decreases with increasing electronegativity of the peripheral atom (ligand).
3) The two electron pairs of a double bond (or the three electron pairs of a triple bond) take up more room on the surface of an atom than the one electron pair of a single bond.

Example 1.4 In H_2O the H–O–H angle is 104.5° and in NH_3 the H–N–H angle is 107.8°. Explain the deviation in bond angles in NH_3 and H_2O with respect to tetrahedral angle.

Answer This is because lone pair-lone pair repulsion in H_2O is greater than lone pair-bond pair repulsion in NH_3.

Example 1.5 Explain the bond angles in F_2O (F–O–F angle: 103.3°) and in H_2O (H–O–H angle: 104.5°).

Answer Fluorine, more electronegative than hydrogen, occupies less space on the surface of the molecule. Hence, the lone pair occupies more space and thereby squeezes the two F's closer.

Let us consider the shape of SF_4 and ClF_3 (Fig. 1.2).

If there are several possible molecular structures involving 90° interactions, the most favored spatial arrangement is the one that minimizes the number of lone pair–lone pair interactions at 90°. Based on this criterion alone if it cannot be differentiated between two possible structures, then the most favored spatial arrangement is the one that minimizes the number of lone pair–bond pair interactions at 90°.

Example 1.6 Compare the structure of SF_4 and SOF_4.

Answer According to VSEPR model, the structures of SF_4 and OSF_4 are as shown below.

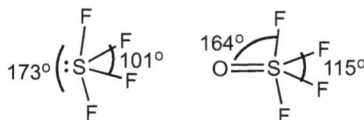

SF₄ (type: AB₄E₁)

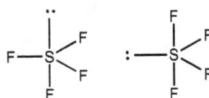

No. of lone pair-lone pair repulsion at 90°	0	**0**
No. of lone pair-bond pair repulsion at 90°	3	**2**

ClF₃ (type: AB₃E₂)

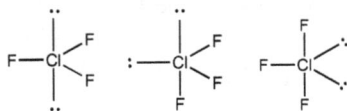

No. of lone pair-lone pair repulsion at 90°	0	1	**0**
No. of lone pair-bond pair repulsion at 90°	6	3	**4**

Fig. 1.2 To arrive at the shape of SF_4 (distorted tetrahedral or see-saw) and ClF_3 (T-shape).

The difference is in the equatorial plane; the presence of a lone pair in SF_4 and a S=O bond in OSF_4. This makes the differences in F_{ax}–S–F_{ax} and F_{eq}–S–F_{eq} angles. Double bonds with greater repulsions than single bonds are situated in the equatorial position.

(b) Trigonal bipyramidal molecules

It is evident that it is not possible to place five equivalent points uniformly on the surface of a sphere apart from forming a regular pentagon. Thus, the distribution of five electron pairs deviates from the regular polyhedra predicted by VSEPR model for other coordination numbers (Table 1.1). A second problem encountered lies in the placement of bonding and nonbonding electron pairs in the nonequivalent positions of the trigonal bipyramid.

For a species with five bond/lone pairs containing one nonbonding electron pair, e.g. SF_4 the nonbonding pair (lone pair) will occupy one of the equatorial positions in the trigonal plane. This can be explained by examining the axial and equatorial positions. The axial position has three neighboring pairs at 90° while the equatorial position has two neighboring pairs at 90° and two more at 120° (Fig. 1.2).

The different bond lengths in trigonal bipyramidal AX_5 molecules can be explained by the difference in the total repulsions between the bonds in the equatorial and the axial positions.

Consider the following molecular geometries.

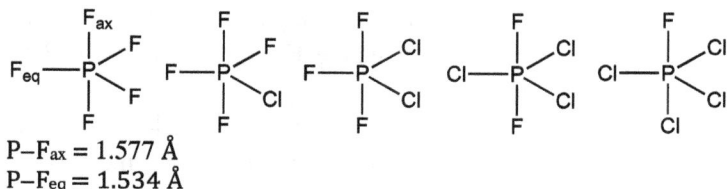

$P-F_{ax} = 1.577$ Å
$P-F_{eq} = 1.534$ Å

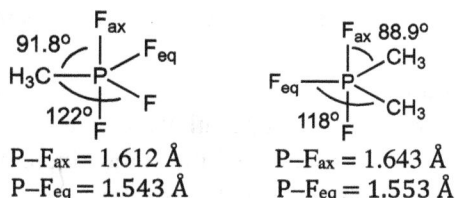

$P-F_{ax} = 1.612$ Å
$P-F_{eq} = 1.543$ Å

$P-F_{ax} = 1.643$ Å
$P-F_{eq} = 1.553$ Å

For any molecule in which a central atom with spherical inner shells has five electron pairs in its valence shell forming single bonds to five equivalent atoms/ligands, the most noteworthy features of five-coordinate molecules are as follows:

1) The molecular geometry is expected to be ideal or slightly distorted trigonal bipyramidal.
2) The axial bonds are longer than the equatorial bonds.
   The axial electron pairs in this arrangement are not equivalent to the equatorial pairs, because the former have three nearest neighbors at 90° while the latter have only two such neighboring pairs, equilibrium can only be attained if the axial pairs are at a greater distance from the nucleus than the equatorial pairs.
3) Methyl groups occupy equatorial positions.
   The smallest electron pairs, which have the smallest interactions with other electron pairs, tend to go into the axial positions. The larger

(Continued)

(Continued)

electron pairs occupy the equatorial positions, where there is more room for them. Thus, the most electronegative ligands, which have the smallest bonding electron pairs, always go into the axial positions and the less electronegative ligands e.g. the methyl groups occupy the equatorial positions.

This is a quite general rule, e.g., in PF_3Cl_2, $PFCl_4$, and $P(CH_3)Cl_4$ the more electronegative F or CF_3 groups in $P(CH_3)_3$-$(CF_3)_2$ occupy the axial positions.

4) All the bond lengths increase and the ratio of the length of the axial bonds to the length of the equatorial bonds r_{ax}/r_{eq} increases as the number of CH_3 substituents increases.

5) Methyl substitution causes the P–F bonds to be bent away from the CH_3 groups.

The substitution of a fluorine by a methyl group in $P(CH_3)F_4$ decreases the effective electronegativity of the P and allows all the bonding pairs to move away from the P thus increasing all the bond lengths. Moreover, the F_{ax} bonds are closer to the large electron pair bonding the methyl group than the F_{eq} bonds, hence they suffer a greater repulsion and increase more in length than the equatorial bonds. They are also pushed away from the electron pair bonding the methyl group so that the F_{ax}–P–F_{ax} bond angle becomes less than $180°$, in just the same way as the large, lone pairs cause the same angle in the SF_4 and ClF_3 molecules to be less than $180°$ (Fig. 1.2).

Table 1.2 collates the predicted geometry and shape of molecules and molecular ions, using VSEPR model.

(c) Summary of VSEPR method: molecular geometry/shape

VSEPR model has been developed into a useful tool to predict molecular structure (geometry/shape), based on the arrangement of bonding and nonbonding electron pairs in the valence shell. The method is fairly easy to apply and gives the correct molecular geometry for an extraordinarily large number of molecules. This means that the predicted shapes for these

Table 1.2 Examples of the predicted geometry and actual geometry (molecular shape), based on VSERPR model of molecules and molecular ions

Number of bonds[a]	Number of lone pair	Type	Predicted geometry/ Arrangement	Actual geometry (molecular shape)	Examples
2	0	AB_2	linear	linear	CO_2, NO_2^+
2	0 but an odd electron	'AB_2E'	equilateral triangular/ triangular planar	angular	NO_2
3	0	AB_3	equilateral triangular/ triangular planar	equilateral triangular/ triangular planar	SO_3, CO_3^{2-}
2	1	AB_2E_1	equilateral triangular/ triangular planar	angular	SO_2, BF_2^-, NO_2^-
4	0	AB_4/ AB_2B_2'	tetrahedral	tetrahedral	$SnCl_4$, XeO_4, BF_4^-, BeF_4^{2-}, ClO_4^-, $IO_2F_2^+$
3	1	AB_3E_1/ $AB_2B'E_1$	tetrahedral	trigonal pyramidal	NH_3, XeO_3, H_3O^+, SO_3^{2-}, $SnPh_3^-$, $(CH_3)_2SO$
2	2	AB_2E_2	tetrahedral	V-shape/bent	H_2O, NH_2^-, ICl_2^+
2	2	$ABB'E_2$	tetrahedral	V-shape/bent	$FClO$, $PhIO$
5	0	AB_5/ AB_3B_2'	trigonal bipyramidal	trigonal bipyramidal	PCl_5, PF_5, $SbCl_5$, SiF_5^-, XeO_3F_2
5	0	AB_4B'	trigonal bipyramidal	trigonal bipyramidal	SOF_4

(Continued)

Table 1.2 (*Continued*)

Number of bonds[a]	Number of lone pair	Type	Predicted geometry/ Arrangement	Actual geometry (molecular shape)	Examples
4	1	AB_4E_1	trigonal bipyramidal	distorted tetrahedral/ see-saw	SF_4, $SeCl_4$, $TeCl_4$, ICl_4^+
4	1	$AB_2B'_2E_1$	trigonal bipyramidal	distorted tetrahedral/ see-saw	XeO_2F_2
3	2	AB_3E_2	trigonal bipyramidal	T-shape	ClF_3, XeF_3^+, $PhICl_2$
2	3	AB_2E_3	trigonal bipyramidal	linear	I_3^-, ICl_2^-, XeF_2
6	0	AB_6	octahedral	octahedral	SF_6, IO_6^{5-}
5	1	AB_5E_1	octahedral	square pyramidal	BrF_5, IF_5, XeF_5^+, TeF_5^-
5	1	$AB_4B'E_1$	octahedral	distorted square pyramidal	$XeOF_4$, F_4BrO^-
4	2	AB_4E_2	octahedral	square planar	XeF_4, BrF_4^-, ICl_4^-

[a]Nature of bond should not be confused with number of bonds.

molecules are in excellent accord with experimentally determined molecular shapes.

The VSEPR model proposed that the geometry of a molecule is determined primarily by the repulsive interactions between the electron pairs in the valence shell of the central atom. The repulsions are at a minimum for the equilibrium geometry. The electron pairs (bond pairs and lone pairs) around a central atom will adopt a spatial arrangement that maximizes their angle of separation.

Deviations from the regular polyhedral shapes are rationalized by the following well-known assumptions or rules:

1. Lone pairs repel more strongly than bond pairs (trend in repulsion: lone-lone-lone pair > lone pair-bond pair > bond pair-bond pair).
2. The repulsion exerted by a bond pair decreases with increasing electronegativity of the ligand.
3. When the central atom is much more electronegative than the peripheral atoms (ligands), then the unshared electron pairs are less stereochemically active than in the presence of a central atom of low electronegativity.
4. Multiple bonds repel more strongly than single bonds.

(d) Some significant limitations of the VSEPR model

- This useful model fails to explain isoelectronic species (i.e. elements having the same number of electrons). The species may vary in shapes despite having the same number of electrons.
- This model is silent on the compounds of transition elements, except d^0, high-spin d^5, and d^{10} electronic distribution. This is because the compounds of other d^n ions have additional non-zero crystal field stabilization energy (see Chapter 6).
- The model cannot predict the molecular shape of molecules, where the geometry depends on molecular conformation. Examples are Al_2Cl_6, S_2F_{10}, NH_2OH, P_4, P_4O_6, P_4O_{10} (Fig. 1.3), B_2H_6, S_8 etc.

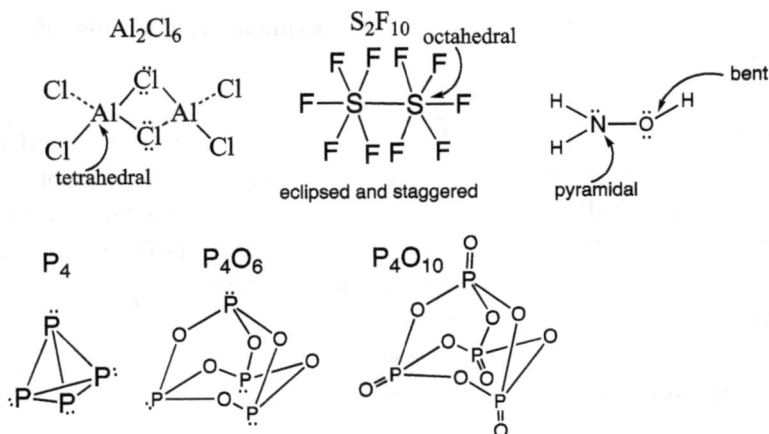

Fig. 1.3   The molecular structure / geometry of selected molecules, including P_4.

1.3 Hybridization

It is important to recognize that the VSEPR model provides an approach to bonding and molecular geometry (shape) based on the Pauli principle that is completely independent of the valence bond (VB) theory or of any orbital description of bonding. The VB theory is often taught at more or less the same time as the VSEPR model, and this can be a source of considerable confusion, as it can lead to the belief that the VSEPR model depends on, or is derived from, the VB theory. In fact, VSEPR model is completely independent of the VB theory, which gives a simple and very approximate description of the bonding in a molecule, but in general it does not predict geometry.[7]

Valence bond theory requires the presence of unpaired electrons in an atom to participate in covalent bonding. This condition in turn limits the number of covalent bonds to the number of unpaired electrons in an atom. Hybridization, based on the mixing of atomic orbitals to generate equivalent hybrid orbitals, is a very useful concept. It is a common misconception that hybridization is the cause of a particular molecular shape. This is not true. It is the energy which decides any particular shape.

---

[7]D. Ebertin and M. Monroe, *J. Chem. Educ.* **1982**, *59*, 285.

An important feature of hybridization is that a hybrid orbital has direc-
tional property, which arises from the constructive interference between
the orbital lobes.

Hybrid orbitals of different compositions of atomic orbitals are considered to match different molecular geometries. For example, sp hybridization is used to justify the electron distribution needed for linear molecules,
and sp^2 and sp^3 hybridizations are used to reproduce the electron distribution necessary for equilateral triangular planar/triangular planar and tetrahedral molecules, respectively, such as on B in BF_3 and C in CH_4. If in
CH_4 there are four equivalent bonds (sp^3 hybridization; four sp^3 hybrid
orbitals on C with one electron each interacts with four s orbitals of H with
one electron each to form four equivalent two-center two-electron ($2c$-$2e^-$)
C–H covalent bonds), then repulsion between electron pairs will be minimum if the four hybrid orbitals point to a tetrahedron, which would give
the observed bond angle of $109°28'$.

Table 1.3 summarizes examples of number of bond pair(s)/lone pair(s),
arrangement of electron pair(s), shape of molecules, composition of atomic
orbitals involved in the formation of hybrid orbitals.

Most hybridizations result in equivalent hybrid orbitals. However, for
trigonal bipyramidal (TBP) geometry, the equatorial bonds (sp^2 orbitals)
are stronger and the two linear axial bonds (dp orbitals) are weaker.

(a) Bent's Rule

More electronegative substituents prefer hybrid orbitals having less s character and the more electropositive substituents prefer hybrid orbitals having
more s character.

In TBP arrangement of PCl_3F_2 (see Section 1.2 (b)), three P–Cl_{eq} bonds
in the equatorial plane can have sp^2 hybridization (s, p_x, and p_y orbitals
of P forms three sp^2 hybrid orbitals). These three hybrid orbitals interact
with three s orbitals of three F's in the equatorial plane. Two P–F_{ax} bonds
can have pd hybridization (comprising p_z and dz^2 orbitals of P). These
two hybrid orbitals interact with two F's in the axial position with their p
orbitals. This results in sp^3d hybridization. Understandably, two pd hybrid
orbitals will have less s character and three sp^2 hybris orbitals will have
more s character. This clarifies Bent's rule.

Table 1.3 The ideal geometry, hybridization, and shape of molecules and molecular ions

Number of bonds/lone pair	Ideal geometry	Number of hybrid orbitals/type of hybridization	Composition of atomic orbitals	Molecular geometry (shape)	Examples
2 / 0	linear	2 / sp	$s + p_x/p_y/p_z$	linear	$BeCl_2$, CO_2
3 / 0	equilateral triangular planar/triangular planar	3 / sp^2	$s + p_x + p_y$	equilateral triangular planar/triangular planar	BCl_3, BF_3, CO_3^{2-}
2 / 1	equilateral triangular planar/triangular planar	3 / sp^2	$s + p_x + p_y$	angular	NO_2^-
4 / 0	tetrahedral	4 / sp^3	$s + p_x + p_y + p_z$	tetrahedral	CH_4, BF_4^-, ClO_4^-, SO_4^{2-}
3 / 1	tetrahedral	4 / sp^3	$s + p_x + p_y + p_z$	pyramidal	NH_3, H_3O^+,
2 / 2	tetrahedral	4 / sp^3	$s + p_x + p_y + p_z$	V-shape	H_2O, ICl_2^+, Cl_2O
4 / 0^a	square planar	4 / dsp^2	$dx^2 - y^2 + s + p_x + p_y$	square planar	$[Ni^{II}(CN)_4]^{2-}$, $[Pt^{II}Cl_4]^{2-}$
5 / 0	trigonal bipyramidal	5 / sp^3d	$s + p_x + p_y + p_z + dz^2$	trigonal bipyramidal	PCl_5
4 / 1	trigonal bipyramidal	5 / sp^3d	$s + p_x + p_y + p_z + dz^2$	see-saw/distorted tetrahedral	SF_4
3 / 2	trigonal bipyramidal	5 / sp^3d	$s + p_x + p_y + p_z + dz^2$	T-shape	ClF_3
2 / 3	trigonal bipyramidal	5 / sp^3d	$s + p_x + p_y + p_z + dz^2$	linear	ICl_2^-, I_3^-

(Continued)

Table 1.3 (*Continued*)

Number of bonds/ lone pair	Ideal geometry	Number of hybrid orbitals/ type of hybridization	Composition of atomic orbitals	Molecular geometry (shape)	Examples
5 / 0	square pyramidal	5 / dsp^3	$dx^2 - y^2 + s + p_x + p_y + p_z$	square pyramidal	$[Ni^{II}(CN)_5]^{3-}$
6 / 0	octahedral	6 / sp^3d^2	$s + p_x + p_y + p_z + dx^2 - y^2 + dz^2$	octahedral	PF_6^-
6 / 0ᵇ	octahedral	6 / d^2sp^3	$dx^2 - y^2 + dz^2 + s + p_x + p_y + p_z$	octahedral	$[Fe^{II}(CN)_6]^{4-}$
5 / 1	octahedral	6 / sp^3d^2	$s + p_x + p_y + p_z + dx^2 - y^2 + dz^2$	square pyramidal	BrF_5
4 / 2	octahedral	6 / sp^3d^2	$s + p_x + p_y + p_z + dx^2 - y^2 + dz^2$	square planar	BrF_4^-

a,bThe dsp^2 and d^2sp^3 hybridizations are considered for spin-paired/low-spin transition metal complexes (see Chapter 6).

(b) Berry pseudorotation

Pentacoordinate compounds are of particular interest due to the low-barrier interconversions (permutational barrier of PF_5: only 3.1 kcal/mol) between the two ideal (i.e., trigonal bipyramidal (TBP) and square pyramidal (SP)) polyhedra as the limiting polytopal structures, which result in their stereochemical non-rigidity (fluxionality). The classical Berry mechanism, illustrated in Fig. 1.4, interconverts axial and equatorial ligands (atoms/groups) without breaking bonds and with minimal structural distortions through an SP transition state (TS) of C_{4v} symmetry featuring a large apical-M-basal angle θ (Fig. 1.4). In equilibrium SP structures, θ is typically ranging from 105° to 125°. The TS connects two chemically-equivalent TBP stereoisomers of D_{3h} symmetry, in which the axial and equatorial ligands interchanged.

> The Berry pseudorotation considers exchange of the axial and equatorial atoms in trigonal bipyramidal compounds.

Substances having the composition XPF_4 have been subjected to particularly detailed study. The combination of ^{31}P and ^{19}F NMR spectroscopy with other physical techniques makes it possible to define the ground-state geometry of these substances as TBP, with the substituent X in an equatorial position. The axial-equatorial exchange reactions of many of these compounds occur at rates that are convenient for study by magnetic resonance techniques. The observation that ^{19}F-^{31}P coupling is preserved during these exchange reactions guarantees their intramolecularity (although not their unimolecularity). For example, the line-broadening and coalescence observed in the ^{19}F NMR spectrum of PF_5 have been rationalized on the basis of a Berry pseudorotation that is fast relative to the experimental NMR timescale.

Figure 1.4 indicates that two basal ligands (1 and 2) are moving towards the *trans*-position pivotal ligand 5, while the other two (3 and 4) are simultaneously moving away from it. The D_{3h} structure provides two different fluorine environments, which are present in a 2 to 3 ratio. Actually, we can determine this through the application of the operations of the D_{3h} point group (See Chapter 2). For example, carrying out a C_3 operation leads to the exchange of F positions in the equatorial plane but does not affect the

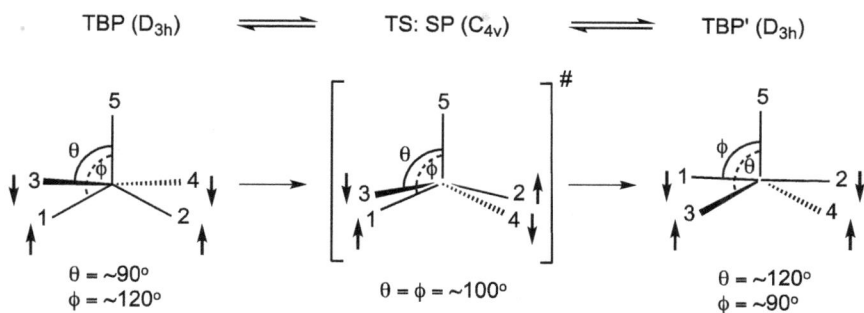

Fig. 1.4 Classical Berry pseudorotation mechanism interconverting two trigonal bipyramidal (TBP) stereoisomers of D_{3h} symmetry via a transition state (TS) of C_{4v} symmetry.

two axial F atoms. On the other hand, it can be seen that if a C_2 operation is carried out, this leads to a pairwise exchange of equatorial F atoms at the same time as it leads to a pairwise exchange of the axial F atoms. What we find is that none of the group operations leads to an exchange between the axial and the equatorial positions.

This is why D_{3h} symmetry for PF_5 mandates the presence of two distinct F environments in a 2 to 3 ratio. In the ^{19}F NMR spectrum of PF_5, a single peak representing all five F's is observed, even at low temperatures. This is due to interconversion between two five-coordinate geometries, which is faster than the NMR timescale.

Example 1.7 Consider that Berry pseudorotation can be slowed down at very low temperatures. Predict the variable-temperature ^{19}F NMR spectra with spin-spin splitting of (a) SF_4 and (b) ClF_3 (I for $^{19}F = 1/2$; S and Cl are not NMR active).

Answer

SF_4: At higher temperature, two axial F_{ax} and two equatorial F_{eq} interconvert rapidly. Hence, only one ^{19}F signal. However, at low temperature the rate of interconversion slows down (frozen) and hence two ^{19}F signals, one for two F_{ax} and one for two F_{eq} with intensity ratio 1:1 (splitting as per $2nI + 1$ rule).

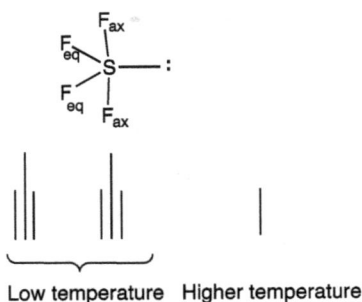

Low temperature   Higher temperature

SF_4: AB_4E_1 (see saw / distorted tetrahedral)

ClF_3: At higher temperature, two F_{ax} and an F_{eq} interconvert rapidly. Hence, only one ^{19}F signal. However, at low temperature the rate of interconversion slows down (frozen) and hence two ^{19}F signals, one doublet for two F_{ax} and one triplet for one F_{eq} with intensity ratio 2:1 (splitting as per $2nI + 1$ rule), respectively.

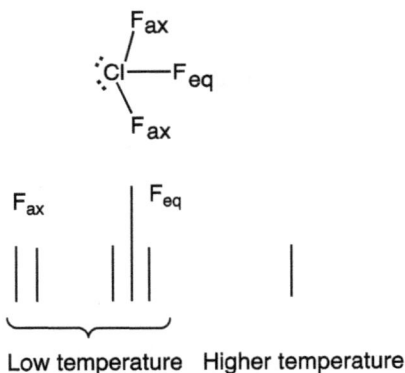

Low temperature   Higher temperature

ClF_3: AB_3E_2 (T shape); The $Cl-F_{ax}$ bonds are larger than $Cl-F_{eq}$ bonds.

(c) Bonding involving $p\pi$-$p\pi$ and $d\pi$-$p\pi$ interactions

Let us consider the structure of BF_3 and BF_4^-. According to VSEPR model, BF_3 belongs to AB_3 type and hence assumes equilateral triangular planar geometry and BF_4^- belongs to AB_4 type and has tetrahedral geometry. Thus, three and four B–F bonds in BF_3 and BF_4^-, respectively, are expected to be purely single bonds. However, B–F bond length in BF_3 is 130 pm and that in BF_4^- it is 143 pm. This can be explained based on partial double bond character in BF_3. Three single B–F bonds in BF_3 results in only six valence electron on B. However, if we invoke electron donation from filled p orbital on F to empty p orbital on B then between B and F in BF_3 partial double bond formation may occur. As a result of this bonding, B increases its valence electron from six towards eight. The orbital interaction with only one B–F bond is shown in Fig. 1.5. Based on energy and symmetry matching (see Chapter 3) the interaction is definitely feasible. The bonding situation that prevails in BF_3 is due to $p\pi$-$p\pi$ interaction.

Let us compare the structure of $(SiH_3)_2O$ and H_2O. According to VSEPR model, both $(SiH_3)_2O$ and H_2O belong to AB_2E_2 type and hence these molecules are expected to have V-shape. Therefore, two Si–O bonds in $(SiH_3)_2O$ and two H–O bonds in H_2O are expected to be purely single bonds. However, it is not so. The shape of $(SiH_3)_2O$ is almost linear. In fact, Si–O–Si bond angle is $\sim 180°$ in $(SiH_3)_2O$ and H–O–H angle is $104.5°$ in H_2O. This can be explained based on partial double bond character in $(SiH_3)_2O$. If we invoke donation of electron from filled p orbital on O to empty d orbital on Si then in $(SiH_3)_2O$ partial double bond formation between Si and O may take place. As a consequence, the molecule attains AB_2 type and assumes linear geometry. The orbital interaction with only one Si–O bond is shown in Fig. 1.6. Based on energy and symmetry matching the interaction is feasible. The bonding in $(SiH_3)_2O$ is due to $d\pi$-$p\pi$

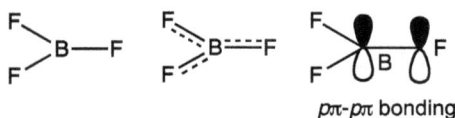

$p\pi$-$p\pi$ bonding

Fig. 1.5   The $p\pi$-$p\pi$ bonding in BF_3.

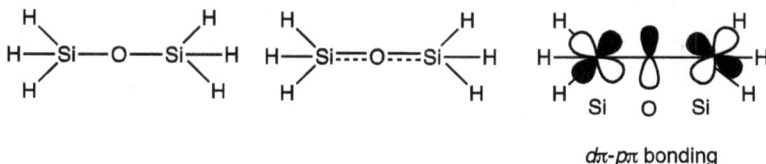

$d\pi$-$p\pi$ bonding

Fig. 1.6   The $d\pi$-$p\pi$ bonding in $(SiH_3)_2O$.

interaction. Also, due to unavailability of electron pairs on O, the basicity of $(SiH_3)_2O$ is much reduced compared to that of H_2O.

Example 1.8 Diethyl ether is more basic than the corresponding disilyl ether. Explain.

Answer The lone pair on O in $(CH_3)_2O$ makes it basic since these electrons are available for donation. However, in case of disilyl ether, the lone pair on O is donated to the empty $3d$ orbital of Si of suitable energy to form a $p\pi$-$d\pi$ bond. Hence, these lone pairs are not available as readily as in case of diethyl ether. Thus disilyl ether is less basic. The bond angle difference justifies this argument.

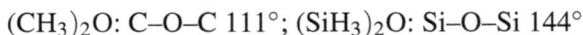

$$(CH_3)_2O: C–O–C\ 111°; (SiH_3)_2O: Si–O–Si\ 144°$$

Example 1.9 The basicity of $(SiH_3)_3N$ is much weaker than that of NH_3.

Answer Exactly that in $(SiH_3)_2O$, in $(SiH_3)_3N$ the bond angle Si–N–Si is close to 120°, revealing partial double bond formation between Si and N in $(SiH_3)_3N$. Due to unavailability of electron pair on N, the basicity of $(SiH_3)_3N$ is much reduced compared to that of H_3N.

(d) Multi-center bond models

In addition to conventional two-center two-electron $(2c\text{-}2e^-)$ covalent bond model, there could also be three-center four-electron $(3c\text{-}4e^-)$ bonding interaction.

Example 1.10 Explain the bonding in (a) SF_4, (b) BrF_5, and (c) IF_5 in terms of three-center four electron $(3c\text{-}4e^-)$ bond model.

Answer SF_4 (AB_4E_1: see-saw/distorted tetrahedral; see Section 1.2): Sulfur uses one of the p orbitals and two electrons and two fluorine uses one p orbital and one-electron each in the formation of a 3c-4e$^-$ F_{ax}–S–F_{ax} linear bond. Sulfur uses two more electrons and two of the p orbitals in the formation of two conventional 2c-2e$^-$ S–F bonds. The lone pair on sulfur remains in the spherical $3s$ orbital.

BrF_5 (AB_5E_1: square pyramidal; see Section 1.2): Bromine uses one of the p orbitals and two electrons and two fluorine uses one p orbital and one-electron each in the formation of two 3c-4e$^-$ F_{eq}–Br–F_{eq} linear bonds. Bromine uses one electron and one of the p orbitals in the formation of conventional 2c-2e$^-$ Br–F bond. The lone pair on bromine remains in the spherical s orbital.

IF$_5$ (AB$_5$E$_1$: square pyramidal; see Section 1.2): Iodine uses two 3c-4e$^-$ linear F_{eq}–I–F_{eq} bonds (two electrons from I and one electron from two F_{eq}; two such bonds) and one 2c-2e$^-$ conventional covalent I–F_{ax} bond (one electron from I and one electron from F_{ax}). The lone pair on iodine remains in the spherical s orbital.

Example 1.11 Explain the bonding in XeF$_2$ using (a) the participation of d orbitals and (b) the three-center four-electron bond model.

Answer According to VSEPR model the geometry of XeF$_2$ is linear (AB$_2$E$_3$ type). Xe can form sp^3d hybrid orbitals. Three of these can be used to accommodate three lone pairs. Two can be used to form two 2c-2e$^-$ covalent Xe–F bonds. Alternatively, the hybridization can be $sp^2 + pd$. Three sp^2 hybrid orbitals can accommodate three lone pairs on Xe and pd hybrid orbitals to be used for two linear bonds with two F atoms.

In sp^2 hybrid orbitals Xe can accommodate its three lone pairs in the equatorial plane. Using a p orbital of Xe with two electrons and a p orbital each from the two F's with one electron each forming a 3c-4e$^-$ linear bond.

(e) Exceptions to the VSEPR rules and possible effects of ligand-ligand repulsions

The VSEPR model assumes that the geometry/shape of a molecule is determined solely by the interaction between the electron pairs in the valence-shell and that ligand-ligand repulsions are generally of lesser importance. In almost every case, the VSEPR model predicts the correct molecular geometry. We have considered so far that lone pairs are stereochemically active, as they usually affect molecular geometry.

However, for high coordination numbers and for sufficiently large ligands or sufficiently small central atoms it is clear that repulsions between ligands could be of importance in determining molecular geometry. For example, the ions SbBr$_6^{3-}$ and TeBr$_6^{2-}$, and some closely related molecules all have seven electron pairs, including one lone pair in their valence shells. Notably, they all have regular octahedral structures, although they should have structures, based on a preferred arrangement of seven electron pairs (Table 1.1). Although it is not possible to predict the most likely arrangement of seven electron pairs with complete certainty, the six ligands would not be expected to have an octahedral arrangement. Therefore, it seems reasonable to assume that the ligand-ligand repulsions dominate the stereochemistry in these cases.

If the lone pair did occupy the valence shell in $TeBr_6^{2-}$, some of the Br···Br separations would be very small, e.g., 2.9 Å in a pentagonal bipyramid arrangement of the seven electron pairs. Such a small Br···Br distance would imply very considerable repulsions between the bromines. Given the fact that it has regular octahedral geometry, it would appear that the lone pair is forced inside the valence shell into a spherical s orbital. The observed bond length of 2.75 Å compared with 2.51 Å predicted from the sum of the covalent radii is consistent with the idea that the lone pair is inside the valence shell and thus contributes considerably to the shielding of the bonding electron pairs from the tellurium nucleus, effectively decreasing the electronegativity of Te and increasing the bond lengths. In these rare cases lone pairs do *not* affect the molecular geometry and these are termed *stereochemically inactive* lone pairs. In VSEPR model, these stereochemically inactive lone pairs are rationalized by placing them in an s-orbital, which is non-directional. An electron pair in an s-orbital has a spherically symmetric electron density, and hence cannot influence the shape of a molecule.

The shape of XeF_6 (AB_6E_1) is distorted octahedral/capped octahedral in which all the six positions are occupied by F atoms and the lone pair of electrons of Xe atom is present at the corner of one of the triangular faces. This is a case of *stereochemically active* lone pair. It is interesting to note that in XeF_8^{2-} (AB_8E_1), which should have 8 bond pairs and 1 lone pair around Xe. The dianion adopts a square antiprismatic structure, without any obvious position in space for the lone pair to occupy. Here the lone pair is *stereochemically inactive*.

The geometry of $XeOF_5^-$ ($AB_5B'E_1$) is distorted pentagonal pyramid. The five Xe−F bonds are in the pentagonal plane. A lone pair resides on one of the axial positions and in the other a double bonded Xe=O.

Further reading

C. E. Housecroft and A. G. Sharpe, *Inorganic Chemistry*, 2nd edition, Pearson Education Limited (2005)

D. F. Shriver and P. W. Atkins, *Inorganic Chemistry*, 3rd edition, Oxford University Press (1999)

J. D. Lee, *Concise Inorganic Chemistry*, 5th edition, Blackwell Science Limited (1996)

J. E. Huheey, E. A. Keiter, and R. L. Keiter, *Inorganic Chemistry: Principles of Structure and Reactivity*, 4th edition, Addison-Wesley Publishing Company (1993)

W. L. Jolly, Modern Inorganic Chemistry, 2nd edition, McGraw-Hill Inc., McGraw-Hill International Editions Chemistry Series (1991)

F. A. Cotton and G. Wilkinson, *Basic Inorganic Chemistry*, John Wiley & Sons, Inc.; Fourth Wiley Eastern Reprint (1988)

K. F. Purcell and J. C. Kotz, Inorganic Chemistry, Saunders Golden Sunburst Series, W. B. Saunders Company, Holt-Saunders Japan (1985)

B. Douglas, D. H. McDaniel, and J. J. Alexander, Conceps and Models of Inorganic Chemistry, 2nd edition, John Wiley & Sons, Inc. (1983)

Exercises

1.1 In SOF_2 the S–O and S–F bond lengths are 142 pm and 158 pm, respectively. The S–O bond length is shorter than a single bond. Draw its Lewis dot structure and justify its Lewis base character.

1.2 Like SO_2, in SO_3 the S–O bond length is 143 pm. Draw Lewis structure of SO_3. Given: in SO_4^{2-} the S–O bond length is 149 pm.

1.3 The S–S and S–O bond lengths in thiosulfate ion are 201 pm and 147 pm, respectively. Draw its Lewis dot structure.

1.4 In Cl_2O_7 the terminal Cl–O bonds are 140.5 pm, compared to 171 pm for the bridging Cl–O–Cl. Draw its Lewis dot structure.

1.5 Arrive at the molecular geometry of (i) $XeOF_2$, (ii) $XeOF_4$, (iii) XeO_2F_2, and (iv) XeO_3.

1.6 Predict the molecular shape of the ions (i) ClO_3^-, (ii) I_3^-, (iii) NH_2^-, and (iv) IO_6^{5-}.

1.7 Justify the trend of O–N–O angle in the series NO_2, NO_2^-, and NO_2^+ are 134°, 115°, and 180°, respectively.

1.8 Comment on the molecular geometry of XeF_6 and $SbCl_6^{3-}$ as distorted octahedral/capped octahedral and perfect octahedral, respectively.

Chapter 2

Molecular Symmetry

The concept of symmetry is important to almost every aspect of life. In chemistry, symmetry is important both at a molecular level (molecular properties; the subject matter of this chapter) and within crystalline systems (crystal structures). Crystal symmetry is beyond the scope of this book. An understanding of symmetry is essential in discussions of bonding (molecular orbital theory treatment; the subject matter of Chapter 3) and spectral properties, and calculations of molecular properties. Thus, familiarity with symmetry simplifies our task and aids our understanding. The basic concepts of molecular symmetry and very basics of necessary group theory knowledge are discussed first. It should be mentioned here that group theory is the mathematical treatment of symmetry.[1,2]

This chapter does not set out to give a comprehensive survey of molecular symmetry, but rather to introduce some common terminology and its meaning. In this chapter, the fundamental language of group theory (symmetry operation, symmetry element, point group, and character table) is introduced.

2.1 Symmetry, point groups, and character tables

During all symmetry operations, which can be applied to a molecule, at least one point in the molecule – the center of gravity – remains unchanged.

---

[1]M. Orchin and H. H. Jaffé, *J. Chem. Educ.* **1970**, *47*, 246; *J. Chem. Educ.* **1970**, *47*, 372; *J. Chem. Educ.* **1970**, *47*, 510.
[2]M. Zeldin, *J. Chem. Educ.* **1966**, *43*, 17.

Symmetry of this kind is therefore called *point symmetry*. Let us consider a set of molecules I–VI.

(a) Symmetry operations and symmetry elements

Proper rotation axis

During a 360° rotation around the z-axis of the molecules I–VI, I repeats itself twice and II–VI repeat themselves, two, three, four, five, and six times, respectively. The z-axes in these molecules are thus called a 2-fold, 3-fold, 4-fold, 5-fold, and 6-fold rotation axes, respectively. If the angle through which the molecules must be rotated to attain the superimposable image is designated θ, the molecules have $(360/\theta)$-fold rotational axes. The rotational axis is denoted as C_n, where n = $360°/\theta$. These symmetry operations are rotation around an axis, while the symmetry element is the rotational axis C_n.

Among molecules I–VI, except I and II, all are planar and the z rotational axis (Fig. 2.1) in each case is perpendicular to the molecular plane (*xy*). In I and II, the z rotational axis is colinear with proper rotation axis.

> The symmetry operation C_n is a proper rotation around an axis, while the symmetry element is the *proper rotational axis* C_n.

All linear molecules such as H_2, CO (homonuclear diatomic and heteronuclear diatomic molecule, respectively), and C_2H_2, CO_2 possess C_∞,

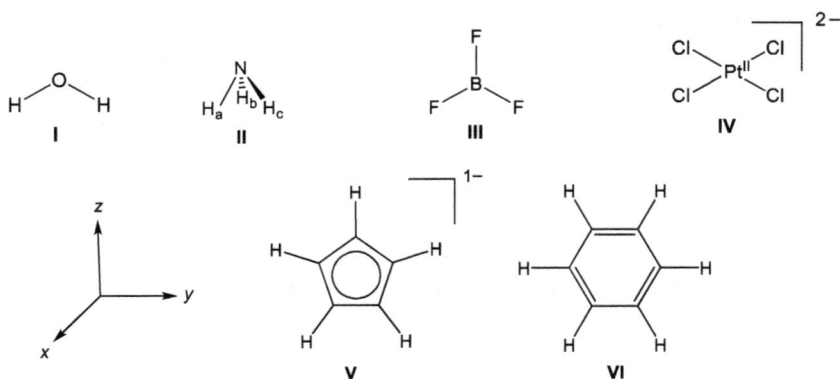

Fig. 2.1 A set of four neutral molecules and two ions, exhibiting symmetry of various kinds.

since rotation about the internuclear axis by any angle gives an orientation identical with the original.

It is important to recognize that all molecules have an infinite number of C_1 axes, since a 360° rotation around any or all axes passing through the center of gravity of the molecule returns it to its original position. An operation which leaves the molecule identical to the original is called the *identity operation*, denoted by E and thus for I $C_2^2 = E$, for II $C_3^3 = E$, and since all these operations, by definition, leave the molecule unchanged, doing nothing to the molecule $(C_1 = E)$ is also an identity operation. A very simple way to return a molecule to its original position, after a symmetry operation, consists of simply reversing the operation. Thus, the C_3 (clockwise rotation) of II by 120° around the z-axis passing through N and the center of the base of the triangle, takes H_a into H_b, H_b into H_c, and H_c into H_a. Now if this were followed by C_3' or C_3^{-1}, the inverse of C_3 (anticlockwise rotation), the molecule would be returned to its original orientation.

Reflection of a plane

If a molecule is bisected by a plane, and each atom in one half of the bisected molecule is reflected (the operation) through the plane and encounters a similar atom in the other half, the molecule is said to possess a mirror plane (the symmetry element). The operation and the element are denoted by σ.

Molecules I and II have two and three vertical planes (σ_v), respectively. Molecule I has $\sigma_v(xz)$ and $\sigma_v(yz)$ and molecule II has three vertical planes $(\sigma_v, \sigma_v', \text{ and } \sigma_v'', \text{ passing through three N–H bonds and bisecting the other sides of the plane of the triangle, respectively})$. Molecule III has three vertical mirror planes (σ_v), each including one B–F bond and bisecting the angle between the other two B–F bonds. Molecule IV has four vertical planes, the planes along the x- and y-axes. Molecule VI has a set of three vertical planes, which pass through opposite atoms, and which may be designated $3\sigma_v$.

All planar molecules/ions (Fig. 2.1: III–VI) have at least one plane of symmetry, the molecular plane. Thus, the xy plane is a horizontal mirror plane (σ_h) in all the molecules/ions.

Molecule IV has, in addition, a set of two mirror planes $(2\sigma_d)$ bisecting opposite Cl–Pt–Cl angles and molecule VI has, in addition, a set of three mirror planes bisecting opposite C–C bonds ($3\sigma_d$), which bisect the angles between the x- and y-axes.

> The reflection operation is a mirror plane. The operation and the element are denoted by σ. There are three kinds of mirror planes: σ_v, σ_h, and σ_d.

Center of symmetry

When a straight line is drawn from any atom in a molecule through the center of the molecule and, if continued in the same direction, encounters an equivalent atom equidistant from the center (the operation), and if the same operation can be performed on all atoms, then the molecule possesses a center of symmetry, designated as i for inversion.

Since each atom is thus reflected through the center into an equivalent atom, atoms must occur in pairs (excepting any atom which may lie on the center itself) with the members of the pair equidistant but in opposite directions from the center. Thus, molecules IV and VI have i, but I–III and V do not.

> For the center of symmetry or inversion center, the operation and the element are denoted by i.

Improper rotational axis

If a molecule is rotated around an axis and the resulting orientation is reflected in a plane perpendicular to this axis (the operation) and if the resulting orientation is superimposable on the original, the molecule is said to possess a rotation-reflection axis (the element).

The axis around which the rotation was performed is the rotation-reflection axis, and it is designated as S_n, where n is the order. This axis is called an improper rotational axis and is thus distinguished from a rotational or proper rotational axis, C_n.

> The rotation-reflection operation and the element is designated as S_n. This axis is called an improper axis of rotation.

A vertical axis (z axis) through the Ni(II) center of tetrahedral $[NiCl_4]^{2-}$ (VII) (see Chapter 6) (Fig. 2.2) is a S_4 axis, taking Cl1 into Cl4, Cl2 into Cl3, Cl3 into Cl1, and Cl4 into Cl2. Thus, the S_4 axis passes through

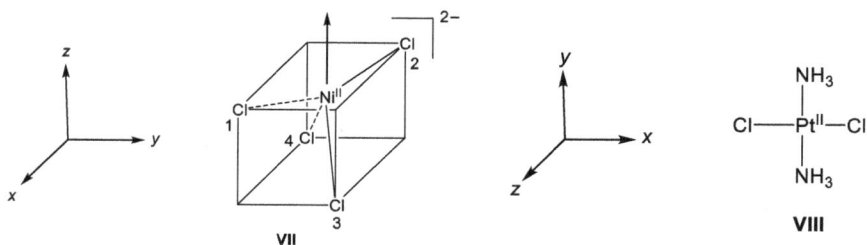

Fig. 2.2 $[Ni^{II}Cl_4]^{2-}$ and *trans*-$[Pt^{II}(NH_3)_2Cl_2]$, exhibiting symmetry of various kinds.

the center of opposite faces of the cube and since there are 6 faces there are three S_4 axes. And because for each S_4 (clockcwise) there is a (counterclockwise) operation, there are thus a total of $6S_4$ operations in a tetrahedral molecule, and all of these S_4 operations transform in a similar manner and hence belong to the same class. Note that these S_4 axes are not C_4 axes. The C_4 axes coincide with C_2 axes.

It should be pointed out that not only is i equivalent to S_2 but σ is equivalent to S_1. Usually it is easier to find the plane of symmetry than the S_1 axis, but the S_1 axis will be any axis perpendicular to the plane of symmetry.

trans-$[Pt^{II}(NH_3)_2Cl_2]$ (VIII) contains a $C_2^{(z)}$ main rotation axis, two C_2 axes $C_2^{(x)}$ and $C_2^{(y)}$ perpendicular to $C_2^{(z)}$ and three σ_v planes ($\sigma_v(xz)$, $\sigma_v(yz)$, and $\sigma_v(xy)$). It also has an inversion center i. It belongs to the D_{2h} point group (see below). This representation ignores the orientation of hydrogens on NH_3.

The principles of point group symmetry and the protocol for determining molecular point groups are summarized below and collated in Table 2.1. This will allow us to identify the basic symmetry elements present in a molecule: principal rotation axis and other rotation axes, planes of symmetry, center of symmetry (or inversion center), rotation-reflection axes, and identity. The difference between the symmetry elements and associated symmetry operations must be understood.

(b) Protocol for determining molecular point groups

A simple approach to classify a molecule in a point group, in which it belongs, is presented below.

Linear molecules
look for $\infty C_2 \perp C_\infty$: Yes $D_{\infty h}$; No $C_{\infty v}$

Proper axis, C_n (n = maximum): No C_s, C_i, C_1
look for σ: Yes C_s (E, σ only)
look for i: Yes C_i (E, i only) ; No C_1 (E only)

look for $C_2 \perp C_n$:
No C_n (E, C_n only)
No but has S_{2n} but no σ: $S_4 \{E, S_4, C_2, S_4^3\}^*$
No but has S_{2n} and i but no σ: $S_6 \{E, C_3, C_3^2, i, S_6^5, S_6\}^*$
*Special nomenclature: $S_1 \equiv C_{1v} \equiv C_{1h} \equiv C_s$; $S_n \equiv C_{nh}$ for odd n; $S_2 \equiv C_i$)
look for σ_v: Yes $C_{nv} \{E, C_n, n\sigma_v$ only$\}$
look for $\sigma_h (\perp C_n)$: Yes C_{nh} (also i for n even)
look for $nC_2 \perp C_n$: Yes $D_n \{E, C_n, nC_2^\perp$ only$\}$
look for σ_h: Yes $D_{nh} \{E, C_n, nC_2^\perp, \sigma_h, S_3, n\sigma_v\}$
Yes D_{nh} (i for n even) $\{E, C_n, nC_2^\perp, \sigma_h, S_3, n\sigma_v\}$
look for σ_d (bisects C_2): Yes D_{nd} (i for n odd)

Special symmetries (other noncoincident C_n with n > 2)

n = 3 $\rightarrow$ T
look for σ_d: Yes T_d

n = 4 $\rightarrow$ O
look for σ or i: Yes O_h

Linear molecules: $C_{\infty v}$ and $D_{\infty h}$
Any molecule:
C_1 (E only); C_s (E, σ_h only); C_i (E, i only); C_n (E, C_n only)
C_{nv} (E, C_n, nσ only); C_{nh} (E, C_n, σ_h, S_n only)
D_n(E, C_n, $nC_2^\perp$ only); D_{nh}(E, C_n, $nC_2^\perp$, σ_h only)

Table 2.1 Examples of the presence of symmetry element(s) and the corresponding point group

Molecules and molecular ions	Symmetry element(s)	Point group
CHClBrI	E	C_1
HOCl, quinoline	E, σ_h	C_s
1,2-dibromo 1,2-dichloroethane	E, i	C_i
N_2, O_2, CO, NO, HCN	E, $2C_\infty^\phi$, $\infty\sigma_v$	$C_{\infty v}$
H_2O_2	E, C_2	C_2
NO_2^-, SF_4, ClF_3, Cl_2O, SO_2,	E, $C_2^{(z)}$, σ_v (xz), σ_v' (yz)	C_{2v}
NH_3, $POCl_3$	E, C_3, C_3^2/C_3^{-1}, σ_v, σ_v', $\sigma_{v''}$	C_{3v}
$XeOF_4$	E, C_4, C_4^3/C_4^{-1}, C_2, σ_v, σ_v', σ_d, σ_d'	C_{4v}
trans-1,2-dichloroethylene	E, C_2, i, σ_h	C_{2h}
C_2H_2, CO_2, BeH_2	E, $2C_\infty^\phi$, $\infty\sigma_v$, i, $2S_\infty^\phi$, ∞C_2	$D_{\infty h}$
Twistane ($C_{10}H_{16}$)	E, $C_2^{(z)}$, $C_2^{(y)}$, $C_2^{(x)}$	D_2
$[Co^{III}(en)_3]^{3+}$ (en = 1,2-diaminoethane)	E, $2C_3$, $3C_2$	D_3
trans-$[Pt^{II}(NH_3)_2Cl_2]$, tetracyanoethylene	E, $C_2^{(z)}$, $C_2^{(y)}$, $C_2^{(x)}$, i, $\sigma_v(xy)$, σ_v' (xz), σ_v'' (yz)	D_{2h}
NO_3^-, BF_3, CO_3^{2-}, SO_3,	E, $2C_3$, $3C_2$, σ_h, $2S_3$, $3\sigma_v$	D_{3h}
$[Pt^{II}Cl_4]^{2-}$	E, $2C_4$, C_2, $2C_2'$, $2C_2''$, i, $2S_4$, σ_h, $2\sigma_v$, $2\sigma_d$	D_{4h}
Benzene (C_6H_6)	E, $2C_6$, $2C_3$, C_2, $3C_2'$, $3C_2''$, i, $2S_3$, $2S_6$, σ_h, $3\sigma_d$, $3\sigma_v$	D_{6h}
Allene (C_3H_4)	E, $2S_4$, C_2, $2C_2'$, $2\sigma_d$	D_{2d}
tetraphenylmethane	E, S_4, C_2, S_4^3	S_4
CH_4, $[Ni^0(CO)_4]$, $[NiCl_4]^{2-}$	E, $8C_3$, $3C_2$, $6S_4$, $6\sigma_d$	T_d
SF_6, $[Ir^{II}Cl_6]^{4-}$	E, $8C_3$, $6C_2$, $6C_4$, $3C_2$ ($= C_4^2$), i, $6S_4$, $8S_6$, $3\sigma_h$, $6\sigma_d$	O_h

(c) Character tables

In the study of the properties of molecules, the motion of the molecule itself and the motion of the atoms relative to one another (vibrations and IR absorptions) as well as the motion of the electrons in the molecule (MO theory and electronic spectra) are of interest. For a molecule like H_2O which belongs to the point group C_{2v}, all motions of the molecule must be either symmetric or antisymmetric with respect to each of the four symmetry operations E, $C_2^{(z)}$, $\sigma_v(xz)$, $\sigma_v(yz)$ of the group. Symmetric behavior is characterized as $+1$ and antisymmetric behavior as -1 and we call the $+1$ and -1 the "character" of the motion with respect to the symmetry operation.

Let us examine the behavior of the p_x, p_y, and p_z orbitals in the H_2O molecule. Under the symmetry operations of point group C_{2v} (Table 2.2), the p_x orbital is $+1$ with respect to E (since identity operation leaves the molecule unchanged, obviously has the character $+1$), -1 with respect to $C_2^{(z)}$ (rotation around the z-axis changes the signs of the two lobes; hence the p_x orbital is antisymmetric and has the character -1), $+1$ with respect to $\sigma_v(xz)$ (reflection on $\sigma_v(xz)$ mirror transforms the orbital into itself and hence has the character $+1$), and -1 with respect to $\sigma_v(yz)$ (reflection on $\sigma_v(yz)$ gives a change in 'sign' and hence is -1). This $+1$, -1, $+1$, -1 behavior is one of the four possible ways that every property of the molecule can be described. The symmetry species is B_1 (see below). Similarly, for p_y orbital and p_z orbital the behaviors are $+1$, -1, -1, $+1$ (the symmetry species B_2) and $+1$, $+1$, $+1$, $+1$ (the symmetry species A_1), respectively. The behavior of p_y orbital is displayed in Fig. 2.3.

The four distinct behavior patterns (A_1, A_2, B_1, and B_2; for A_2 (see below, R_z)) are called *symmetry species* or *irreducible representations*,

Table 2.2   Character table for point group C_{2v}

C_{2v}	E	$C_2^{(z)}$	$\sigma_v(xz)$	$\sigma_v(yz)$		
A_1	$+1$	$+1$	$+1$	$+1$	z	x^2, y^2, z^2
A_2	$+1$	$+1$	-1	-1	R_z	xy
B_1	$+1$	-1	$+1$	-1	x, R_y	xz
B_2	$+1$	-1	-1	$+1$	y, R_x	yz

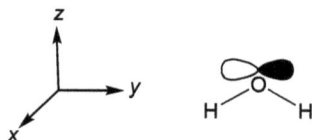

Fig. 2.3 The p_y orbital in H_2O: $+1, -1, -1, +1$.

and their number is equal to the order of the group. The total number of symmetry operations in a point group is called the *order of the group*; in C_{2v} the order is four.

The symmetry species in the case of C_{2v} are shown in the first column of Table 2.2. All symmetry species that are symmetric with respect to the highest rotational axis are designated by A, and those antisymmetric are designated by B. This table shows that under the identity operation E, every property is symmetric, as expected, since the E operation does nothing. Table 2.2 is called the character table of the point group C_{2v}. In the character table for C_{2v} it should be noted that in the last column of the row of characters in the representation B_2 we have the symbol y, which tells us that motion in the y direction of the water molecule transforms as B_2. The behavior of rotations about x, y, z axes (and R_x, R_y, R_z operations; see below) are also included.

Protocol for assigning symmetry representations:

1. All one-dimensional representations are designated either A or B; two-dimensional are designated E; three-dimensional species are designated T.
2. One-dimensional species which are symmetric with respect to rotation by $360°/n$ about the principal C_n axis [symmetric meaning: $\chi(C_n) = 1$; χ represents the character (see below)] are designated A, which those antisymmetric in this respect [$\chi(C_n) = -1$] are designated B.
3. Subscripts 1 and 2 are usually attached to A's and B's to designate those which are, respectively, symmetric and antisymmetric with respect to a C_2 perpendicular to the proper rotation axis or, if such a C_2 axis is lacking, to a vertical plane of symmetry.

(Continued)

(Continued)

> 4. Primes and double primes are attached to all letters, when appropriate, to indicate those which are, respectively, symmetric and antisymmetric with respect to σ_h.
> 5. In groups with a center of inversion, the subscript g and u are attached to symbols for representations which are symmetric and antisymmetric, respectively, with respect to inversion.

Let us examine the translational motion of the water molecule in the direction of each of the cartesian axes. First let us ascertain how a translation along the positive y-axis (T_y) (Fig. 2.3) transforms under the symmetry operations of C_{2v}. To simplify the analysis, let us imagine an arrow from the center of the molecule along the y-axis. Under E the arrow is unchanged, so the motion is $+1$ with respect to E. Under $C_2^{(z)}$ we visualize that now the arrow is pointing in a direction opposite to the original, i.e., the vector has changed sign and the motion in the y direction is thus antisymmetric with respect to $C_2^{(z)}$ and hence has the character -1. On reflection in the xz-plane, the arrow would again be pointing in the direction opposite to the original, as in $C_2^{(z)}$, and again the character would be -1. Finally, reflection in the yz-plane leaves the arrow unchanged and hence under this operation, the character is $+1$. Thus, the symmetry species is B_2. Motion in the x and z directions can be similarly analyzed. For the behavior of translations along x (T_x) and z (T_z) axes the representations are B_1 and A_1, respectively.

> Since the p_x, p_y, and p_z orbitals behave like translations in these directions, this notation also tells us how these orbitals transform.

Similar transformations of rotational motion around the three cartesian axes can be assigned to symmetry species. Let us analyze how rotation around the z-axis (R_z) transforms. Again, to help us analyze the situation, we employ rotating arrows and visualize whether or not the direction of the arrows is getting reversed under the symmetry operations of C_{2v}. The molecule is now rotating around the z-axis in a clockwise direction. The behavior of rotation about z axis for E, $C_2^{(z)}$, $\sigma_v(xz)$, $\sigma_v(yz)$ operations in point group C_{2v} is represented as $+1$ (the arrow direction does not change; do nothing operation), $+1$ (the arrow direction does not change, as it is along the direction of motion), -1 (the arrow direction changes), and -1 (the arrow direction changes). The symmetry species is A_2. Similarly, for R_x and R_y the behaviors are $+1, -1, -1, +1$ and $+1, -1, +1, -1$, respectively, and hence the symmetry representations are B_2 and B_1, respectively.

> The behavior of translations along x, y, z axes (T_x, T_y, and T_z operations, respectively) are similar to that of R_y, R_x, and p_z, respectively. The four symmetry species (A_1, A_2, B_1, B_2) represent every property of the molecule, which belongs to point group C_{2v}.

It should be noted that the character tables such as those shown in Table 2.2 (and Table 2.3 and Table 2.4; see below) all include in a last column the notations x, y, z and R_x, R_y, R_z. These symbols are assigned to particular symmetry species in each point group. They inform us to which symmetry species the translations (of a molecule, for example) along the x-, y-, and z-axes belong, and to which symmetry species the rotations R_x, R_y, R_z, around the x-, y-, and z-axes, respectively, belong.

Let us consider the $[Pt^{II}Cl_4]^{2-}$ ion IV (Fig. 2.1) and examine how the three p orbitals of the Pt(II) ion transform under the symmetry operations for point group D_{4h}. The x- and y-axes (Fig. 2.4) are C_2' axes; the C_2 axes which bisect the angles between bonds are designated C_2'' axes. The two

Table 2.3 Character table for point group D_{4h}

D_{4h}	E	$2C_4$	C_2	$2C_2'$	C_2''	i	$2S_4$	σ_h	$2\sigma_v$	$2\sigma_d$	
A_{1g}	1	1	1	1	1	1	1	1	1	1	x^2+y^2, z^2
A_{2g}	1	1	1	-1	-1	1	1	1	-1	-1	
B_{1g}	1	-1	1	1	-1	1	-1	1	1	-1	x^2-y^2
B_{2g}	1	-1	1	-1	1	1	-1	1	-1	1	xy
E_g	2	0	-2	0	0	2	0	-2	0	0	(xz, yz)
A_{1u}	1	1	1	1	1	-1	-1	-1	-1	-1	
A_{2u}	1	1	1	-1	-1	-1	-1	-1	1	1	z
B_{1u}	1	-1	1	1	-1	-1	1	-1	-1	1	
B_{2u}	1	-1	1	-1	1	-1	1	-1	1	-1	
E_u	2	0	-2	0	0	-2	0	2	0	0	(x, y)

Table 2.4   Character table for point group D_{2h}

D_{2h}	E	$C_2^{(z)}$	$C_2^{(y)}$	$C_2^{(x)}$	i	$\sigma_v(xy)$	$\sigma_v(xz)$	$\sigma_v(yz)$		
A_g	1	1	1	1	1	1	1	1		x^2, y^2, z^2
B_{1g}	1	1	−1	−1	1	1	−1	−1	R_z	xy
B_{2g}	1	−1	1	−1	1	−1	1	−1	R_y	xz
B_{3g}	1	−1	−1	1	1	−1	−1	1	R_x	yz
A_u	1	1	1	1	−1	−1	−1	−1		
B_{1u}	1	1	−1	−1	−1	−1	1	1	z	
B_{2u}	1	−1	1	−1	−1	1	−1	1	y	
B_{3u}	1	−1	−1	1	−1	1	1	−1	x	

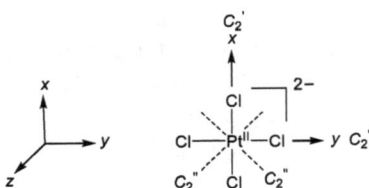

Fig. 2.4   The coordinate system and the $[Pt^{II}Cl_4]^{2-}$ ion.

vertical planes of symmetry that include the x and y axes are σ_v and those that include the C_2'' axes are called σ_d in the character table for point group D_{4h} (Fig. 2.4; Table 2.3). Applying all the symmetry operations on the p_z orbital in turn as listed in the character table of D_{4h} (E, $2C_4$, C_2, $2C_2'$, $2C_2''$, i, $2S_4$, σ_h, $2\sigma_v$, $2\sigma_d$), we get, $+1, +1, +1, -1, -1, -1, -1, -1, +1, +1$. The result tells us that p_z orbital belongs to symmetry label or symmetry species A_{2u}.

Now let us examine the behavior of the p_x orbital. If we perform a 90° clockwise rotation, $C_4^{(z)}$, we see that p_x is transformed to p_y, and hence p_x is neither symmetric nor antisymmetric under the operation. However, the $C_4^{(z)}$ operation simultaneously transforms p_y into $-p_x$. Obviously, the transformations are related and the two orbitals transform together. Thus, when we perform a clockwise rotation of 90° on the p_x orbital, we get a new orbital which has none of the old orbital in it and is exactly equal to the p_y orbital. If we call the new orbital p_x', we may state the fact mathematically,

$$p_x' = 0p_x + 1p_y$$

Similarly, the transformation of the old p_y before the $C_4^{(z)}$ into the new p'_y, may be written,

$$p'_y = -1px + 0p_y$$

A special method of writing these equations is possible,

$$\begin{pmatrix} p'_x \\ p'_y \end{pmatrix} = \begin{pmatrix} 0 & 1 \\ -1 & 0 \end{pmatrix} \begin{pmatrix} p_x \\ p_y \end{pmatrix}$$

The set of numbers in the central parenthesis is called the *transformation matrix*. This matrix transforms the old set of p_x, p_y on a 90° clockwise rotation to a new set of p_x, p_y orbitals. If the multiplication of the 2 × 2 matrix by the column vector were carried out by the rules of the matrix multiplication, the abovementioned two equations would be obtained. *The numbers appearing in the diagonal from upper left to lower right of the transformation matrix is called the trace of the matrix and the actual number found by the addition is called the character of the transformation matrix.* In the present example the character is 0. Accordingly, if we refer to the character table of D_{4h} (Table 2.3) we must look for a symmetry species which under the C_4 operation has a character of 0 (E_g and E_u).

Instead of rotating the p_x, p_y orbitals 90° in the clockwise direction, if we rotate in the counter-clockwise direction, i.e., instead of C_4 to C'_4 (C_4^{-1}), the transformation may be written as,

$$\begin{pmatrix} p'_x \\ p'_y \end{pmatrix} = \begin{pmatrix} 0 & -1 \\ 1 & 0 \end{pmatrix} \begin{pmatrix} p_x \\ p_y \end{pmatrix}$$

The character of the transformation matrix is again zero. We now see why C_4 and C'_4 belong to the same "class" ($2C_4$), with the same character (zero).

Here we can use a short-cut to determine the correct symmetry species. When we are dealing with a pair of orbitals of equal energy that transform together we have a so-called degenerate set. A set of two degenerate orbitals always belongs to an E species (a set of three degenerate orbitals to a T species). Hence, our orbitals belong to one of the E species in D_{4h}, and it remains only to specify the behavior under the operation i. A p orbital is always antisymmetric to a center of symmetry, and hence our p_x, p_y orbitals together belong to species E_u in point group D_{4h}, the characters of which are found in the last row of the character table.

As an exercise let us determine the transformation matrix of the p'_x, p'_y orbitals under the operation i to confirm that the character is -2 as shown in the character table. Referring to above coordinate system, we see that under i,

$$p'_x = -1p_x + 0p_y$$
$$p'_y = 0px - 1p_y$$

and the character of the transformation matrix is -2.

$$\begin{pmatrix} p'_x \\ p'_y \end{pmatrix} = \begin{pmatrix} -1 & 0 \\ 0 & -1 \end{pmatrix} \begin{pmatrix} p_x \\ p_y \end{pmatrix}$$

The C_4 and C'_4 rotations give, respectively, the transformation matrices,

$$\begin{pmatrix} 0 & 1 \\ -1 & 0 \end{pmatrix} \quad \text{and} \quad \begin{pmatrix} 0 & -1 \\ 1 & 0 \end{pmatrix}$$

These numbers correspond to the values of the sin and cos of $\pm 90°$ and indeed that is their origin. We may generalize the 2×2 transformation matrix for any degree of rotation by using the matrices,

$$\begin{pmatrix} \cos\theta & \sin\theta \\ -\sin\theta & \cos\theta \end{pmatrix} \quad \text{and} \quad \begin{pmatrix} \cos\theta & -\sin\theta \\ \sin\theta & \cos\theta \end{pmatrix}$$

and in either ease $2\cos\theta$ corresponds to the character. Thus, in any doubly degenerate symmetry species, E, the character for the appropriate rotation of $\theta°$ either clockwise or counter-clockwise around the C_n axis is $2\cos\theta$.

In the non-degenerate point groups (A, B symmetry species), the transformation matrices are all 1×1 matrices and so we could immediately assign $+1$ to symmetric and -1 to antisymmetric behavior.

In the case of point groups like D_{2h} (example, *trans*-$[Pt^{II}(NH_3)_2Cl_2]$; Fig. 2.2, Table 2.4) where there are three two-fold axes, and therefore no rotational axis that is of highest order, only the symmetry species that is symmetric to all three C_2 axes is designated A. Where more than one species or representation is symmetric with respect to the highest rotational axis, as in C_{2v}, they are distinguished in the subscripts (or sometimes by primes) and the totally-symmetric species, i.e., the species as in C_{2v} which

Table 2.5   Character table for point group C_{4v}

C_{4v}	E	$2C_4$	C_2	$2\sigma_v$	$2\sigma_d$		
A_1	1	1	1	1	1	z	x^2+y^2, z^2
A_2	1	1	1	-1	-1	R_z	
B_1	1	-1	1	1	-1		x^2-y^2
B_2	1	-1	1	-1	1		xy
E	2	0	-2	0	0	$(x,y)\ (R_x, R_y)$	(xz, yz)

is $+1$ with respect to every operation, is always the A_1 species. Subscripting of the B species is more arbitrary. In the case of molecules belonging to C_{2v}, the rules for orientation given earlier is usually unambiguous and after setting up the coordinate system, the B species that is symmetric to $\sigma_v(xz)$, is called B_1.

2.2   Orbital symmetries

A very important result of molecular symmetry is that it conveys on all orbitals of all atoms in a molecule the property of symmetry, which dictates whether or not orbitals can interact to form bonds. This information is contained in character tables; formal knowledge of group theory is necessary to derive these tables. We shall concentrate on how to use them. Let us take $[Ni^{II}(CN)_5]^{3-}$ as an example, which belongs to the point group C_{4v}. The character table of C_{4v} is displayed in Table 2.5.

Character tables contain two basic types of information:

(1) Symmetry operations (grouped by class) of a point group
(2) Orbital symmetries (defined by rows of characters)

To pursue (2), let us follow a coordinate system for square pyramidal (tetragonal) $[Ni^{II}(CN)_5]^{3-}$ with Ni(II) at origin, z along C_4 (conventional to take z along highest C_n axis), and x, y along equatorial M–L (Ni^{II}–CN) bonds.

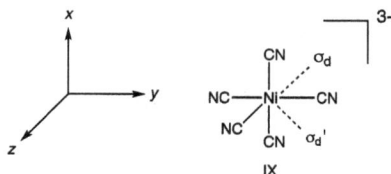

IX

Now let us determine how the valence orbitals of Ni behave, or transform, under the symmetry operations of the point group C_{4v}.

C_{4v}	E	C_4^+	C_4^-	C_2	$\sigma_v(xz)$	$\sigma_v(yz)$	σ_d	σ_d'
s	s	s	s	s	s	s	s	s
p_z	z	z	z	z	z	z	z	z
p_x	x	y	$-y$	$-x$	x	$-x$	y	$-y$
p_y	y	$-x$	x	$-y$	$-y$	y	x	$-x$
z^2	z^2	z^2	z^2	z^2	z^2	z^2	z^2	z^2
x^2-y^2	x^2-y^2	$-(x^2-y^2)$	$-(x^2-y^2)$	x^2-y^2	x^2-y^2	x^2-y^2	$-(x^2-y^2)$	$-(x^2-y^2)$
xy	xy	$-xy$	$-xy$	xy	$-xy$	$-xy$	xy	xy
xz	xz	yz	$-yz$	$-xz$	xz	$-xz$	yz	$-yz$
yz	yz	$-xz$	xz	$-yz$	$-yz$	yz	xz	$-xz$

Two types of behavior are observed:

(i) Result of the symmetry operations $= \pm1$, orbital is transformed into itself or –itself; sets of ±1 in rows define orbital symmetry.
s, dz^2, pz: A_1 totally symmetric; dx^2-y^2: B_1; d_{xy}: B_2 (cf. Table 2.5)

The orbitals dz^2, dx^2-y^2, and dxy are non-degenerate orbitals. Even those with the same l value cannot have the same energy in ML_5, except accidentally.

(ii) Orbital is transformed into itself and another orbital (i.e. orbitals are "mixed up" by symmetry operations of the group). This property is expressible by simple matrix formation (Fig. 2.5).

From this matrix analysis of symmetry operations, the transformation matrix (character χ) is the addition of diagonal traces from upper left to lower right. The characters for (xz, yz) orbital transformation properties, which are doubly degenerate (xz transforms to yz or yz transforms to xz), for the operations E, $2C_4$ (C_4^+ and C_4^-), C_2, $2\sigma_v$, and $2\sigma_d$ of the point group C_{4v} are 2, 0, -2, 0, and 0 (see Table 2.5). The symmetry representation is E.

The discussion of character tables up to this point can be applied to point groups C_1, C_s, C_i, C_{2v}, C_{2h}, D_2, and D_{2h}. All of these point groups do not involve a symmetry axis C or S greater than two-fold (D_{2d} has an S_4 axis). As soon as an axis greater than two-fold arises, the problems of symmetry species become much more difficult.

$$\begin{array}{cc} & xz \quad yz \quad E \\ \begin{array}{c} xz \\ yz \end{array} & \begin{pmatrix} 1 & 0 \\ 0 & 1 \end{pmatrix} \begin{pmatrix} xz \\ yz \end{pmatrix} = \begin{pmatrix} xz \\ yz \end{pmatrix} \end{array} \qquad \begin{array}{cc} & xz \quad yz \quad C_4^{+} \\ \begin{array}{c} xz \\ yz \end{array} & \begin{pmatrix} 0 & 1 \\ -1 & 0 \end{pmatrix} \begin{pmatrix} xz \\ yz \end{pmatrix} = \begin{pmatrix} yz \\ -xz \end{pmatrix} \end{array}$$

$$\chi(E) = 2 \qquad\qquad\qquad \chi(C_4) = 0$$

$$\begin{array}{cc} & xz \quad yz \quad C_2 \\ \begin{array}{c} xz \\ yz \end{array} & \begin{pmatrix} -1 & 0 \\ 0 & -1 \end{pmatrix} \begin{pmatrix} xz \\ yz \end{pmatrix} = \begin{pmatrix} -xz \\ -yz \end{pmatrix} \end{array} \qquad \begin{array}{cc} & xz \quad yz \quad C_4^{-} \\ \begin{array}{c} xz \\ yz \end{array} & \begin{pmatrix} 0 & -1 \\ 1 & 0 \end{pmatrix} \begin{pmatrix} xz \\ yz \end{pmatrix} = \begin{pmatrix} -yz \\ xz \end{pmatrix} \end{array}$$

$$\chi(C_2) = -2 \qquad\qquad\qquad \chi(C_4) = 0$$

$$\begin{array}{cc} & xz \quad yz \quad \sigma_v(xz) \\ \begin{array}{c} xz \\ yz \end{array} & \begin{pmatrix} 1 & 0 \\ 0 & -1 \end{pmatrix} \begin{pmatrix} xz \\ yz \end{pmatrix} = \begin{pmatrix} xz \\ -yz \end{pmatrix} \end{array} \qquad \begin{array}{cc} & xz \quad yz \quad \sigma_v'(yz) \\ \begin{array}{c} xz \\ yz \end{array} & \begin{pmatrix} -1 & 0 \\ 0 & 1 \end{pmatrix} \begin{pmatrix} xz \\ yz \end{pmatrix} = \begin{pmatrix} -xz \\ yz \end{pmatrix} \end{array}$$

$$\chi(\sigma_v) = 0 \qquad\qquad\qquad \chi(\sigma_v') = 0$$

$$\begin{array}{cc} & xz \quad yz \quad \sigma_d \\ \begin{array}{c} xz \\ yz \end{array} & \begin{pmatrix} 1 & 0 \\ 0 & -1 \end{pmatrix} \begin{pmatrix} xz \\ yz \end{pmatrix} = \begin{pmatrix} yz \\ xz \end{pmatrix} \end{array} \qquad \begin{array}{cc} & xz \quad yz \quad \sigma_d' \\ \begin{array}{c} xz \\ yz \end{array} & \begin{pmatrix} -1 & 0 \\ 0 & 1 \end{pmatrix} \begin{pmatrix} xz \\ yz \end{pmatrix} = \begin{pmatrix} -yz \\ -xz \end{pmatrix} \end{array}$$

$$\chi(\sigma_d) = 0 \qquad\qquad\qquad \chi(\sigma_d') = 0$$

Fig. 2.5 Transformation matrix notation for symmetry operations (geometric transformations) for *xz*, *yz* orbital pair in C_{4v} point group.

Orbital degeneracy in octahedral symmetry:
Orbitals are degenerate if interconverted by one or more operations.
$O_h/E\ 8C_3\ 6C_2\ 6C_4\ 3C_2\ i\ldots$

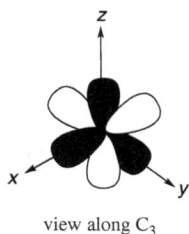

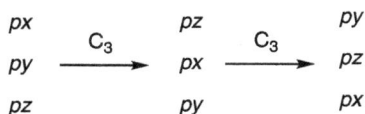

view along C_3

$$\begin{array}{c} px \\ py \\ pz \end{array} \xrightarrow{C_3} \begin{array}{c} pz \\ px \\ py \end{array} \xrightarrow{C_3} \begin{array}{c} py \\ pz \\ px \end{array}$$

Certain other operations have similar consequences.

Hence, *px*, *py*, and *pz* are triply degenerate (t_{1u}).

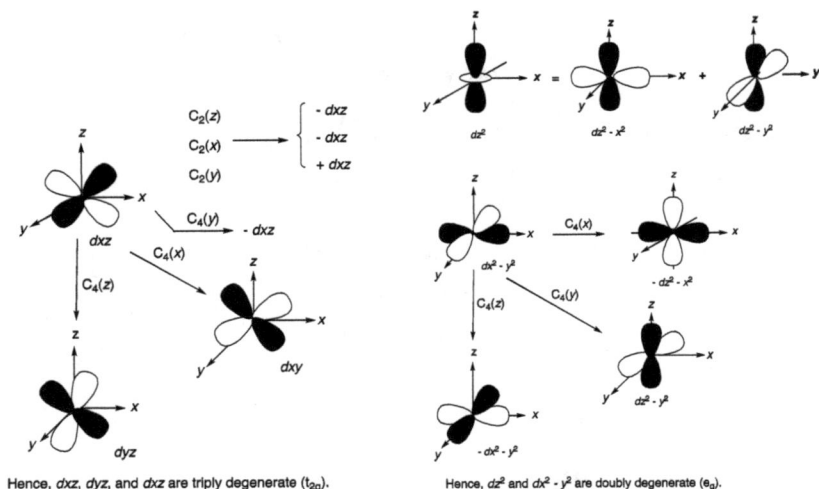

Hence, dxz, dyz, and dxz are triply degenerate (t_{2g}).

Hence, dz^2 and dx^2-y^2 are doubly degenerate (e_g).

Further reading

C. E. Housecroft and A. G. Sharpe, *Inorganic Chemistry*, 2nd edition, Pearson Education Limited (2005)

D. F. Shriver and P. W. Atkins, *Inorganic Chemistry*, 3rd edition, Oxford University Press (1999)

J. E. Huheey, E. A. Keiter, and R. L. Keiter, *Inorganic Chemistry: Principles of Structure and Reactivity*, 4th edition, Addison-Wesley Publishing Company (1993)

W. L. Jolly, *Modern Inorganic Chemistry*, 2nd edition, McGraw-Hill Inc., McGraw-Hill International Editions Chemistry Series (1991)

K. F. Purcell and J. C. Kotz, *Inorganic Chemistry*, Saunders Golden Sunburst Series, W. B. Saunders Company, Holt-Saunders Japan (1985)

B. Douglas, M. H. McDaniel, and J. J. Alexander, *Concepts and Models of Inorganic Chemistry*, 2nd edition, John Wiley & Sons, Inc. (1983).

F. A. Cotton, *Chemical Applications of Group Theory*, second edition, Wiley Eastern Limited, Fifth Wiley Eastern Reprint (1986) (U.S. edition (1971)).

Exercises

2.1 Determine the shape and then assign the point group for the following:
(i) Cl_2O, (ii) NH_2^-, (iii) $POCl_3$, (iv) XeO_2F_2, (v) ICl_2^-.

2.2 Consider a molecule which belongs to the point group C_{2v}. (a) If the molecule translates along the x-axis, represent the observed motion with appropriate symmetry species (irreducible representation) and (b) If the molecule rotates about the z-axis, represent the observed motion with appropriate symmetry representation.

Chapter 3

Molecular Orbital Theory

The valence-bond model (see Chapter 1 and Chapter 6) cannot adequately explain the fact that some molecules contain equivalent bonds with a bond order between that of a single bond and a double bond (the structures of BF_3 (this chapter), $(SiH_3)_2O$, $(SiH_3)_3N$ (Chapter 1)). The best it can do is to suggest that these molecules are hybrids of the two or more Lewis structures that can be written for these molecules. Notably, the paramagnetic character of O_2 cannot be explained by valence-bond model. These problems, and many others, can be overcome by using a model of bonding based on *molecular orbitals*. Molecular orbital theory (a delocalized bonding approach) is more powerful than valence-bond theory (a localized bonding approach) because the orbitals reflect the geometry of the molecule to which they are applied (see below).

Molecules are built from two or more bound atoms. It is possible to combine the known orbitals of constituent atoms in a molecule to describe its electron orbitals. The molecular orbitals (MO's) represent regions in a molecule where an electron is likely to be found. They are obtained by combining atomic orbitals (AO's). An MO can specify a molecule's electron configuration, and most commonly, it is represented as a *linear combination of atomic orbitals* (the LCAO-MO method), especially in qualitative or approximate usage. These considerations provide a simple model of bonding in a molecule, understood through MO theory. This powerful bonding model was proposed by American physicist and chemist R. S. Mulliken (1896–1986; Nobel Prize in Chemistry: 1966).

Although MO theory is computationally demanding, the principles on which it is based are similar with those we use to write electron configurations for atoms. The key difference is that in MO's, the electrons are allowed to interact with more than one atomic nucleus at a time. Just as with AO's, an energy-level diagram is created by listing the MO's in order of increasing energy. The orbitals are then filled with the required number of valence electrons according to the Pauli principle.

> Each MO can accommodate a maximum of two electrons with opposite spins.

3.1 LCAO-MO theory

Electron configurations of atoms are described as wave functions, the basic set of functions that describe a given atom's electrons. The wave functions (Ψ) which are solutions of the Schrödinger equation (Austrian and naturalized Irish physicist E. Schrödinger (1887–1961); Nobel Prize in Physics: 1933) are commonly called *orbitals*. This function can be used to calculate the probability of finding any electron in any specific region around an atom's nucleus. An orbital may also refer to the physical region where the electron can be calculated to exist, as defined by the orbital's particular mathematical form. The basic functions are one-electron functions centered on the nuclei of component atoms in a molecule.

The MO description of a molecule treats the orbitals as the property of the entire molecule, considering the influence of the various nuclei, and the interactions among all the electrons. The usual simplification of this formidable task is to recognize that while electrons are near a particular nucleus, they should be influenced primarily by that nucleus. This permits us to use the familiar descriptions of AO's. The MO's are obtained as LCAO's. The allowed combinations of AO's are limited by the symmetry of the molecule and the symmetry properties of the AO's. The MO's must transform as symmetry species (irreducible representations) of the point group in which the molecule belongs (see below). Irreducible representations represent the symmetries of specific molecular properties.

> The MO theory assumes that the valence electrons of the atoms within a molecule become the valence electrons of the entire molecule.

The bonds in polyatomic molecules are built in the same way as in diatomic molecules, the only difference being that we use more AO's to construct the MO's, and these MO's spread over the entire molecule. Thus, the MO wave function for two interacting AO's can be written as follows,

$$\Psi = c_1 \phi_1 \pm c_2 \phi_2$$

In deciding which AO's may be combined to form MO's, the following conditions should be satisfied:

1. The energy of the AO's must have closely similar energy (*'energy matching'*).
2. The symmetry of the AO's in which the molecule belongs will decide whether it will be a bonding, an antibonding or a nonbonding situation (*'symmetry matching'*).
3. The overlap of the AO's must be very effective, judged by the magnitude of the *overlap integral*.

Minimizing the total energy of the system determines an appropriate set of coefficients (c_1, c_2 etc.) for linear combinations.

The shape of the MO's and their respective energies are approximated by comparing the energies of the individual atoms' AO's – or molecular fragments – and applying known values for repulsion and other similar factors. The MO's are constructed by taking linear combinations of the valence orbitals of atoms within the molecule. Symmetry will allow us to treat more complex molecules by helping us to determine which AO's combine to make MO's.

> MO's are obtained from LCAO's. It should be remembered that orbitals are wave functions. These linear combinations occur when the symmetries of the AO's are the same and the orbitals are close in energy. This allows understanding the differences between bonding, antibonding, and nonbonding MO's.

(a) Basic rules of MO theory

Rule 1: Orbitals must have the same symmetry (same symmetry species/irreducible representation) to have non-zero overlap.

Rule 2: The interaction of n AO's combine to form n MO's. If n = 2, one MO is bonding and one antibonding. The bonding orbital is more stable than the lower-energy AO. The antibonding orbital is less stable than the higher-energy AO. The bonding orbital is stabilized less than the anti-bonding orbital is destabilized (see below).

One of LCAO's initial assumptions is that the number of MO's is equal to the number of AO's included in the linear expansion.

Rule 3: If the AO's are degenerate, their interaction is proportional to their overlap integral, S.

Rule 4: If the AO's are non-degenerate, their interaction is proportional to $S^2/\Delta E$, where ΔE is the energy separation between the AO's. In this case the bonding orbital is mostly localized on the atom with the lower energy AO, usually the more electronegative atom. The antibonding orbital is mostly localized on the atom with the higher energy AO.

(b) The overlap criterion of bonding

Let us start with the familiar notion that the overlap of two or more orbitals can lead to a bonding interaction and that the stronger the overlap the stronger the bond. To have physical meaning, each orbital function must be normalized to unity, corresponding to unit probability that the electron will be found in all of space.

The mathematical condition is $\int \Psi_i^2 d\tau = 1$, where $\Psi_i^2(x, y, z)$ is the point probability for an electron in Ψ_i at the point (x, y, z).

The MO function is given by,

$$\Psi_i = c_1\phi_1 + c_2\phi_2 + \cdots$$

where ϕ's are the valence AO's of the atoms that constitute the molecule and c's are weighting coefficients that express how much of each AO is in the MO. The probability function is given by (if we use real ϕ's, then Ψ_i will be real),

$$\Psi_i^2 = c_1^2\phi_1^2 + c_2^2\phi_2^2 + 2c_1c_2\phi_1\phi_2 + \cdots$$

and the integrated probability is

$$\int \Psi_i^2 d\tau = c_1^2 \int \phi_1^2 d\tau_1 + c_2^2 \int \phi_2^2 d\tau_2 + 2c_1c_2 \int \phi_1\phi_2 d\tau_1 d\tau_2 + \cdots$$
$$= c_1^2 + c_2^2 + 2c_1c_2 S_{12} + \cdots$$

The last equation follows, if the individual AO's are normalized ($\int \phi_1^2 d\tau_1 = 1$, $\int \phi_2^2 d\tau_2 = 1$) and S_{12} is the *overlap integral* ($= \int \phi_1 \phi_2 d\tau_1 d\tau_2$) between AO's, ϕ_1 and ϕ_2.

Since the sum of terms on the right side of the integrated probability equation constitutes the total probability of the electron in space, it is logical to view the terms c_1^2 or c_2^2 represent as the probability that the electron in a MO behaves like an electron does in atomic orbitals ϕ_1 or ϕ_2, and thus it is said that c_1^2 or c_2^2 is the probability that the electron is to be found in AO's, ϕ_1 or ϕ_2. The term $2c_1 c_2 S_{12}$ represents the *overlap probability* between AO's, ϕ_1 and ϕ_2. The greater the magnitude of this term, the greater is the probability that the electron is in the overlap region of the AO's, ϕ_1 and ϕ_2. Understandably, this quantity should be related to the strength of the chemical bond between the atoms for which ϕ_1 and ϕ_2 are AO's.

Before proceeding to a more in-depth examination of the significance of these terms in the probability function, it is necessary to examine the types of orbital overlap in MO functions, the significance of the relative phases of the overlapping AO's, and the signs of overlap integrals. The usual overlap situations are shown in Fig. 3.1. The overlap integral S between two AO's assumes a value in the range $-1 < S < +1$.

Now let us illustrate some common types of orbital interactions:

Relative to the non-interacting situation, electron density flows out of the atomic orbitals into the internuclear region for a bonding interaction. For the antibonding interaction, electron density flows in opposite directions out of the internuclear region. In other words, the sign of $c_1 c_2 S$ identifies the interaction of two AO's in a MO as bonding or antibonding. Based on these simple arguments of electron probability, one expects the electron to be stabilized (lower energy) in the bonding situation and destabilized (higher energy) in the antibonding situation.

Relative to two atoms at the same distance:

$S = +1$ identifies bonding; build-up of electron charge between nuclei

$S = -1$ identifies antibonding; decrease in electron charge between nuclei

Depending upon whether the interfering lobes have the same or different signs (the signs indicate the sign of the wave function; in Fig. 3.1 and in the figures to follow, blackened lobes represent + sign and un-blackened

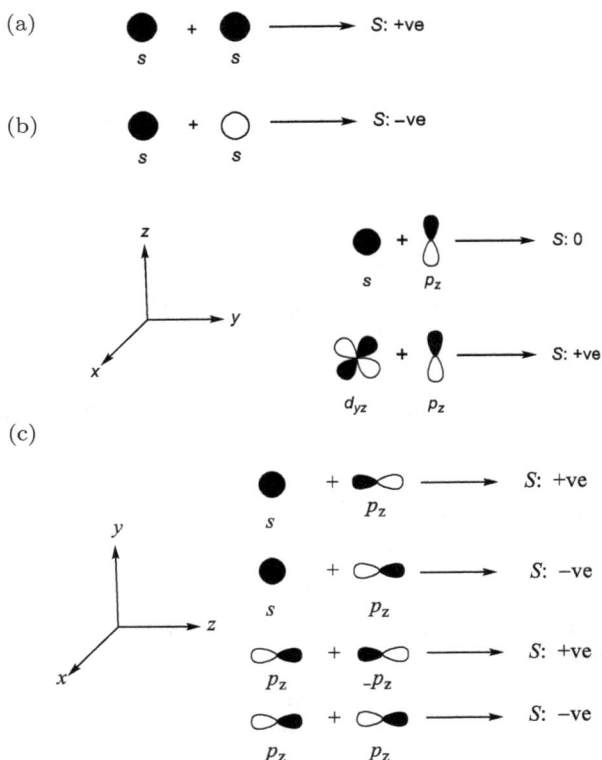

Fig. 3.1   The orbital interactions and the overlap integral.

lobes represent − sign), the overlap integral would have a positive value or a negative value.

$S = 0$ (nonbonding; no electron charge between nuclei): if the symmetry between two AO's ensures a net zero value.

Although Ψ's include both radial and angular wave functions, the sign of S is usually dictated by angular wave functions.

Using the specific case of overlapping atomic orbitals s and p_z in which the nucleus of the atom with the s orbital lies in the nodal plane of the p_z orbital of the second atom (Fig. 3.1(c)), the overlap product $\phi_s \phi_p$ has positive values at all points above the nodal plane and negative values at all points below this plane. Furthermore, because the nodal plane is a reflection plane, for each point above the plane (with a positive value for $\phi_s \phi_p$) there is a corresponding point below the plane with the same magnitude of the $\phi_s \phi_p$ product but with a negative sign. Consequently, when the integration is performed (the $\phi_s \phi_p$ products are summed over all points), a value for S identically equal to zero results.

Let us consider illustrative MO formation. A MO is formed by interaction of ≥ 2 AO's centered on their respective atoms. It is appropriate to answer at this point the question, "What is the physical significance of the sign of an overlap integral (S)? To do so, let us consider a linear combination of two atomic orbitals s and p_z (Fig. 3.1(c)), and that the nucleus of the atom with the s orbital (ϕ_1) lies on the z axis and in the negative z direction with respect to the atom with the p_z orbital (ϕ_2). Two situations can arise from the linear combinations (Fig. 3.2):

$$\Psi = c_1 \phi_1 + c_2 \phi_2 \quad \text{or} \quad \Psi = c_1 \phi_1 - c_2 \phi_2,$$

where c_1 and c_2 have positive values.

In the situation of destructive interference $(\Psi = c_1 \phi_1 + c_2 \phi_2)$ in the internuclear region, a node is created in that region. This is antibonding interaction between the two orbitals. This is characterized by the MO (a). In the situation of constructive interference $(\Psi = c_1 \phi_1 - c_2 \phi_2)$ in the internuclear region, no node is created in that region. This is characterized by the MO (b). This is bonding interaction between the two orbitals.

The integrated probability functions for the two MO's differ only by the sign of the overlap term,

$$\int \Psi^2 d\tau = c_1^2 + c_2^2 \pm 2c_1 c_2 S_{12} = 1$$

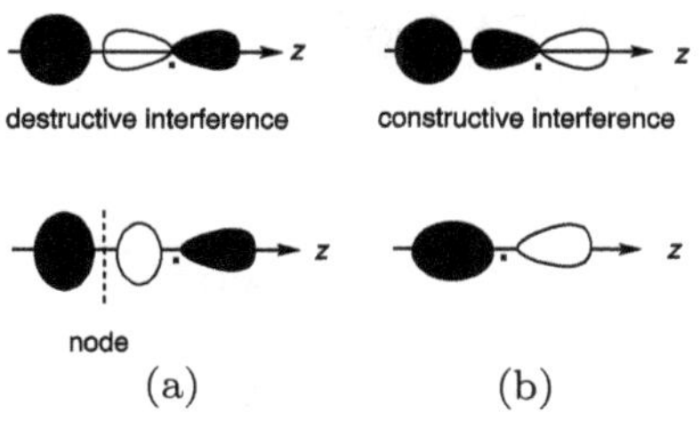

destructive interference   constructive interference

node
(a)   (b)

Fig. 3.2 The orbital interactions between s and p_z orbitals.

S_{12} is the same for both MO's and is inherently negative in sign because the s orbital overlaps the negative p_z lobe. Now, for the antibonding MO the overlap term gives a negative contribution to the probability, whereas for the bonding MO the overlap contribution is positive. Because the sum of c_1, c_2, and $2c_1c_2S_{12}$ must be equal to 1 in both cases, we find that,

$$c_1^2 + c_2^2 \text{ (antibonding case)} > 1 > c_1^2 + c_2^2 \text{ (bonding case)}$$

For the bonding situation the electron probability in the AO's is reduced below the value of unity that is appropriate to the case in which the orbitals do not interact at all (the nonbonding case, $S_{12} = 0$), whereas the antibonding situation results in greater electron probability in the AO's than for the nonbonding case.

---

1) Two non-interacting orbitals have $S = 0$.
2) Two bonding orbitals have $c_1c_2S > 0$ and $c_1^2 + c_2^2 < 1$ by the amount $2c_1c_2S$.
3) Two antibonding orbitals have $c_1c_2S < 0$ and $c_1^2 + c_2^2 > 1$ by the amount $2c_1c_2S$.

Let us consider LCAO involving two *basis* orbitals d_{yz} (ϕ_a) and p_z (ϕ_b) (Fig. 3.1(c)). Here we provide arbitrary illustration involving two AO's $\phi_a + \phi_b$, which have the correct symmetry to overlap. Their interaction is expressed by a linear combination.

Then, Ψ_{MO} (bonding orbital) $= N (\phi_a + \lambda\phi_b)$ and Ψ^*_{MO} (antibonding orbital) $= N^*(\lambda\phi_a - \phi_b)$, where N and N^* are normalization constants, and λ represents admixture coefficient of the AO's in the MO's, which could assume a value in the range $1 \geq \lambda \geq 0$.

When $\lambda = 0$, $\Psi_{MO} \approx \phi_a$ and $\Psi^*_{MO} \approx -\phi_b$. It means no orbital interaction i.e. ionic bonding, no covalence.

When $\lambda = 1$,

$$\Psi_{MO} = N(\phi_a + \phi_b) \text{ and } \Psi^*_{MO} = N^*(\phi_a - \phi_b) \tag{1}$$

It means maximum covalence. These two cases of MO formation are represented in Fig. 3.3.

If $\lambda = 0.5$ (see below),

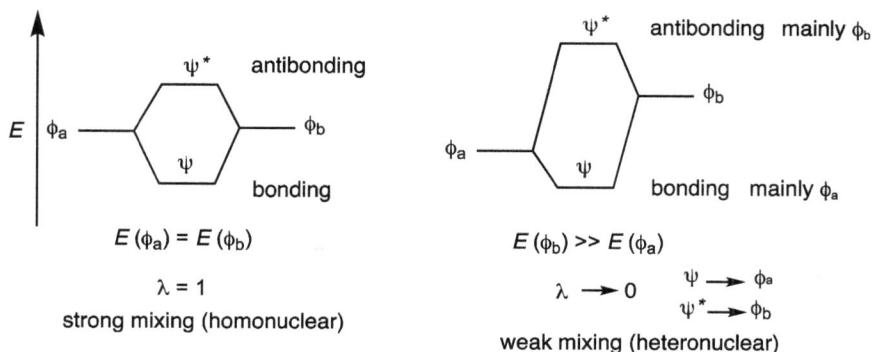

Fig. 3.3 The two cases of MO formation due to orbital interactions.

$\Psi_{MO} = \{1/(1.25)^{1/2}\}(\phi_a + 0.5\phi_b); \phi_a = 80\% \text{ and } \phi_b = 20\%$
$\Psi^*_{MO} = \{1/(1.25)^{1/2}\}(0.5\phi_a - \phi_b); \phi_a = 20\% \text{ and } \phi_b = 80\%$

It should be kept in mind that the square of a normalized wave functions (MO's) and normalized AO's, both are unity.

Thus,

$$\int \phi_a^2 d\tau = 1, \quad \int \phi_b^2 d\tau = 1, \quad \int \Psi^2_{MO} d\tau = 1, \quad \text{and} \quad \int \Psi^{*2}_{MO} d\tau = 1 \tag{2}$$

For 'in-phase' (constructive overlap between two AO's) combination,

From eq 1 and eq 2, $\Psi^2_{MO} d\tau = N^2(\int \phi_a + \lambda \phi_b)^2 d\tau = N^2(\int \phi_a^2 d\tau + 2\lambda \int \phi_a \phi_b d\tau + \lambda^2 \int \phi_b^2 d\tau) = 1$

$\Psi^2_{MO} d\tau = N^2(1 + 2\lambda S + \lambda^2) = 1$, where $S = \int \phi_a \phi_b \, d\tau$

Then,

$$N = 1/(1 + 2\lambda S + \lambda^2)^{1/2} = 1/(2 + 2S)^{1/2} \text{ (for } \lambda = 1) = 1/\{2(1 + S)\}^{1/2} \tag{3}$$

Then,

$$\Psi_{MO} = \{1/(1 + 2\lambda S + \lambda^2)^{1/2}\}(\phi_a + \lambda \phi_b) \tag{4}$$

S is $+$ (Fig. 3.1(c)); eq 4 signifies build-up of electron density between the nuclei. This implies bonding situation.

Then eq 4 simplifies to (from eq 3),

$$\Psi_{MO} = 1/\{2(1+S)\}^{1/2}(\phi_a + \phi_b) \tag{5}$$

Neglecting S,

$$\Psi_{MO} = (1/\sqrt{2})(\phi_a + \phi_b) \tag{6}$$

Similarly, for 'out-of phase' (destructive overlap between two AO's) combination,

$$\Psi_{MO}^{*2}d\tau = N^{*2}(\lambda\phi_a - \phi_b)^2 d\tau$$

$$= N^{*2}\left(\lambda^2\int\phi_a^2 d\tau - 2\lambda\int\phi_a\phi_b d\tau + \int\phi_b^2 d\tau\right) = 1$$

$$= N^{*2}(\lambda^2 - 2\lambda S + 1) = 1$$

$$N^* = 1/(1 - 2\lambda S + \lambda^2)^{1/2} = 1/\{(2 - 2S)^{1/2}\}$$

$$(\text{for } \lambda = 1) = 1/\{2(1-S)\}^{1/2} \tag{7}$$

Then,

$$\Psi_{MO}^* = \{1/(1 - 2\lambda S + \lambda^2)^{1/2}\}(\lambda\phi_a - \phi_b) \tag{8}$$

S is still $+$; eq 8 signifies decrease in electron density between the nuclei. This implies antibonding situation.

From eq 4 and eq 8, including S, N $[= 1/(1 + 2\lambda S + \lambda^2)^{1/2}]$ is less than N^* $[= 1/(1 - 2\lambda S + \lambda^2)^{1/2}]$. It means that the antibonding orbital is raised more than the bonding orbital is lowered.

From eq 7, eq 8 simplifies to,

$$\Psi_{MO}^* = 1/\{2(1-S)\}^{1/2}(\phi_a - \phi_b) \tag{9}$$

Neglecting S,

$$\Psi_{MO}^* = (1/\sqrt{2})(\phi_a - \phi_b) \tag{10}$$

The energies of the MO's are given by, $E = \int\Psi H\Psi d\tau$

Now in any MO wave function the square of the coefficient of any combined (basis) orbital represents the fractional contribution of that orbital to the MO. Neglecting the overlap integral (S), which is numerically small, from eq 4,

ϕ_a contributes $1/(1+\lambda^2)$ to Ψ_{MO} and from eq 8, ϕ_a contributes $\lambda^2/(1+\lambda^2)$ to Ψ_{MO}^*

Thus, total contribution of ϕ_a to Ψ_{MO} and $\Psi_{MO}^* = (1+\lambda^2)/(1+\lambda^2) = 1$

Similarly, total contribution of ϕ_b to Ψ_{MO} and $\Psi_{MO}^* = 1$

In the applications to follow, the S will be neglected so that for an LCAO, the normalization constant is the square root of the sum of the squares of the orbital coefficients (c_1, c_2 etc.).

In the hydrogen atom, the $1s$ atomic orbital has the lowest energy, while the remaining orbitals ($2s$, $2p_x$, $2p_y$, and $2p_z$) are of equal energy (i.e., degenerate). For all other atoms, the $2s$ atomic orbital is of lower energy than the degenerate $2p_x$, $2p_y$, and $2p_z$ orbitals. In atoms, electrons occupy AO's, but in molecules they occupy MO's, which surround the molecule.

(c) Examples of MO approach to diatomic molecules

Any diatomic/triatomic/tetratomic molecule can be considered to have a generic form AB_n, in which A represents central atom, B's are the peripheral atom(s) and n denotes the number of such atom(s). This would help to build up the concept of ligand group orbitals (LGO's), which in turn will help to understand structure (molecular and electronic) and bonding from MO approach. To better appreciate LGO's, it is important to understand that the orbitals of the peripheral atoms (ligands) of a central atom cannot be treated independently, because they transform under the symmetry operations of the molecular point group concerned. A relatively straightforward, stepwise method to better understand the usefulness of MO theory in inorganic chemistry is described in this chapter.

Let us start the MO approach with the simplest molecule H_2. Two $1s$ AO's of two H atoms and hence there are two MO's for H_2, the lower energy orbital has greater electron density between the two nuclei. This is the bonding MO and is of lower energy than the two $1s$ atomic orbitals of H atoms. This makes the bonding MO more stable than two separated AO's. The upper MO has a node in the electronic wave function and the electron density is low between the two positively charged nuclei. The energy of the upper orbital is greater than that of the $1s$ atomic orbitals, and such an orbital is called an antibonding MO. The energy levels in a H_2 molecule

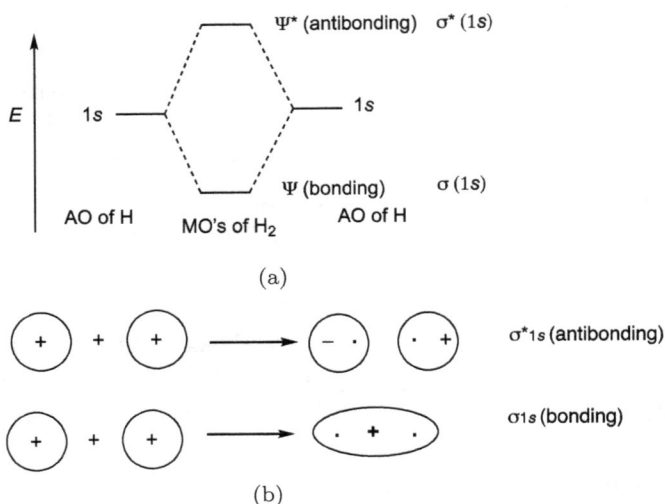

Fig. 3.4 (a) MO energy level diagram of H_2, showing bonding and antibonding interaction of atomic orbitals of two hydrogens, and (b) forms of MO's (. represents nuclei).

can be represented in a diagram (Fig. 3.4), showing how the two $1s$ atomic orbitals combine to form two MO's, one bonding (σ) and one antibonding (σ^*).

The next step in constructing an MO diagram is filling the newly formed MO's with electrons. Three general rules apply:

- The Aufbau principle states that orbitals are filled starting with the lowest energy.
- The Pauli exclusion principle states that the maximum number of electrons occupying an orbital is two, with opposite spins.
- Hund's rule states that when there are several MO's with equal energy, the electrons occupy the MO's one at a time before two occupy the same MO.

In H_2 the two electrons in H_2 occupy the bonding MO with anti-parallel spins (Pauli's exclusion principle).

(i) Homonuclear diatomic molecules

It has been discussed that atomic orbitals can interact with each other in-phase, out-of-phase, and no electron density between the two nuclei. In the bonding MO, with energy much lower than the original AO's, the electrons are present in between the interacting nuclei to minimize internuclear repulsion. In the antibonding MO, with energy much higher than the original AO's, any electron present is in lobes pointing away from the central internuclear axis. For a corresponding σ-bonding orbital, such an orbital would be symmetrical, but are differentiated from it by an asterisk, as in σ^*. For a π-bond (the interacting orbitals are perpendicular to the internuclear axis), corresponding bonding and antibonding orbitals would not have such symmetry around the bond axis and are designated π and π^*, respectively.

The filled MO that is highest in energy is called the *highest occupied molecular orbital* (HOMO) and the empty MO just above it is the *lowest unoccupied molecular orbital* (LUMO). The electrons in the bonding MO's are called bonding electrons, and any electrons in the antibonding orbital are called antibonding electrons.

Let us note the following. (i) The σ bonding orbital is lowest in energy due to the greater end-on-end overlap. (ii) Electrons preferentially occupy MO's that are lower in energy. (iii) If two electrons occupy the same MO, they must be spin-paired. (iv) When occupying degenerate MO's, electrons occupy separate orbitals with parallel spins, before pairing. (iv) Bond order (BO) is defined as half the difference between the number of bonding and antibonding electrons. Stable bonds have a positive bond order. Bond order is an index of bond strength and is used extensively in valence bond theory.

According to Fig. 3.5, electron distribution in diatomic molecules from H_2 to F_2:

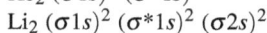

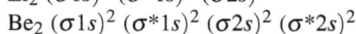

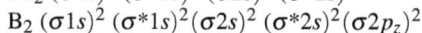

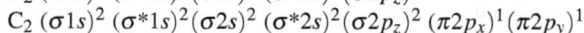

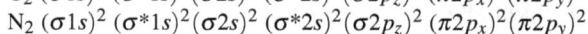

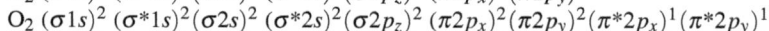

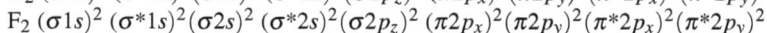

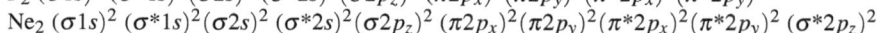

H_2 $(\sigma 1s)^2$

He_2 $(\sigma 1s)^2$ $(\sigma^* 1s)^2$

Li_2 $(\sigma 1s)^2$ $(\sigma^* 1s)^2$ $(\sigma 2s)^2$

Be_2 $(\sigma 1s)^2$ $(\sigma^* 1s)^2$ $(\sigma 2s)^2$ $(\sigma^* 2s)^2$

B_2 $(\sigma 1s)^2$ $(\sigma^* 1s)^2 (\sigma 2s)^2$ $(\sigma^* 2s)^2 (\sigma 2p_z)^2$

C_2 $(\sigma 1s)^2$ $(\sigma^* 1s)^2 (\sigma 2s)^2$ $(\sigma^* 2s)^2 (\sigma 2p_z)^2$ $(\pi 2p_x)^1 (\pi 2p_y)^1$

N_2 $(\sigma 1s)^2$ $(\sigma^* 1s)^2 (\sigma 2s)^2$ $(\sigma^* 2s)^2 (\sigma 2p_z)^2$ $(\pi 2p_x)^2 (\pi 2p_y)^2$

O_2 $(\sigma 1s)^2$ $(\sigma^* 1s)^2 (\sigma 2s)^2$ $(\sigma^* 2s)^2 (\sigma 2p_z)^2$ $(\pi 2p_x)^2 (\pi 2p_y)^2 (\pi^* 2p_x)^1 (\pi^* 2p_y)^1$

F_2 $(\sigma 1s)^2$ $(\sigma^* 1s)^2 (\sigma 2s)^2$ $(\sigma^* 2s)^2 (\sigma 2p_z)^2$ $(\pi 2p_x)^2 (\pi 2p_y)^2 (\pi^* 2p_x)^2 (\pi^* 2p_y)^2$

Ne_2 $(\sigma 1s)^2$ $(\sigma^* 1s)^2 (\sigma 2s)^2$ $(\sigma^* 2s)^2 (\sigma 2p_z)^2$ $(\pi 2p_x)^2 (\pi 2p_y)^2 (\pi^* 2p_x)^2 (\pi^* 2p_y)^2$ $(\sigma^* 2p_z)^2$

(a)

(b)

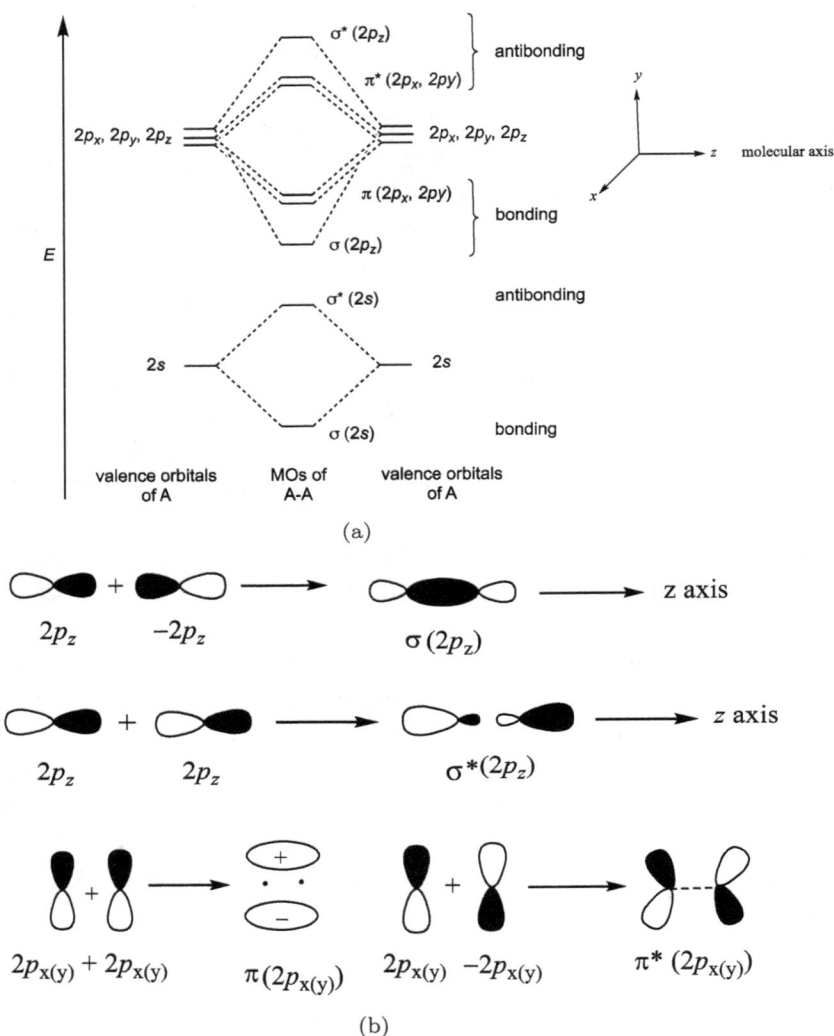

Fig. 3.5 (a) MO energy level diagram of homonuclear diatomic molecules of the first period of the periodic table, showing bonding and antibonding interaction of atomic orbitals of two atomic orbitals, and (b) forms of MO's (. represents nuclei).

Bond order:

H_2 (2–0)/2 = 1; He_2 (2–2)/2 = 0; Li_2 (2–0)/2 = 1; Be_2(2–2)/2 = 0; B_2(2–0)/2 = 1; C_2 (4–0)/2 = 2; N_2 (6–0)/2 = 3; O_2 (6–2)/2 = 2; F_2(6–4)/2 = 1; Ne_2 (6–6)/2 = 0.

The experimental result is that B_2 is paramagnetic with two unpaired electrons (according to Fig. 3.5 it should be diamagnetic) and BO of 1, and C_2 is diamagnetic (according to Fig. 3.5 it should be paramagnetic with two unpaired electrons) and BO of 2. These experimental facts suggest that we should consider the mixing of σ orbitals belonging to the same representation. The net effect of the mixing is that the lower-energy σ orbital $(s+s)$ is lowered still further and the higher-energy σ orbital $(2pz+2pz)$ is raised above the level of the π orbitals. In other words, s-p interaction results in interchange of the ordering between $\sigma 2pz$ and $\pi 2px,y$ in Fig. 3.5.

Let us consider two relevant molecular orbitals $\sigma 2s$ and $\sigma 2pz$ for s-p interaction. Keeping in mind that A_2 molecules belong to $D_{\infty h}$ point group, we should examine the effect of symmetry operations on these two orbitals. If the energy difference between the two orbital sets is small and if their symmetry match (for $\sigma 2s$ and $\sigma 2pz$ molecular orbitals all symmetry operations of $D_{\infty h}$ point group give rise to identical characters), then they can interact giving rise to a modified MO diagram for homonuclear diatomic molecules (Fig. 3.6).[1]

The extent of orbital mixing between orbitals is *inversely* proportional to their energy separation. The energy gap between the $2s$ and $2p$ orbital in the elements becomes larger as the atomic number increases. In H (or in a H-like ion), there is no energy difference between $2s$ and $2p$. The energy of an orbital is dependent only on the principal quantum number n. In atoms with more than one electron, $2s$ is lower in energy than $2p$. As the nuclear charge increases across the Li – F period, both $2s$ and $2p$ orbitals lower in energy due to the increased nuclear charge, but the $2s$ orbital is affected to a larger degree. An electron in a $2s$ orbital is less shielded by the other electrons than an electron in a $2p$ orbital. The $2s$ electron experiences a higher nuclear charge and drops to lower energy. Thus, the difference between $2s$ and $2p$ orbitals is larger in F than in Li and consequently the importance of s-p mixing is smaller in F.

---

[1] A. Haim, *J. Chem. Educ.* **1991**, *68*, 737.

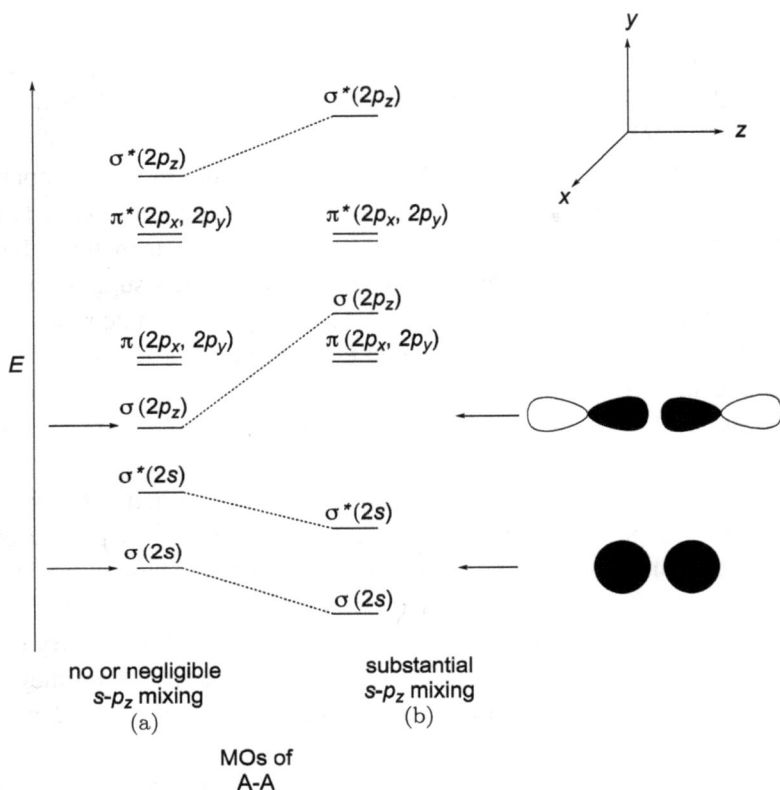

Fig. 3.6   Modified MO energy-level diagram for B_2, C_2, and N_2.

According to Fig. 3.6, the electron distribution in B_2, C_2, and N_2 are:

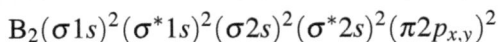

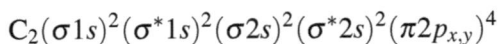

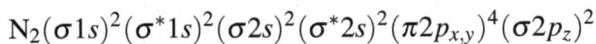

$$B_2(\sigma 1s)^2(\sigma^* 1s)^2(\sigma 2s)^2(\sigma^* 2s)^2(\pi 2p_{x,y})^2$$

$$C_2(\sigma 1s)^2(\sigma^* 1s)^2(\sigma 2s)^2(\sigma^* 2s)^2(\pi 2p_{x,y})^4$$

$$N_2(\sigma 1s)^2(\sigma^* 1s)^2(\sigma 2s)^2(\sigma^* 2s)^2(\pi 2p_{x,y})^4(\sigma 2p_z)^2$$

The electron distribution in B_2, C_2, and N_2 is in conformity with experimental result. According to both with (Fig. 3.6) and without (Fig. 3.5) s-p mixing, the N_2 molecule is diamagnetic with BO $= 3$. The significant experimental result is that N_2^+ has a single σ electron, so for N_2^+, and presumably for N_2, the energy-level ordering involving s-p mixing in Fig. 3.6 justifies.

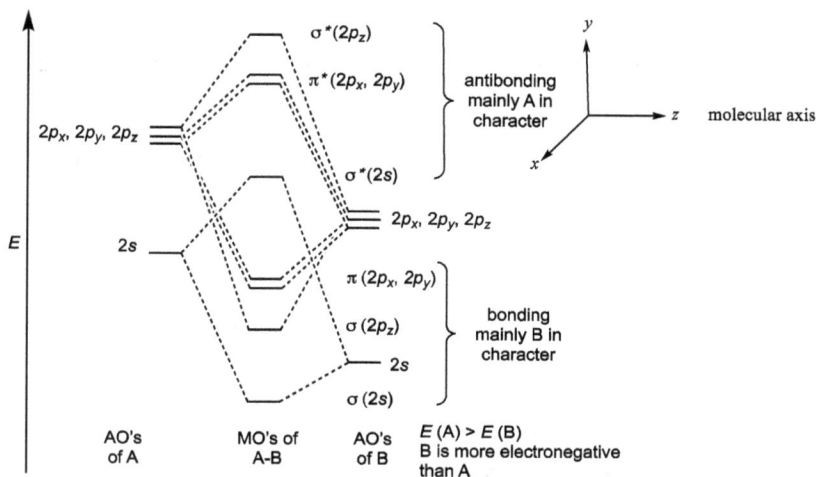

Fig. 3.7 MO energy-level diagram for CO, NO, CN⁻.

(ii) Heteronuclear diatomic molecule

Due to difference in the electronegativity between A and B, the two MO's $\sigma 2pz$ and degenerate $\pi 2px$ and $\pi 2py$ orbital set are below $\sigma*2s$ MO. The corresponding antibonding orbitals are destabilized accordingly (more stabilized bonding MO is expected to result in more destabilized antibonding MO).

According to Fig. 3.7, the electron distribution in CO, NO, and CN⁻ are:

CO (10 valence electrons) : $(\sigma 2s)^2(\sigma 2p_z)^2(\pi 2p_{x,y})^4(\sigma^*2s)^2$

$\quad$ (BO $= (8-2)/2 = 3$)

NO (11 valence electrons) : $(\sigma 2s)^2(\sigma 2p_z)^2(\pi 2p_{x,y})^4(\sigma^*2s)^2(\pi*2p_{x,y})^1$

$\quad$ (BO $= (8-3)/2 = 2.5$)

CN⁻ (10 valence electrons) : $(\sigma 2s)^2(\sigma 2p_z)^2(\pi 2p_{x,y})^4(\sigma^*2s)^2$

$\quad$ (BO $= (8-2)/2 = 3$)

Example 3.1 The bond lengths (Å) in B_2, C_2, N_2, O_2, and F_2 are 1.59, 1.24, 1.10, 1.21, 1.44, respectively. Rationalize the experimental result.

Answer

According to Fig. 3.6 (*s-p* mixing) the electron distribution in B_2, C_2, and N_2 are:

$$B_2 \cdots (\pi 2p_{x,y})^2; \quad BO = 1; \quad \text{Bond length} = 1.59\,\text{Å}$$

$$C_2 \cdots (\pi 2p_{x,y})^4; \quad BO = 2; \quad \text{Bond length} = 1.24\,\text{Å}$$

$$N_2 \cdots (\pi 2p_{x,y})^4 (\sigma 2p_z)^2; \quad BO = 3; \quad \text{Bond length} = 1.10\,\text{Å}$$

According to Fig. 3.5 (no *s-p* mixing) the electron distribution in O_2 and F_2 are:

$$N_2 \cdots (\sigma 2p_z)^2 (\pi 2p_{x,y})^4; \quad BO = 3; \quad \text{Bond length} = 1.10\,\text{Å}$$

$$O_2 \cdots (\sigma 2p_z)^2 (\pi 2p_{x,y})^4 (\pi^* 2p_{x,y})^2; \quad BO = 2; \quad \text{Bond length} = 1.21\,\text{Å}$$

$$F_2 \cdots (\sigma 2p_z)^2 (\pi 2p_{x,y})^4 (\pi^* 2p_{x,y})^4; \quad BO = 1; \quad \text{Bond length} = 1.44\,\text{Å}$$

Example 3.2 The bond lengths (Å) in O_2, O_2^+, and O_2^- are 1.21, 1.12, and 1.28, respectively. Rationalize the experimental result.

Answer

According to Fig. 3.5 the electron distribution in O_2, O_2^+, and O_2^- are:

$$O_2 \cdots (\sigma 2p_z)^2 (\pi 2p_{x,y})^4 (\pi^* 2p_{x,y})^2; \quad BO = 2; \quad \text{Bond length} = 1.21\,\text{Å}$$

$$O_2^+ \cdots (\sigma 2p_z)^2 (\pi 2p_{x,y})^4 (\pi^* 2p_{x,y})^1; \quad BO = 2.5; \quad \text{Bond length} = 1.12\,\text{Å}$$

$$O_2^- \cdots (\sigma 2p_z)^2 (\pi 2p_{x,y})^4 (\pi^* 2p_{x,y})^3; \quad BO = 1.5; \quad \text{Bond length} = 1.28\,\text{Å}$$

The number of electron(s) in the antibonding orbital follows the trend: O_2^+ < O_2 < O_2^-. This in turn increases BO and hence bond length.

Example 3.3 The bond energies (kcal/mol) in B_2, C_2, N_2, O_2, and F_2 are 69, 144, 225, 118, and 36, respectively. Rationalize the experimental result.

Answer

According to Answer of Example 3.1 the BO in B_2, C_2, and N_2 are 1, 2, and 3, respectively. This data is in conformity with bond energy trend.

Similarly, the BO in O_2 and F_2 are 2 and 1, respectively. This data also rationalizes the bond energy trend.

3.2 The electronic structure by MO theory

For molecular orbital treatment it is the valence orbitals, which are important.

To understand the electronic structure by MO theory the following steps are to be followed.

- Identify the point group of the molecule concerned (homonuclear diatomics: H_2, N_2 O_2 etc $D_{\infty h}$; heteronuclear diatomics: CO, NO etc. $C_{\infty v}$; triatomic/tetratomic molecules: BeH_2 $D_{\infty h}$, H_2O C_{2v}, NH_3 C_{3v}, BF_3 D_{3h} etc.)
- Identify the atomic orbitals which are available to form bonds (the valence AO's)
- Consider symmetry elements of the point group, transform the atomic orbitals and label the symmetry operations
- When the symmetries of the atomic orbitals of the central atom and the peripheral atoms are the same and the orbitals are close in energy, construct the MO diagram
- Number of AO's = Number of MO's
- The strength of the bond depends upon the degree of orbital overlap
- Draw forms of all the molecular orbitals of the molecule

(i) Diatomic molecule: AB type (HF; linear, $C_{\infty v}$)

In HF, the F $2s$ and $2p_z$ (internuclear axis: z) orbitals have the proper symmetry for forming σ bonds with the H $1s$ orbital. The F $2s$ orbital is so much lower in energy than the H $1s$ orbital that this orbital is effectively nonbonding. The F $2pz$ orbital has the proper symmetry and reasonable energy for σ bonding interaction. Since the F $2pz$ orbital is considerably lower in energy than the H $1s$ orbital due to difference in electronegativity, an unsymmetrical energy-level diagram results (Fig. 3.8). The bonding σ MO is closer in energy to the F $2pz$ orbital (more F character) and the σ^* MO is closer in energy to the H $1s$ orbital. It means that in HF$^-$, the electron density for σ^* MO orbital would be greater near H. The F $2px$ and $2py$ orbitals are nonbonding.

(ii) Triatomic molecule: AB$_2$ type, σ bonding (BeH$_2$; linear, $D_{\infty h}$)

Let us consider the character table (partial) for $D_{\infty h}$ (Table 3.1).

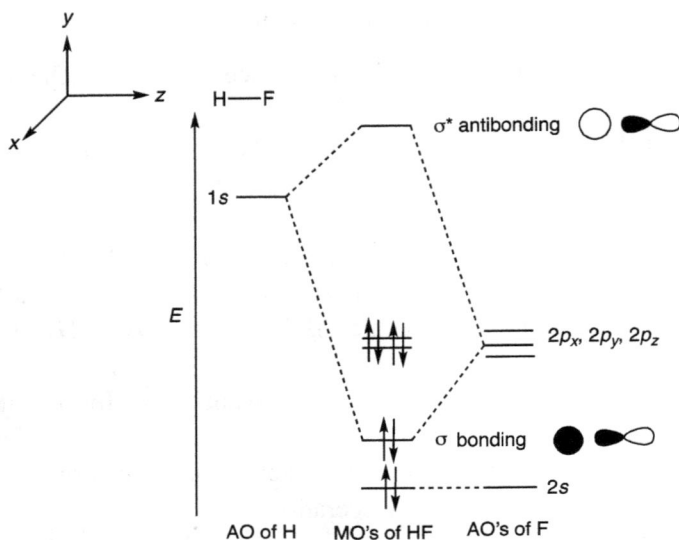

Fig. 3.8   Qualitative MO energy-level diagram for HF.

Table 3.1   Partial character table for point group $D_{\infty h}$

$D_{\infty h}$/symmetry	E	$2C_{\infty}^{\phi}$	$\infty\sigma_v$	i	$2S_{\infty}^{\phi}$	∞C_2	
Σ_g^+	+1	+1	+1	+1	+1	+1	s
Σ_u^+	+1	+1	+1	−1	−1	−1	z
π_u	2	$2\cos\phi$	0	−2	$2\cos\phi$	0	(x,y)

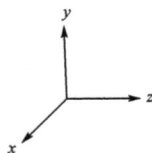

H—Be—H

The AO's (valence orbitals) on the Be central atom are $2s$, $2p_x$, $2p_y$, and $2p_z$. It should be appreciated that the orbitals $2p_x$ and $2p_y$ are perpendicular to the internuclear axis (z axis) and symmetry operations $2C_{\infty}^{\phi}$ (two operations: one clockwise rotation and the other anticlockwise) on $2p_x$ and $2p_y$ will transform to themselves (symmetry operation on p_x will transform to p_y and symmetry operation on p_y will transform to p_x) and hence will be degenerate. The symmetry representations of $2s$ and $2p_z$ orbitals are Σ_g^+ and Σ_u^+, respectively [sign of the lobe (s orbital is spherically symmetric) of the p orbital (wave function) does not change with respect to $2C_\infty^\phi$ and is represented by Σ; changes to itself $+/-$ with respect to $\infty\sigma_v$, represented by superscript '+'; sign of the lobe of the orbital changes $+/-$ to itself with respect to i, represented by g/u, respectively].

> The convention is the use of lowercase letters for the symmetry species of orbitals and uppercase letters are used for representations in the general sense and for energy states.

Using the concept of LGO's, the two linear combinations of H $1s$ orbitals, $+/+$ combination and $+/-$ combinations are represented as Φ_1 and Φ_2, respectively.

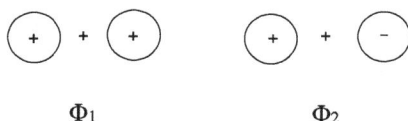

$$\Phi_1 \qquad\qquad \Phi_2$$

The next step is to apply symmetry operations of $D_{\infty h}$ on Φ_1 and Φ_2 and assign the symmetry representations (representation as that of valence orbital; no change: $+1$, sign of the wave function changes to itself: -1).

$D_{\infty h}$/symmetry	E	$2C_\infty^\phi$	$\infty\sigma_v$	i	$2S_\infty^\phi$	∞C_2	
Σ_g^+	$+1$	$+1$	$+1$	$+1$	$+1$	$+1$	Φ_1
Σ_u^+	$+1$	$+1$	$+1$	-1	-1	-1	Φ_2

For the two LGO's Φ_1 and Φ_2, the symmetry representations are Σ_g^+ and Σ_u^+, respectively.

Now considering the two basic requirements of formation of MO energy-level diagram, energy matching and symmetry matching, the LGO's are combined with orbitals on Be that have the same symmetry. Since $2p_x$, $2p_y$ orbitals on Be do not match the symmetry of the LGO's of two H's, it will remain nonbonding and degenerate. The other orbitals on Be combine with the appropriate LGO's to form bonding and antibonding MO's. The MO diagram of BeH_2 is shown in Fig. 3.9.

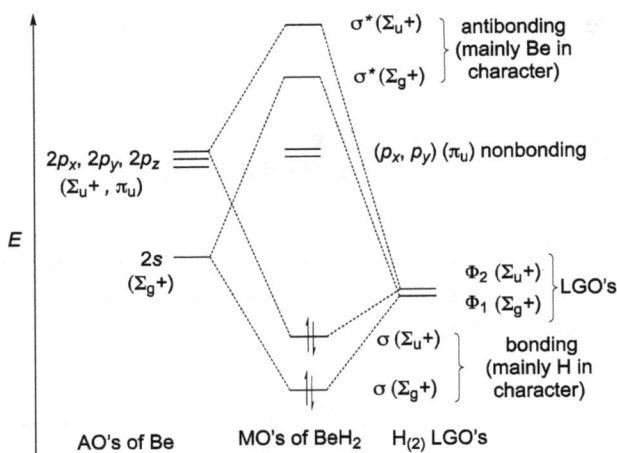

Fig. 3.9   Qualitative MO energy-level diagram for BeH_2 ($D_{\infty h}$).

(iii) Triatomic molecule: AB_2 type, σ and π bonding (CO_2; linear, $D_{\infty h}$)

In the case of CO_2, the O $2s$ and $2p_z$ orbitals have the proper symmetry for forming σ bonds with the C $2s$ and $2p_z$ orbitals. The O $2s$ orbital is so much lower in energy than the C $2s$ orbital that this orbital is effectively remains nonbonding. The orbitals on two O's can combine with orbitals on C that have the same symmetry (C $2s$ with O $2p_z$ and C $2p_z$ with O $2p_z$) and we can identify the σ bonding interactions. For π bonding interactions, the $2p_x$, $2p_y$ orbitals on C can combine with the appropriate orbitals on two O's to form bonding and antibonding MO's. The MO diagram of CO_2 is shown in Fig. 3.10(a) and the orbital interactions are displayed in Fig. 3.10(b)).

Total number of valence electron = 4 (from central C atom) + 2×6 (from two O atoms) = 16

The C–O bonds (two σ and two π): $(s\sigma)^2 \ (p\sigma)^2 \ (p\pi)^4 \ [\approx sp$ hybrids] and two pairs of nonbonding electrons

Hence, the bond order for C–O bond

$\qquad = 2\,$pair of electrons$/2\,$bonds$(\sigma$ bonds$)$

$\qquad +2$ pair of electrons$/2$ bonds $(\pi$ bonds$)$

$\qquad = 1\,$pair (for σ bond) $+ 1$ pair (for π bond) $= 2$

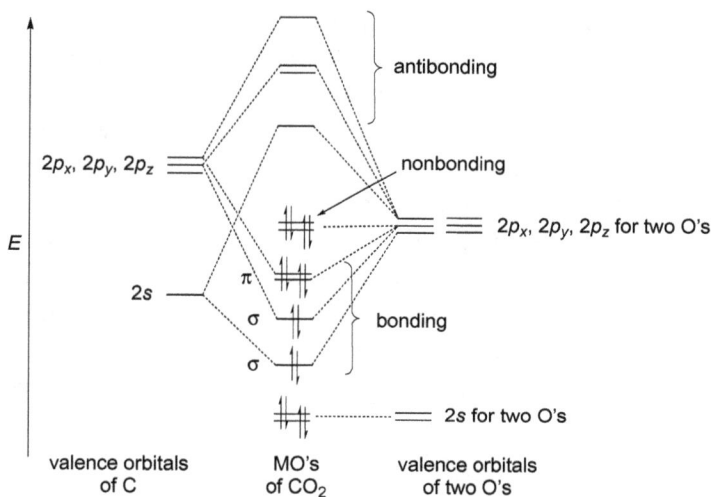

(a)

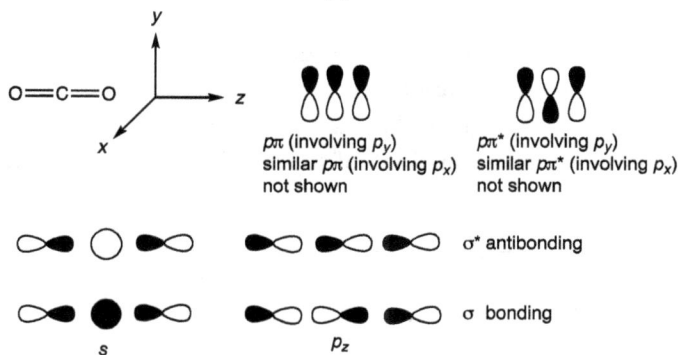

(b)

Fig. 3.10   (a) Qualitative MO energy-level diagram for CO_2 ($D_{\infty h}$) and (b) the forms of MO's.

(iv) Triatomic molecule: AB_2E_2 type (H_2O; V-shape/bent, C_{2v})

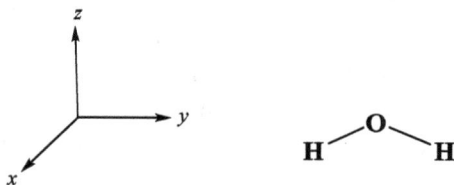

Approach 1: Symmetry considerations

Let us consider the character table for C_{2v} (Table 2.2). The AO's (valence orbitals) on the O central atom are $2s$, $2p_x$, $2p_y$, and $2p_z$. The symmetry representations of these orbitals are a_1, b_1, b_2, and a_1, respectively (no change is represented by $+1$, sign of the lobe of the orbital (wave function) changes to itself, represented by -1; sign of the lobe of the orbital (wave function) changes to itself $+/-$ with respect to $C_2^{(z)}$, represented by a/b, respectively; sign of the lobe of the orbital (wave function) changes to itself $+/-$ with respect to $\sigma_v(xz)$, represented by 1/2, respectively.

Using the concept of LGO's, the two linear combinations of H 1s orbitals, $+/+$ combination and $+/-$ combinations, are represented as Φ_1 and Φ_2, respectively:

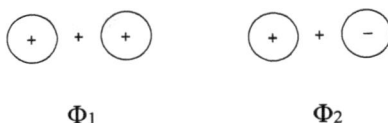

$$\Phi_1 \qquad\qquad\qquad \Phi_2$$

The next step is to apply symmetry operations of C_{2v} on Φ_1 and Φ_2 and assign the symmetry representations (representation as that of valence orbital; no change: $+1$, sign of the wave function changes to itself: -1). Thus, the symmetry representations of the two LGO's Φ_1 and Φ_2, are a_1 and b_2, respectively.

C_{2v}	E	$C_2^{(z)}$	$\sigma_v(xz)$	$\sigma_v(yz)$	
a_1	$+1$	$+1$	$+1$	$+1$	Φ_1
b_2	$+1$	-1	-1	$+1$	Φ_2

Now considering energy matching and symmetry matching, these LGO's are combined with orbitals on O that have the same symmetry. Since $2p_x$ orbital on O does not match the symmetry of the LGO's of two H's, it will remain nonbonding. The other orbitals on O will combine with the appropriate LGO's to form bonding and antibonding MO's. The qualitative MO diagram of H_2O and the forms of MO's are shown in Fig. 3.11.

We note that the number of AO's, including LGO's, is six and it is equal to the number of MO's, which is six. The HOMO ($1b_1$) is nonbonding and mainly localized on the O atom. The next lowest MO ($2a_1$) can be thought of primarily a nonbonding orbital, as it has a lobe pointing away from the

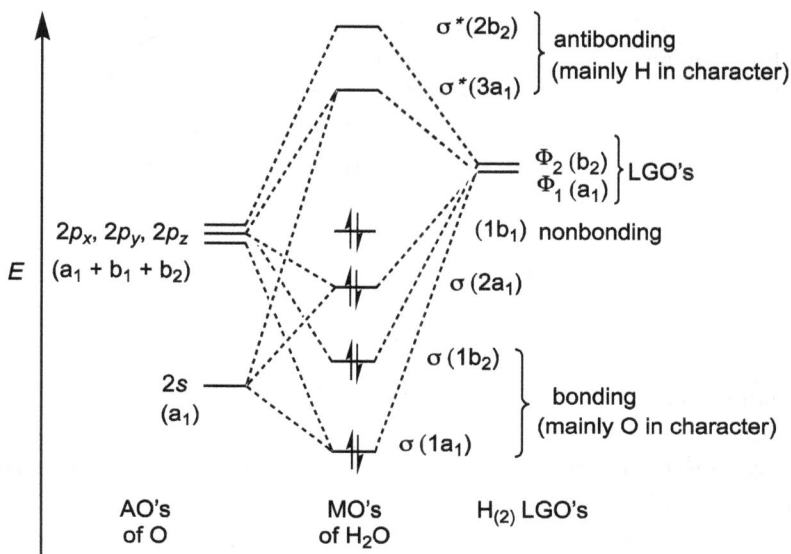

(a)

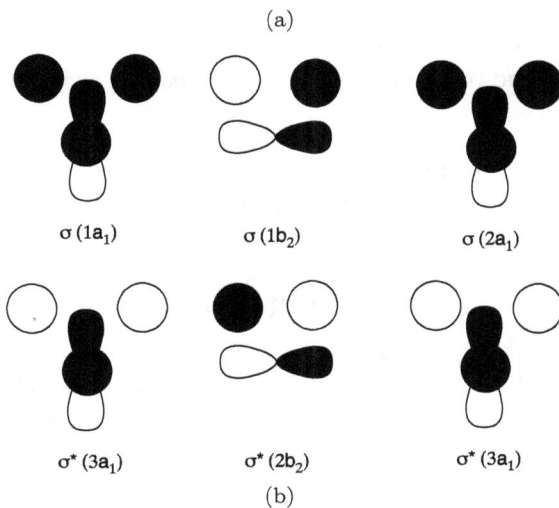

(b)

Fig. 3.11 (a) Qualitative MO energy-level diagram of H_2O (C_{2v}) and (b) forms of MO's.

two H's (see Fig. 3.11). There are two next lower in energy MO's which are bonding in character. It should be kept in mind that H_2O can have only two bonding interactions with two H's. From the lower energy bonding MOs, it is understandable that σ ($1a_1$) and σ ($1b_2$) have more O character.

Now, $\sigma(1a_1) = c_1(2s) + c_2(2p_z) + c_3(\Phi_1)$, but based on energy consideration this MO is mainly O ($2s$). Therefore, $\sigma(1a_1) = \{1/(1+\lambda^2)\}(2s + \lambda\Phi_1)$.

Similarly, $\sigma(1b_2) = \{1/(1+\lambda^2)\}(2p_y + \lambda\Phi_2)$.

$$\sigma(2a_1) = c_4(2s) + c_5(2p_z) + c_6(\Phi_1) \approx \{1/(1+\lambda^2)\}(2p_z + \lambda\Phi_1).$$

The admixture coefficient (λ) is obtained from quantum mechanical calculations.

Total number of valence electron $= 6$ (from central O atom) $+ 2$ (from two peripheral H atoms) $= 8$

Two O–H bonds: σ ($1a_1$)^2 $+ \sigma$ ($1b_2$)^2; two lone-pairs: b_1 (pure 'p') $+\sigma$ ($2a_1$) $[\approx (s+p)]$.

Total number of O–H bonds $= 2$

Hence, the bond order of O–H bonds $= 2$ pair of electrons$/2$ bonds $= 1$ (pair)$/1$ bond $= 1$

Approach 2: Group theoretical considerations

To find the reducible representations for the LGO's, the following well-known group theoretical reduction formula is used.

$$a_i = 1/h \sum_R g\chi_i(R)\chi_t(R)$$

where a_i is the number of times the i-th irreducible representation appears in the reducible representation, h is the order of the group, R represents the symmetry operation, g is the order of the class, $\chi_i(R)$ is the irreducible representation character in R and $\chi_t(R)$ is the reducible representation or total representation character in R. The orbitals that are going to combine linearly must have the same symmetry and be close in energy.

To identify the symmetries of the LGO's for the H atoms of H_2O, the internal coordinates for the σ bonds (σ_1 and σ_2) in Fig. 3.12 must be considered. The reducible representations for the LGO's are given by the

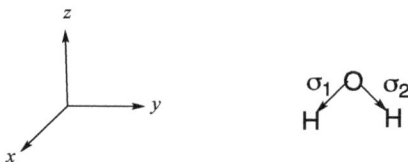

Fig. 3.12 Internal coordinates of H_2O.

number of internal coordinates (σ) that do not move when each of the different symmetry operations is applied. Let us consider the character table for C_{2v} (Table 2.2).

C_{2v}	E	$C_2^{(z)}$	$\sigma_v(xz)$	$\sigma_v(yz)$
Γ_{red}	2	0	0	2

$$a_{A1} = 1/4[(1 \times 2 \times 1) + (1 \times 0 \times 1) + (1 \times 0 \times 1) + (1 \times 2 \times 1)] = 1$$

$$a_{A2} = 1/4[(1 \times 2 \times 1) + (1 \times 0 \times 1) + (1 \times 0 \times -1) + (1 \times 2 \times -1)] = 0$$

$$a_{B1} = 1/4[(1 \times 2 \times 1) + (1 \times 0 \times -1) + (1 \times 0 \times 1) + (1 \times 2 \times -1)] = 0$$

$$a_{B2} = 1/4[(1 \times 2 \times 1) + (1 \times 0 \times -1) + (1 \times 0 \times -1) + (1 \times 2 \times 1)] = 1$$

[Remember for C_{2v}: Number of symmetry operations $= 4$. Entry within bracket: 1 (one E) $\times$ 2 ($\Gamma_{red} = 2$ for E) $\times$ 1 (for E operation, character $= 1$ from character table) (for a_1), 1 (one E) $\times$ 0 ($\Gamma_{red} = 0$ for $C_2^{(z)}$) $\times$ 1 (for $C_2^{(z)}$ operation, character $= 1$ from character table) (for a_1), 1 (one E) $\times$ 0 ($\Gamma_{red} = 0$ for $\sigma_v(xz)$) $\times$ 1 (for $\sigma_v(xz)$ operation character $= 1$ from character table) (for a_1), 1 (one E) $\times$ 2 ($\Gamma_{red} = 0$ for $\sigma_v(yz)$ $\times$ 1 (for $\sigma_v(yz)$ operation character $= 1$ from character table) (for a_1)]

Thus, the symmetries of the two LGO's $(+/+$ and $+/-$ combinations; see above) for H_2O are a_1 and b_2, respectively. It is evident from Table 2.2 that we can ignore the situation with a_2 and b_1, based on σ bonding interactions (entries of the last column). The results will be '0'.

Rest of the procedure is as discussed above.

(v) Tetratomic molecule: AB_3 type, σ bonding (BH_3; equilateral triangular planar, D_{3h})

Table 3.2   Character table for point group D_{3h}

D_{3h}	E	$2C_3$	$3C_2$	σ_h	$2S_3$	$3\sigma_v$		
A_1'	1	1	1	1	1	1		x^2+y^2, z^2
A_2'	1	1	−1	1	1	−1	R_z	
E'	2	−1	0	2	−1	0	(x,y)	(x^2-y^2, xy)
A_1''	1	1	1	−1	−1	−1		
A_2''	1	1	−1	−1	−1	1	z	
E''	2	−1	0	−2	1	0	(R_x, R_y)	(xz, yz)

D_{3h}	E	$2C_3$	$3C_2$	σ_h	$2S_3$	$3\sigma_v$
$\Gamma\sigma$	3	0	1	3	0	1

Although chemically BH_3 is a dimer B_2H_6, but we shall discuss the general approach for the simpler BH_3. Let us consider the molecular plane as xy plane. From the character table of D_{3h} (Table 3.2) and $\Gamma\sigma$,

$$a_{A1'} = 1/12[(1 \times 3 \times 1) + (2 \times 0 \times 1) + (3 \times 1 \times 1) + (1 \times 3 \times 1)$$
$$+(2 \times 0 \times 1) + (3 \times 1 \times 1)] = 1$$

$$a_{A2'} = 1/12[(1 \times 3 \times 1) + (2 \times 0 \times 1) + (3 \times 1 \times -1) + (1 \times 3 \times 1)$$
$$+(2 \times 0 \times 1) + (3 \times 1 \times -1)] = 0$$

$$a_{E'} = 1/12[(1 \times 3 \times 2) + (2 \times 0 \times -1) + (3 \times 1 \times 0) + (1 \times 3 \times 2)$$
$$+(2 \times 0 \times -1) + (3 \times 1 \times 0)] = 1$$

$$a_{A1''} = 1/12[(1 \times 3 \times 1) + (2 \times 0 \times 1) + (3 \times 1 \times 1) + (1 \times 3 \times -1)$$
$$+(2 \times 0 \times -1) + (3 \times 1 \times -1)] = 0$$

$$a_{A2''} = 1/12[(1 \times 3 \times 1) + (2 \times 0 \times 1) + (3 \times 1 \times -1) + (1 \times 3 \times -1)$$
$$+(2 \times 0 \times -1) + (3 \times 1 \times 1)] = 0$$

$$a_{E''} = 1/12[(1 \times 3 \times 2) + (2 \times 0 \times -1) + (3 \times 1 \times 0) + (1 \times 3 \times -2)$$
$$+(2 \times 0 \times 1) + (3 \times 1 \times 0)] = 0$$

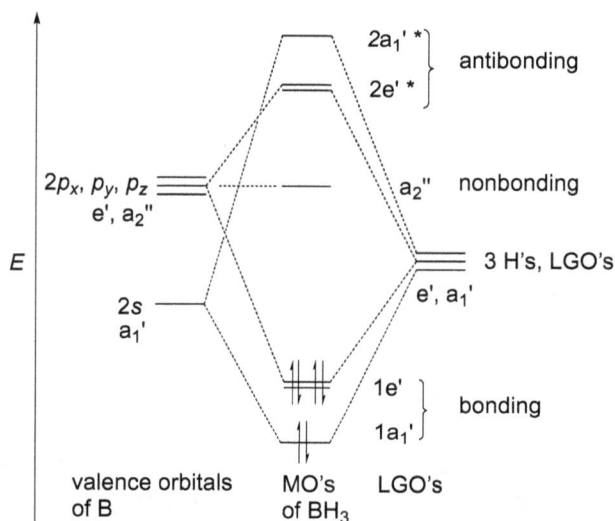

Fig. 3.13 Qualitative MO energy-level diagram for BH_3 (D_{3h}).

Thus, the symmetries of the LGO's for BH_3 are $a_1' + e'$. It is evident from Table 3.2 that we can ignore the situation with A_2', A_1'', A_2'', and E'', based on σ bonding interactions (entries of the last column). The results will be '0'. The qualitative MO diagram of BH_3 is shown in Fig. 3.13.

Total number of valence electron = 3 (from central B) + 3 (from three peripheral H's) = 6

Three B–H bonds: $\sigma (1a_1')^2 + \sigma (1e')^4$ [$\approx (s+2p)$, sp^2 hybridization]

Total number of B–H bonds = 3

Hence, the bond order for B–H bond = 3 pair of electrons/3 bonds = 1 (pair)/1 bond = 1

(vi) Tetratomic molecule: AB_3 type, σ and π bonding (BF_3; triangular planar, D_{3h})

This is a much more involved system than BH_3, solely because the basis set for three F's is more extensive ($2s$, $2p_x$, $2p_y$, $2p_z$ orbitals) than that for three H's ($1s$ orbital). However, the problem can be approached in exactly the same way.

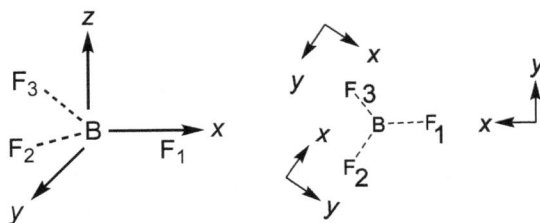

Fig. 3.14 (*a*) Coordinate axes and (b) local coordinate axes, considering symmetry.

All orbitals are referenced to the coordinate system shown in Fig. 3.14. All four atoms are in xy plane and the z axis is perpendicular to the xy plane.

According to D_{3h} character table (Table 3.2), the following are the valence orbitals (basis orbitals) of B to interact with three F $2s$, $2p_x$, $2p_y$, $2p_z$ orbitals.

$$2s(A_1')   \quad \sigma \text{ bonding}$$
$$2p_x, 2p_y(E')   \quad \sigma + \pi_{||}(\text{in-plane}) \text{ bonding}$$
$$2p_z(A_2'')   \quad \pi_{\perp} (\text{out-of-plane}) \text{ bonding}$$

Now we have to obtain the LGO's for the F basis set. First, we consider σ bonding. Let us consider the $2s$ orbitals on the F atoms as a symmetry-related set of three (Fig. 3.15(a)). Next we will run it through all operations of the group, and reduce the resultant reducible representation to arrive at the symmetries of the LGO's.

Let us appreciate the concept of local axes at the three F atoms of BF_3. It is shown for three LGO's for three F's $2s$ orbitals. Also note that these LGO's will effectively remain as nonbonding (see below), as they are tightly bound to the core and do not interact significantly with B $2s$ orbital. As that for BH_3, here also $\Gamma_\sigma = a_1' + e'$. These combinations of s orbitals look like as displayed in Fig. 3.15(a). Two mutually perpendicular nodal planes could readily be seen in e'.

For B–F σ-bonding, the LGO's (Φ_1 to Φ_3) are as follows.

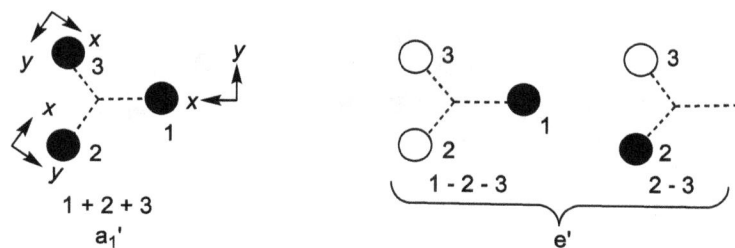

(a)

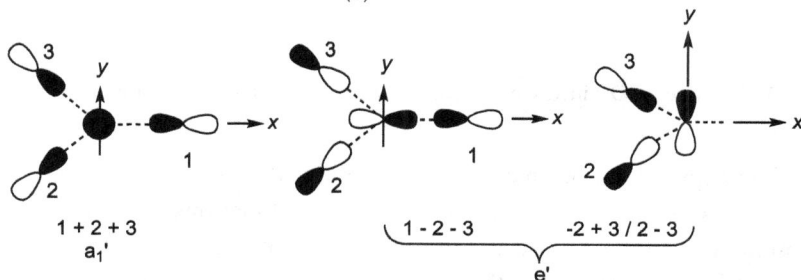

$\Phi_1 (a_1') = 1 / \sqrt{3} (p_{x1} + p_{x2} + p_{x3})$ $\Phi_2 (e') = 1 / \sqrt{6} (2p_{x1} - p_{x2} - p_{x3})$;
$\Phi_3 (e') = 1 / \sqrt{2} (p_{x2} - p_{x3})$

(b)

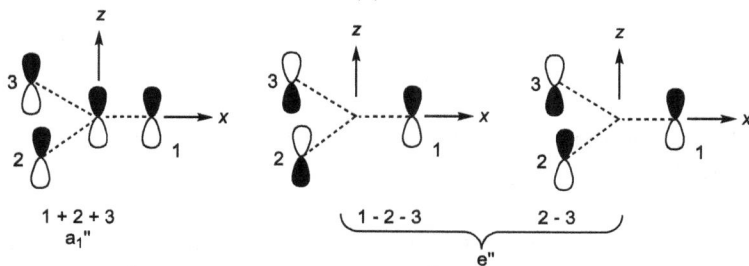

$\Phi_4 (a_2'') = 1 / \sqrt{3} (p_{z1} + p_{z2} + p_{z3})$ $\Phi_5 (e'') = 1 / \sqrt{6} (2p_{z1} - p_{z2} - p_{z3})$;
$\Phi_6 (e'') = 1 / \sqrt{2} (p_{z2} - p_{z3})$

(c)

Fig. 3.15   (*Continued*)

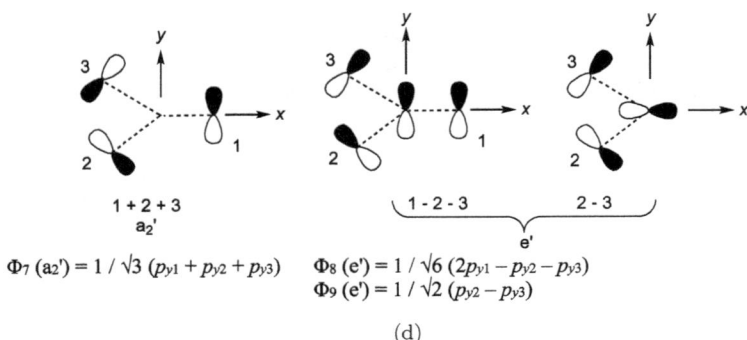

$\Phi_7 (a_2') = 1 / \sqrt{3} \, (p_{y1} + p_{y2} + p_{y3})$ $\Phi_8 (e') = 1 / \sqrt{6} \, (2p_{y1} - p_{y2} - p_{y3})$

$\Phi_9 (e') = 1 / \sqrt{2} \, (p_{y2} - p_{y3})$

(d)

Fig. 3.15   Nine LGO's involving (a) $2s$, (b) $2px$, (c) $2pz$, and (d) $2py$ orbitals of three F's.

The expressions of nine LGO's (normalized to 1 and orthogonal; each set of LGO's is a set of linearly independent functions) for three F's are presented in Fig. 3.15 and the familiar orbital presentations are shown in Fig. 3.15(b)–(d). Note that the expressions for nine LGO's (three F's with each $2p_x$, $2p_y$, $2p_z$ orbitals) are written, considering local symmetry at the three F's.

Φ_1 LGO: each p_x orbital contributes equally. All AO's are normalized to unity. This LGO has same symmetry as B $2s$.

Φ_2 LGO: p_{x1} contributes to 2 LGO's, because it does not overlap with B2py in Φ_3. This LGO matches with B $2p_x$.

Φ_3 LGO: Considering symmetry, the coefficients of px_2 and px_3 must be same.

Total contribution of p_{x1} must be equal to 1. The contribution of p_{x1} to Φ_1 LGO = 1/3. Therefore, the contribution of p_{x1} to Φ_2 LGO = 1 − (1/3 + 0) = 2/3 (square of $2/\sqrt{6} = 4/6 = 2/3$).

This LGO matches with B $2p_y$.

Considering best interaction between orbitals, the construction of LGO's is based on the concept of "matching symmetry". Hence, the symmetries of LGO's are set by the procedure followed. The symmetries of

LGO's are confirmed by orbital transformations under D_{3h} symmetry operations. They are $a_1' + e'$ (see the discussion on BH_3).

For B–F π-bonding ($\pi_\perp$ out-of-plane bonding; xy is the nodal plane), the LGO's are Φ_4 to Φ_6. It is π as it is side-on type bonding.

One LGO Φ_4 is obtainable by "matching symmetry". Other two LGO's Φ_5 and Φ_6 can be obtained by orbital transformations under D_{3h} symmetry operations.

D_{3h}	E	$2C_3$	$3C_2'$	σ_h	$2S_3$	$3\sigma_v$
$\Gamma\pi_\perp$ (red)	3	0	-1	-3	0	1

$$a_{A1'} = 1/12[(1 \times 3 \times 1) + (2 \times 0 \times 1) + (3 \times -1 \times 1) + (1 \times -3 \times 1)$$
$$+ (2 \times 0 \times 1) + (3 \times 1 \times 1)] = 0$$

$$a_{A2'} = 1/12[(1 \times 3 \times 1) + (2 \times 0 \times 1) + (3 \times -1 \times -1)$$
$$+ (1 \times -3 \times 1) + (2 \times 0 \times 1) + (3 \times 1 \times -1)] = 0$$

$$a_{E'} = 1/12[(1 \times 3 \times 2) + (2 \times 0 \times -1) + (3 \times -1 \times 0) + (1 \times -3 \times 2)$$
$$+ (2 \times 0 \times -1) + (3 \times 1 \times 0)] = 0$$

$$a_{A1''} = 1/12[(1 \times 3 \times 1) + (2 \times 0 \times 1) + (3 \times -1 \times 1) + (1 \times -3 \times -1)$$
$$+ (2 \times 0 \times -1) + (3 \times 1 \times -1)] = 0$$

$$a_{A2''} = 1/12[(1 \times 3 \times 1) + (2 \times 0 \times 1) + (3 \times -1 \times -1) + (1 \times -3 \times -1)$$
$$+ (2 \times 0 \times -1) + (3 \times 1 \times 1)] = 1$$

$$a_{E''} = 1/12[(1 \times 3 \times 2) + (2 \times 0 \times -1) + (3 \times -1 \times 0)$$
$$+ (1 \times -3 \times -2) + (2 \times 0 \times 1) + (3 \times 1 \times 0)] = 1$$

Thus, the symmetries of the $\pi_\perp$ LGO's for BF_3 are $a_2'' + e''$. It is evident from Table 3.2 and $\pi_\perp$ (out-of-plane) bonding interactions (Fig. 3.15(c)) that we should consider only a_2'' and e''.

For B–F $\pi_\parallel$-bonding (in-plane), the LGO's are Φ_7 to Φ_9. It is π as it is side-on type bonding.

D_{3h}	E	$2C_3$	$3C_2'$	σ_h	$2S_3$	$3\sigma_v$
$\Gamma\pi_\parallel$ (red)	3	0	-1	3	0	-1

For $\pi_\parallel$ (in plane) bonding:

$$a_{A1'} = 1/12[(1 \times 3 \times 1) + (2 \times 0 \times 1) + (3 \times -1 \times 1) + (1 \times 3 \times 1)$$
$$+ (2 \times 0 \times 1) + (3 \times -1 \times 1)] = 0$$

$$a_{A2'} = 1/12[(1 \times 3 \times 1) + (2 \times 0 \times 1) + (3 \times -1 \times -1) + (1 \times 3 \times 1)$$
$$+ (2 \times 0 \times 1) + (3 \times -1 \times -1)] = 1$$

$$a_{E'} = 1/12[(1 \times 3 \times 2) + (2 \times 0 \times -1) + (3 \times -1 \times 0) + (1 \times 3 \times 2)$$
$$+ (2 \times 0 \times -1) + (3 \times -1 \times 0)] = 1$$

$$a_{A1''} = 1/12[(1 \times 3 \times 1) + (2 \times 0 \times 1) + (3 \times -1 \times 1) + (1 \times 3 \times -1)$$
$$+ (2 \times 0 \times -1) + (3 \times -1 \times -1)] = 0$$

$$a_{A2''} = 1/12[(1 \times 3 \times 1) + (2 \times 0 \times 1) + (3 \times -1 \times -1) + (1 \times 3 \times -1)$$
$$+ (2 \times 0 \times -1) + (3 \times -1 \times 1)] = 0$$

$$a_{E''} = 1/12[(1 \times 3 \times 2) + (2 \times 0 \times -1) + (3 \times -1 \times 0) + (1 \times 3 \times -2)$$
$$+ (2 \times 0 \times 1) + (3 \times -1 \times 0)] = 0$$

Thus, the symmetries of the $\pi_\parallel$ LGO's for BF_3 are $a_2' + e'$. It is again evident from Table 3.3 and $\pi_\parallel$ (in plane) bonding interactions (Fig. 3.15(d)) that we should consider only a_2' and e'.

The LGO's Φ_1, Φ_2, and Φ_3, comprising p_x orbitals, are involved in σ bonding interaction (matches with B $2s$, B $2p_x$, and B $2p_y$, respectively). While the LGO Φ_4, comprising p_z orbitals, is involved in $\pi_\perp$ (out of plane) interaction (matches with B $2p_z$), Φ_5 and Φ_6, comprising p_z orbitals, will remain nonbonding. The LGO's Φ_7–Φ_9 comprise p_y orbitals. While Φ_7 will remain strictly nonbonding, Φ_8 and Φ_9 are involved in $\pi_\parallel$ (in plane) interactions with B $2p_y$ and $2p_x$, respectively. However, LGO's Φ_8 and Φ_9 will remain essentially nonbonding, as $2p_x$ and $2p_y$ are involved in stronger

Table 3.3 Character table for C_{3v}

C_{3v}	E	$2C_3$	$3\sigma_v$		
A_1	1	1	1	z, z^2	
A_2	1	1	-1	R_z	
E	2	-1	0	$(x, y), (R_x, R_y)$	$(x^2-y^2, xy) (xz, yz)$

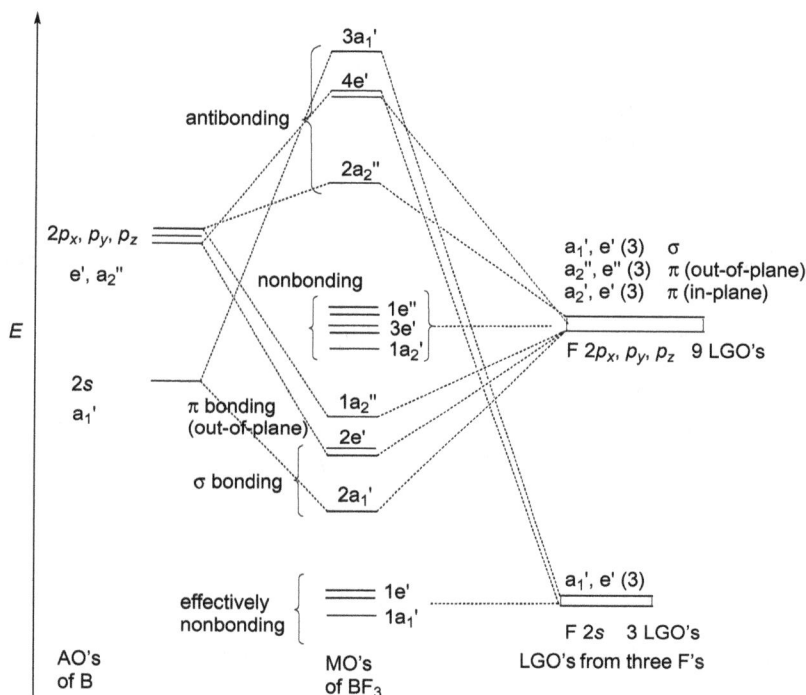

Fig. 3.16 Qualitative MO energy-level diagram for BF_3.

σ-bonding interaction. The qualitative MO energy-level diagram for BF_3 is presented in Fig. 3.16.

Total number of MO's $= 4$ (B) $+ 12$ (3F's) $= 16$

Total number of valence electron $= 3$ (B) $+ 3 \times 7$ (F) $= 24$

Ground state electronic distribution:

$(1a_1')^2(1e')^4 \, (2a_1')^2(2e')^4 \, (1a_2'')^2 \, (1a_2')^2 \, (3e')^4(1e'')^4$.

$(1a_1')^2(1e')^4$: Only slightly perturbed and interacting weakly with B AO's. Thus, it is effectively F $2s$ electrons (nonbonding).

$(2a_1')^2(2e')^4$: σ MO's ($\approx sp^2$ hybrids)

$(1a_2'')^2$: π MO

$(1a_2')^2(3e')^4(1e'')^4$: five lone pairs

Total number of B–F bonds $= 3$

Three B–F σ-bonds: $\sigma \, (2a_1')^2 + \sigma(2e_1')^4$; bond order $= 3$ pair of electrons/3 bonds $= 1$ (pair) / 1 bond $= 1$

One π-bond distributed over three B–F bonds: $(1a_2'')^2$; bond order = 1 pair of electron / 3 bonds = 1 (pair) / 3 bonds = 1/3

Hence, the bond order for B–F bonds = 1 1/3

This corresponds to the resonating Lewis structures, as shown below:

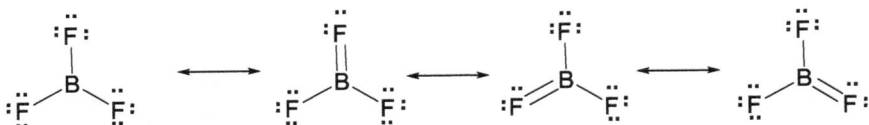

3σ and 1π and 8 lone pairs (nonbonding electrons)

(vii) Tetratomic molecule: AB_3E_1 type (NH_3; pyramidal, C_{3v})

Let us consider the character table for C_{3v} (Table 3.3).

Symmetries of N AO's: s A_1; p_z A_1; p_x, p_y E

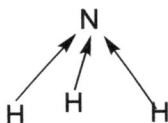

The symmetries of the LGO's are represented as three vectors along the NH bond axes.

C_{3v}	E	$2C_3$	$3\sigma_v$
Γ_{red}	3	0	1

$$a_{A1} = 1/6[(1 \times 3 \times 1) + (2 \times 0 \times 1) + (3 \times 1 \times 1)] = 1$$

$$a_{A2} = 1/6[(1 \times 3 \times 1) + (2 \times 0 \times 1) + (3 \times 1 \times -1)] = 0$$

$$a_E = 1/6[(1 \times 3 \times 2) + (2 \times 0 \times -1) + (3 \times 1 \times 0)] = 1$$

Thus, the symmetries of the LGO's for NH_3 are $a_1 + e$.

Disregarding the $1s$ AO on N, we have seven AO's. All AO's match with LGO's and in this case no nonbonding level. Both s and p_z AO's match the a_1 LGO, so s-p mixing is expected. We must make only three MO's from the s and p_z AO's on N and the a_1 LGO for three H's. For simplicity, we will assume that the s and p_z AO's each form essentially separate bonding MO's, but that together they form a single mixed antibonding MO $3a^*$ (remembering that number of AO's = number of MO's). The resulting MO scheme is shown in Fig. 3.17.

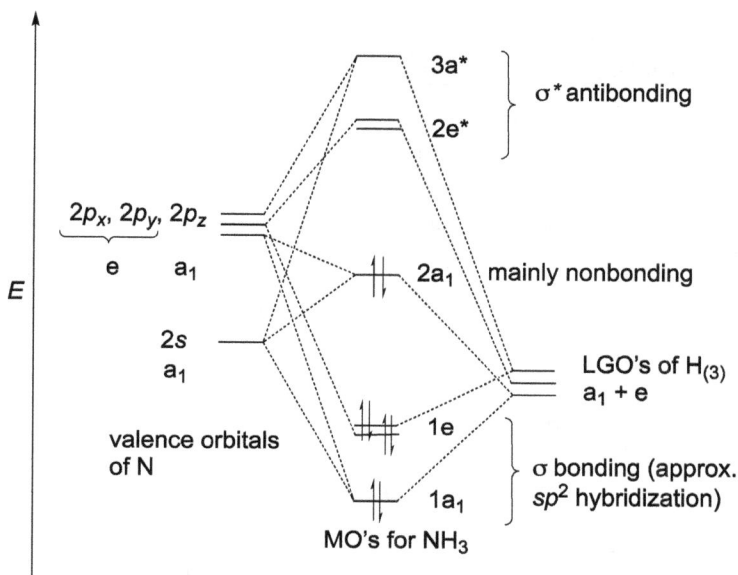

Fig. 3.17 Qualitative MO energy level diagram of NH_3 (C_{3v}).

Total number of valence electron = 5 (from central N atom) + 3 (from three peripheral H atoms) = 8

Total number of N–H bonds = 3

Three N–H bonds: σ $(1a_1)^2 + \sigma$ $(1e)^4$; one lone-pair: $2a_1$ $[\approx (s+p)]$

Hence, the bond order for N–H bonds = 3 pair of electrons/3 bonds = 1 (pair) / 1 bond = 1

The bond order for NH_3 = 3

(viii) Pentatomic molecule: AB_4 type (CH_4; tetrahedral, T_d)

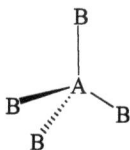

The number of bonds that remain unshifted by the operations are represented by the following:

T_d	E	$8C_3$	$3C_2$	$6S_4$	$6\sigma_d$
$\Gamma\sigma$	4	1	0	0	2

Table 3.4 Character table for T_d

T_d	E	$8C_3$	$3C_2$	$6S_4$	$6\sigma_d$		
A_1	1	1	1	1	1		$x^2+y^2+z^2$
A_2	1	1	1	-1	-1		
E	2	-1	2	0	0		$(2z^2-x^2-y^2, x^2-y^2)$
T_1	3	0	-1	1	-1	(R_x, R_y, R_z)	
T_2	3	0	-1	-1	1	(x, y, z)	(xy, xz, yz)

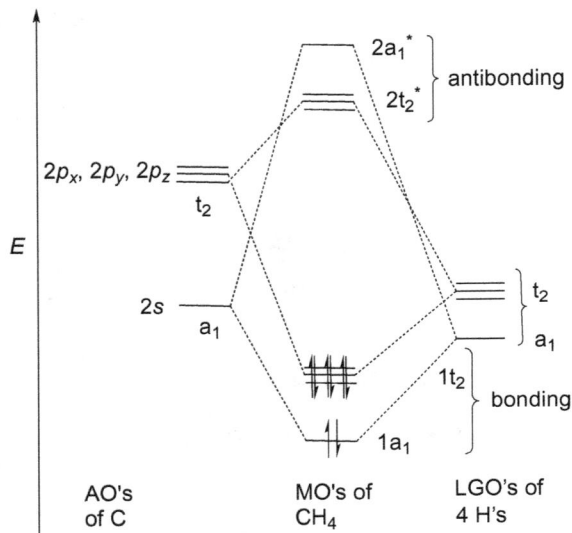

Fig. 3.18 Qualitative MO energy-level diagram of CH_4 (T_d).

The total number of symmetry operation is 24 (Table 3.4). There is one E element, eight C_3, three C_2, six S_4, and six σ_d planes.

The number of times the a_1 and t_2 occur in the representation:

$$a_{A1} = 1/24[(1 \times 4 \times 1) + (8 \times 1 \times 1) + (3 \times 0 \times 1)$$
$$+(6 \times 0 \times 1) + (6 \times 2 \times 1)] = 1$$
$$a_{T2} = 1/24[(1 \times 4 \times 3) + (8 \times 1 \times 0) + (3 \times 0 \times -1)$$
$$+(6 \times 0 \times -1) + (6 \times 2 \times 1)] = 1$$

Thus, the symmetries of the LGO's ($\Gamma\sigma$) for CH_4 are $a_1 + t_2$.

Total number of valence electron = 4 (from central C atom) + 4 (from four peripheral H atoms) = 8. The qualitative MO energy-level diagram for CH_4 (T_d) is displayed in Fig. 3.18.

Total number of C–H bonds $= 4$

Four C–H bonds: $\sigma\,(1a_1)^2 + \sigma\,(1t_2)^6$

Hence, the bond order for C–H bonds $= 4$ pair of electrons/4 bonds $= 1$ (pair)/1 bond $= 1$

(ix) Pentatomic molecule: AB$_4$E$_2$ type (XeF$_4$; square planar, D$_{4h}$)

Let us now consider AB$_4$E$_2$ type XeF$_4$ (point group D$_{4h}$; Table 2.3). As it is a square planar molecule, so we orient the XeF$_4$ in the xy plane with the σ bonds along the x and y axes. To describe the MO energy-level diagram for XeF$_4$ we must find the representations for the combinations of the vectors along the bonds. The σ orbitals of the four ligands can be represented by a set of vectors orientated as shown below. We start with reducible representation ($\Gamma\sigma$) associated with only σ-bonding. The result of operating on the four σ vectors for D$_{4h}$ is presented below.

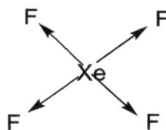

D$_{4h}$	E	2C$_4$	C$_2$	2C$_2'$	2C$_2''$	i	2S$_4$	σ_h	2σ_v	2σ_d
$\Gamma\sigma$	4	0	0	2	0	0	0	4	2	0

We can reduce $\Gamma\sigma$ following the same procedure as done for C$_{2v}$ and C$_{3v}$ or using the characters for σ_h, σ_v, and i as useful clues to obtain $\Gamma\sigma = A_{1g} + B_{1g} + E_u$. The σ bonds for XeF$_4$ transform as A$_{1g}$, B$_{1g}$, and E$_u$.

Let us interpret this information in terms of familiar AO's. We identify the AO's of Xe belonging to the presentations $a_{1g}\,(s, dz^2)$, $b_{1g}\,(dx^2-y^2)$, and $e_u\,(p_x, p_y)$ from the character table (Table 2.3). It is obvious that only these AO's of Xe can participate in σ bonding. Now we have five AO's and we need only four to form four Xe–F bonds. To get an answer, let us consider the criteria of orbital overlap and relative energy of orbitals. The energy of dz^2 orbital is higher than that of s. Thus, the s orbital is more favorable than dz^2. Moreover, the dz^2 orbital has the major lobes along z, orthogonal to the Xe–F bonds, and only the small donut in the xy plane.

This also precludes the possibility of mixing (hybridization) of s and dz^2. Thus, the favored Xe orbitals for four σ bonds are dx^2-y^2, s, and p_x, p_y (p_z is orthogonal to the bonding plane). This combination leads to dsp^2 hybridization (valence bond theory). Total number of LGO's is 16 ($4 \times 3\,p$ orbital and $4 \times 1\,s$ orbital).

We recognize that the F valence orbitals are lower in energy than the Xe orbitals because F is much more electronegative than Xe. Considering the relative energies of the orbitals, a qualitative energy-level MO diagram is presented in Fig. 3.19. The energy of the Xe orbitals increases in the order $5s < 5p \ll 5d$. The b_{1g} (dx^2-y^2 orbital of Xe) and e_u (p_x, p_y orbitals of Xe) MO's involve ideal overlap for σ-bonding, making full use of the lobes directed along the bonds. The Xe d orbitals other than dx^2-y^2 are non-bonding but in D_{4h} point group (Table 2.3) they split as a_{1g} (dz^2), e_g (dxz, dyz), and b_{2g} (dxy). Of these, the dz^2 should have lowest energy because the major lobe is along the z axis and, hence, farthest from the F orbitals in the xy plane. The d_{xy} orbital should be highest in energy because it is in the xy plane.

Total number of valence electron = 8 (from central Xe atom) + 28 (from four peripheral F atoms: s orbital 4×2 and p orbital 4×5) = 36

Number of nonbonding electrons = 28

Total number of Xe–F bonds = 4

Four Xe–F bonds: $\sigma\,(1a_{1g})^2 + \sigma\,(1e_u)^4 + \sigma\,(1b_{1g})^2$

The bond order for Xe–F bonds = 4 pair of electrons/4 bonds = 1 (pair)/1 bond = 1

(x) Hexatomic molecule: AB$_5$ type (BrF$_5$, ML$_5$ (see Chapter 6); square pyramidal, C_{4v})

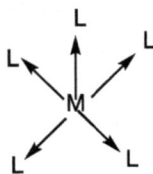

Valence orbitals of M: s, p_z, dz^2 (a_1); dx^2-y^2 (b_1); (p_x, p_y) (d_{xz}, d_{yz}) (e); d_{xy} (b_2) (see Chapter 2, Table 2.5)

(a)

(b)

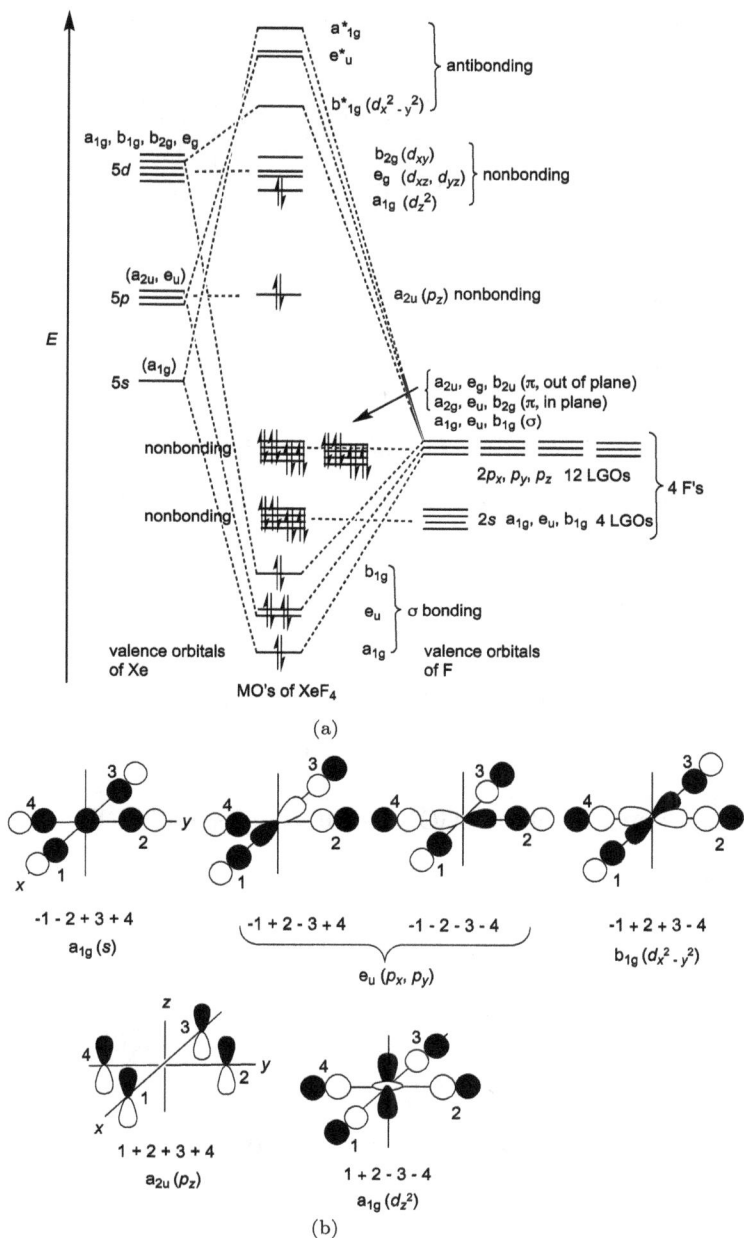

Fig. 3.19 (a) Qualitative MO energy-level diagram for σ-bonding in XeF_4 (D_{4h}) and (b) the forms of MO's.

The number of bonds that remain unshifted by the operations are represented by the following:

C_{4v}	E	$2C_4$	C_2	$2\sigma_v$	$2\sigma_d$
$\Gamma\sigma$	5	1	1	3	1
$\Gamma\sigma$ (basal)	4	0	0	2	0
$\Gamma\sigma$ (axial)	1	1	1	1	1

$$a_{A1} = 1/8[(1 \times 1 \times 5) + (2 \times 1 \times 1) + (1 \times 1 \times 1)$$
$$+ (2 \times 1 \times 3) + (2 \times 1 \times 1)] = 2$$
$$a_{A2} = 1/8[(1 \times 1 \times 5) + (2 \times 1 \times 1) + (1 \times 1 \times 1)$$
$$+ (2 \times -1 \times 3) + (2 \times -1 \times 1)] = 0$$
$$a_{B1} = 1/8[(1 \times 1 \times 5) + (2 \times -1 \times 1) + (1 \times 1 \times 1)$$
$$+ (2 \times 1 \times 3) + (2 \times -1 \times 1)] = 1$$
$$a_{B2} = 1/8[(1 \times 1 \times 5) + (2 \times -1 \times 1) + (1 \times 1 \times 1)$$
$$+ (2 \times -1 \times 3) + (2 \times 1 \times 1)] = 0$$
$$a_{E} = 1/8[(1 \times 2 \times 5) + (2 \times 0 \times 1) + (1 \times -2 \times 1)$$
$$+ (2 \times 0 \times 3) + (2 \times 0 \times 1)] = 1$$

Thus, the symmetries of the LGO's for ML_5 are $2A_1 + B_1 + E$.

$\Gamma\sigma$ (basal):

$$a_{A1} = 1/8[(1 \times 1 \times 4) + (2 \times 1 \times 0) + (1 \times 1 \times 0)$$
$$+ (2 \times 1 \times 2) + (2 \times 1 \times 0)] = 1$$
$$a_{A2} = 1/8[(1 \times 1 \times 4) + (2 \times 1 \times 0) + (1 \times 1 \times 0)$$
$$+ (2 \times -1 \times 2) + (2 \times -1 \times 0)] = 0$$
$$a_{B1} = 1/8[(1 \times 1 \times 4) + (2 \times -1 \times 0) + (1 \times 1 \times 0)$$
$$+ (2 \times 1 \times 2) + (2 \times -1 \times 0)] = 1$$
$$a_{B2} = 1/8[(1 \times 1 \times 4) + (2 \times -1 \times 0) + (1 \times 1 \times 0)$$
$$+ (2 \times -1 \times 2) + (2 \times 1 \times 0)] = 0$$

$$a_E = 1/8[(1 \times 2 \times 4) + (2 \times 0 \times 0) + (1 \times -2 \times 0)$$
$$+ (2 \times 0 \times 2) + (2 \times 0 \times 0)] = 1$$

Thus, the symmetries of the LGO's for ML_5 (basal ligands) are $a_1 + b_1 + e$.

$\Gamma\sigma$ (axial):

$$a_{A1} = 1/8[(1 \times 1 \times 1) + (2 \times 1 \times 1) + (1 \times 1 \times 1) + (2 \times 1 \times 1)$$
$$+ (2 \times 1 \times 1)] = 1$$

$$a_{A2} = 1/8[(1 \times 1 \times 1) + (2 \times 1 \times 1) + (1 \times 1 \times 1) + (2 \times -1 \times 1)$$
$$+ (2 \times -1 \times 1)] = 0$$

$$a_{B1} = 1/8[(1 \times 1 \times 1) + (2 \times -1 \times 1) + (1 \times 1 \times 1) + (2 \times 1 \times 1)$$
$$+ (2 \times -1 \times 1)] = 0$$

$$a_{B2} = 1/8[(1 \times 1 \times 1) + (2 \times -1 \times 1) + (1 \times 1 \times 1) + (2 \times -1 \times 1)$$
$$+ (2 \times 1 \times 1)] = 0$$

$$a_E = 1/8[(1 \times 2 \times 1) + (2 \times 0 \times 1) + (1 \times -2 \times 1) + (2 \times 0 \times 1)$$
$$+ (2 \times 0 \times 1)] = 0$$

Thus, the symmetry of the LGO for ML_5 (axial ligand) is a_1.

Now a qualitative MO energy-level diagram for C_{4v} could be constructed.

3.3 Three-center bonding: Electron-deficient compound B_2H_6

Up to this point we have dealt molecules whose σ-bonding could be described in terms of electron-pair bonds using a localized (Chapter 1) or MO treatment. However, there are many stable molecules for which the total number of valence electrons is insufficient to form a complement of electron-pair bonds amongst nearest neighboring atoms. Such compounds constitute a significant portion of main-group chemistry. The best known examples are the boron hydrides. We consider here the MO treatment for B_2H_6.

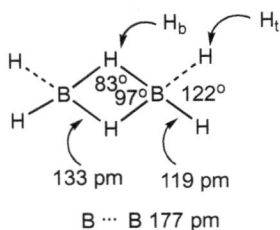

$$B \cdots B \ 177 \ pm$$

Two B atoms: 8 AO's; valence electron = 6
Six H atoms: 6 AO's; valence electron = 6
So, a total of 14 AO's and 12 valence electrons
Important structural features:

Each B atom is $\sim$ tetrahedrally bonded; B–H_b > B–H_t, the latter being consistent with a classical 2c-2e$^-$ bond. The central B–H–B bridge bonds are the unusual feature of B_2H_6 (diborane).

We shall treat these by MO theory, separating them from the four B–H_t bonds. A starting point analogous to previous MO treatments would be as follows:

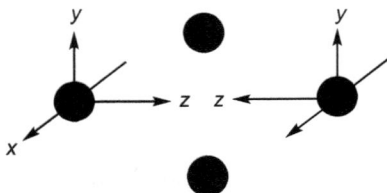

Potential B $\cdots$ B interactions: p_z-p_z (σ), s-s (σ), p_y-p_y(π), p_x-p_x(π)
Potential B $\cdots$ H $\cdots$ B interactions: p_z-s_H-p_z, s_B-s_H-s_B, p_y-s_H-p_y
Therefore, all valence AO's orbitals of two B and two H_b can interact.

These two 'BH_2' units are structurally analogous to H_2O. The two lone pair orbitals of H_2O are vacant in the H_t-B-H_t bridging unit and are oriented so as to overlap well with the two bridging H AO's.

Four sp^3 hybridized orbitals are associated with each B atom. Two LGO's are $\phi_1 + \phi_2$ and $\phi_1 - \phi_2$.

MO treatment for each B–H–B unit:

Two B sp^3 (ϕ_1, ϕ_2) orbitals interact with a H_b $1s$ giving rise to three MO's in C_{2v} symmetry. Let us treat H_b as central atom and two B atoms as

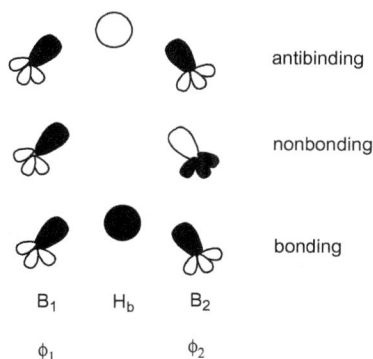

Fig. 3.20 (a) Qualitative MO energy-level diagram for σ-bonding in XeF_4 (D_{4h}) and (b) the forms of MO's.

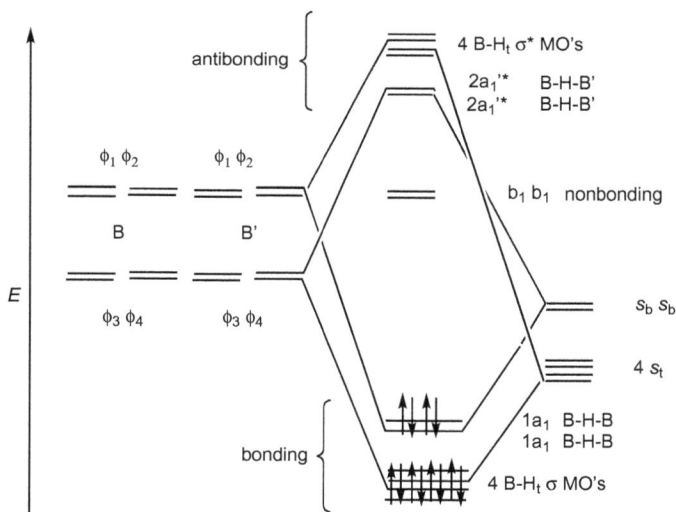

Fig. 3.21 Qualitative MO energy-level diagram for bonding in B_2H_6, with emphasis on three center-two electron bond formation in each BH_2 units.

exterior L atoms, assuming complete covalency.

$$\Psi(1a_1) = 1/2(\phi_1 + \phi_2 + \sqrt{2}s) \quad \text{bonding}$$
$$\Psi^*(2a_1) = 1/2(\phi_1 + \phi_2 - \sqrt{2}s) \quad \text{antibonding}$$
$$\Psi(b_1) = 1/\sqrt{2}(\phi_1 - \phi_2) \quad \text{nonbonding}$$

The coefficients can be easily arrived at, as it can interact only with a_1 LGO. Its coefficient must be,

$(\sqrt{2}/2)^2 = 1/2$

B–H–B bond: electron deficient, $2e^-$ delocalized over three nuclei (3c-$2e^-$ bond model)

B–H_t bond: 2c-$2e^-$ bond model. A qualitative MO energy-level diagram for bonding in B_2H_6 is presented in Fig. 3.21.

The dimerization of BH_3 to form B_2H_6 could be viewed as a donor-acceptor interaction, whereby each B atom attains an octet of electrons via bridge bonding.

Further reading

C. E. Housecroft and A. G. Sharpe, *Inorganic Chemistry*, 2nd edition, Pearson Education Limited (2005)

D. F. Shriver and P. W. Atkins, *Inorganic Chemistry*, 3rd edition, Oxford University Press (1999)

J. D. Lee, *Concise Inorganic Chemistry*, 5th edition, Blackwell Science Limited (1996)

J. E. Huheey, E. A. Keiter, and R. L. Keiter, *Inorganic Chemistry: Principles of Structure and Reactivity*, 4th edition, Addison-Wesley Publishing Company (1993)

W. L. Jolly, *Modern Inorganic Chemistry*, 2nd edition, McGraw-Hill Inc., McGraw-Hill International Editions Chemistry Series (1991)

F. A. Cotton and G. Wilkinson, *Basic Inorganic Chemistry*, John Wiley & Sons, Inc.; Fourth Wiley Eastern Reprint (1988)

K. F. Purcell and J. C. Kotz, *Inorganic Chemistry*, Saunders Golden Sunburst Series, W. B. Saunders Company, Holt-Saunders Japan (1985)

B. E. Douglas, D. H. McDaniel, and J. J. Alexander, *Concepts and Models of Inorganic Chemistry*, 2nd edition, John Wiley & Sons, Inc. (1983)

Exercises

3.1 Construct a qualitative MO energy-level diagram for CO_3^{2-}.

3.2 Construct a qualitative MO energy-level diagram for PF_5.

3.3 Write reducible representations for SF_6 (octahedral point group), considering σ-bonding and also π-bonding.

3.4 Predict the structure of the following compounds/ions and then proceed to treat their bonding aspects using molecular orbital theory.

(i) HF_2^- and (ii) BrF_5.

Chapter 4

Redox Reactions

Chemical reactions in which electrons are transferred from one species to another, oxidation and reduction takes place. Equilibria involving oxidation and reduction processes is the focus of this chapter. The concepts involved are familiar to most readers. In an elementary level, oxidation is defined as the process pertaining to gaining oxygen or losing hydrogen, and reduction is defined as the process involved in losing oxygen or gaining hydrogen. More precisely, oxidation refers to losing one or more electrons and reduction is gaining one or more electrons.

Redox reactions are a fundamental type of chemical reaction that is critical to the bond-making and bond-breaking events that drive nearly all chemical transformations. It should be remembered that a covalent bond involves two electrons. In nature, many reactions that are crucial to life are multielectron redox reactions. For example, sunlight is stored in chemical bonds by photosynthesis, which splits H_2O into O_2 and H_2, as NADPH (nicotinamide adenine dinucleotide phosphate) via a four-electron process. The reverse process, respiration, uses O_2 and four electrons to synthesize ATP as an energy source in cells. Nitrogen fixation (ammonia synthesis), C–H bond functionalization, and CO_2 reduction are all multielectron redox reactions carried out in biology, and each of these processes is facilitated by enzymes that contain first-row transition metals such as Mn, Fe, Co, Ni, and Cu, and second-row transition metals such as Mo and W. Because these metal ions typically react by one-electron pathways, nature often includes redox-active cofactors to expand their reactivity to allow multielectron processes to occur. In synthetic inorganic/organometallic chemistry, redox-active ligands can be used to achieve the same goal and afford multielectron reactivity at a metal ion that is prone to one-electron redox reactions (see Chapter 8).

Metal ions of biological relevance typically react by one-electron pathways. Therefore, nature often includes redox-active cofactors to allow multielectron processes to occur. Redox-active ligands can be used to achieve multielectron reactivity at a metal ion.[1]

4.1 Redox half-cell reactions

The oxidized and reduced species in a half-cell reaction constitute a *redox couple*. A couple is written with the oxidized species before the reduced, as in H^+/H_2, Zn^{2+}/Zn, and $Cr_2^{VI}O_7^{2-}/Cr^{3+}$. A redox reaction is regarded as the outcome of a reduction and an oxidation half-cell reaction.

In a half-cell reaction, the oxidized and reduced species constitute a redox couple. All half-cell reactions are written as reductions (Ox/Red i.e. $Ox + e^- \rightarrow Red$).

Standard potentials

The overall redox reaction is the difference of two half-cell reactions. Hence, the standard Gibbs energy (G°) of the overall reaction is the difference of the standard Gibbs energies of the two half-cell reactions (ΔG°). The overall reaction is favorable (in the sense that $K > 1$) in the direction that corresponds to a negative value of the resulting ΔG°.

Because reduction half-cell reactions always occur in pairs in any redox reaction, only the difference in ΔG° value is of significance. Hence,

---

[1] (a) Forum Issue on Redox-Active Ligands: *Inorg. Chem.* **2011**, *50* (20), 9737. (b) Forum Issue on Applications of Metal Complexes with Ligand-Centered Radicals: *Inorg. Chem.* **2018**, *57* (16) 9577. (c) J. Jacquet, M. D.-E. Murr, and L. Fensterbank, *ChemCatChem* **2016**, *8*, 3310. (d) D. L. J. Broere, R. Plessius, and J. I. van der Vlugt, *Chem. Soc. Rev.* **2015**, *44*, 6886. (e) R. H. Crabtree, *Chem. Soc. Rev.*, **2013**, *42*, 1440. (f) V. Lyaskovskyy and B. de Bruin, *ACS Catal.* **2012**, *2*, 270. (g) V. K. K. Praneeth, M. R. Ringenberg, and T. R. Ward, *Angew. Chem. Int. Ed.* **2012**, *51*, 10228. (h) W. I. Dzik, J. I. van der Vlugt, J. N. H. Reek, and B. de Bruin, *Angew. Chem. Int. Ed.* **2011**, *50*, 3356.

a half-cell reaction was chosen to have $\Delta G° = 0$ and report all other values referenced to it. By convention, the specially chosen half-cell reaction is

$$2H^+(aq) + 2e^- \rightarrow H_2(g)$$

and for reduction of $H^+(aq)$ ions (1M) at 1 bar pressure, $\Delta G° = 0$ at all temperatures. Standard reaction Gibbs energies may be measured by setting up a galvanic cell (an electrochemical cell in which an electric current is generated due to a chemical reaction) in which the reaction driving the electric current through the external circuit is the reaction of interest. The potential difference between the electrodes is then measured and can be converted to a Gibbs energy by using $\Delta G = -nFE$. The potential that corresponds to the $\Delta G°$ of a half-cell reaction is written $E°$, by using $\Delta G° = -nFE°$, where $n =$ number of moles of electrons (equiv) involved in the redox process, $F =$ Faraday constant $= 23.06$ kcal mol^{-1} V^{-1}, $E° =$ the cell potential in V at standard state.

Polarography is an electroanalytical technique that employs mercury drop as a working electrode to measure the reduction potential ($E°$ value) of electroactive species. The technique was discovered by J. Heyrovský, a Czech chemist (1890–1967; Nobel prize in chemistry: 1959).

A redox reaction is considered as the sum of two half-cell reactions. In any redox reaction, only the difference in their standard Gibbs energies ($\Delta G°$) is of significance. An overall reaction is favorable in the direction that corresponds to the resulting overall $\Delta G°$ value to become negative.

The well-known electrochemical series is a list of $E°$ values of half-cell reactions at 25°C.

The potential $E°$ is called the standard reduction potential.

Ox/Red couple with very positive $E°$ value signifies that Ox is a powerful oxidizing agent.

Ox/Red couple with very negative $E°$ value signifies that Red is a powerful reducing agent.

As $\Delta G°$ for the reduction of H^+ is arbitrarily set at zero, the standard reduction potential of the $H^+(aq)/H_2(g)$ couple is also zero at all temperatures.

$$2H^+(aq) + 2e^- \rightarrow H_2(g); \ E°(H^+/H_2) = 0\,V$$

Example 4.1

Given:

The half-cell reactions

$$2H^+(aq) + 2e^- \rightarrow H_2(g); \ E^\circ = 0 \text{ V}$$

$$Zn^{2+}(aq) + 2e^- \rightarrow Zn(s); \ E^\circ = -0.76 \text{ V}$$

Calculate the E° for the net reaction, $2H^+(aq) + Zn(s) \rightarrow Zn^{2+}(aq) + H_2(g)$.

Answer

For half-cell,

$$2H^+(aq) + 2e^- \rightarrow H_2(g); \ \Delta G^\circ = -2F \times 0 \text{ kcal mol}^{-1}$$

For half-cell,

$$Zn^{2+}(aq) + 2e^- \rightarrow Zn(s); \ \Delta G^\circ = -2F \times -0.76 = 2F \times -0.76 \text{ kcal mol}^{-1}$$

Then, for

$$Zn(s) \rightarrow Zn^{2+}(aq) + 2e^-; \ \Delta G^\circ = -2F \times 0.76 \text{ kcal mol}^{-1}$$

Adding two half-cell reactions,

$$\Delta G^\circ(\text{net reaction}) = -2F \times (0 + 0.76) = -2F \times 0.76 \text{ kcal mol}^{-1}$$

Therefore,

$$E^\circ(\text{net reaction}) = 0.76 \text{ V } (\text{since } \Delta G^\circ = -nFE^\circ)$$

Example 4.2

For a Daniel cell,

Given:

$$Zn^{2+}(aq) + 2e^- \rightarrow Zn(s); \ E^\circ = -0.76 \text{ V}$$

$$Cu^{2+}(aq) + 2e^- \rightarrow Cu(s); \ E^\circ = 0.34 \text{ V}$$

Calculate the E° for the net reaction, $Zn(s) + Cu^{2+}(aq) \rightarrow Zn^{2+}(aq) + Cu(s)$.

Answer

For half-cell,

$$Zn^{2+}(aq) + 2e^- \rightarrow Zn(s); \Delta G^\circ = -2F \times -0.76 = 2F \times 0.76$$

Then, for

$$Zn(s) \rightarrow Zn^{2+}(aq) + 2e^-; \Delta G^\circ = -2F \times 0.76$$

For half-cell,

$$Cu^{2+}(aq) + 2e^- \rightarrow Cu(s); \Delta G^\circ = -2F \times 0.34$$

Adding two half-cell reactions, the net reaction

$$Zn(s) + Cu^{2+}(aq) \rightarrow Zn^{2+}(aq) + Cu(s)$$

$$\Delta G^\circ = -2F \times (0.76 + 0.34) \text{ kcal mol}^{-1}$$

Therefore,

$$E^\circ \text{ (net reaction)} = 1.10 \text{ V}$$

Example 4.3 Consider the following standard electrode potential (E°) values to decide what redox reaction, if any, could occur when the following species are mixed: Sn^{2+} and Fe^{2+}; Sn^{2+} and Fe^{3+}; Fe and Sn^{4+}; Fe and Sn^{2+}; Sn and Sn^{4+}.[2]

$$E^\circ(Fe^{3+}(aq)/Fe^{2+}(aq)) = +0.77 \text{ V}$$

$$E^\circ(Sn^{4+}(aq)/Sn^{2+}(aq)) = +0.15 \text{ V}$$

$$E^\circ(Sn^{2+}(aq)/Sn(s)) = -0.14 \text{ V}$$

$$E^\circ(Fe^{2+}(aq)/Fe(s)) = -0.44 \text{ V}$$

Answer

Two possibilities arise for the mixtures of Sn^{2+} and Fe^{2+}. Thus,

$$Fe^{2+} \rightarrow Fe^{3+} \text{ coupled to } Sn^{2+} \rightarrow Sn$$

or

$$Sn^{2+} \rightarrow Sn^{4+} \text{ coupled to } Fe^{2+} \rightarrow Fe$$

---

[2] A. J. Vella, *J. Chem. Educ.* **1990**, *67*, 479.

These reactions are represented, respectively, by the following:

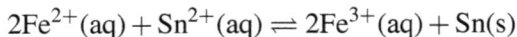

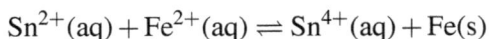

$$2Fe^{2+}(aq) + Sn^{2+}(aq) \rightleftharpoons 2Fe^{3+}(aq) + Sn(s)$$

$$Sn^{2+}(aq) + Fe^{2+}(aq) \rightleftharpoons Sn^{4+}(aq) + Fe(s)$$

Given the E° values associated with the redox couple, it is obvious that Sn^{2+} can act as an electron source (reducing agent) only if Fe^{3+} is present as an electron sink. This corresponds to the redox reaction:

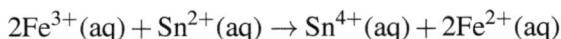

$$2Fe^{3+}(aq) + Sn^{2+}(aq) \rightarrow Sn^{4+}(aq) + 2Fe^{2+}(aq)$$

Fe^{2+} does not act as an electron sink (oxidizing agent) toward Sn^{2+}. Thus, no redox reaction should occur between the ions Sn^{2+} and Fe^{2+}. Moreover, the E° values associated with the redox couple reveals that Fe(s) should be able to react with Sn^{4+} in the following way:

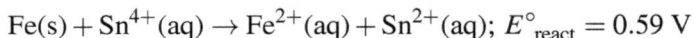

$$Fe(s) + Sn^{4+}(aq) \rightarrow Fe^{2+}(aq) + Sn^{2+}(aq); \; E^\circ_{react} = 0.59 \text{ V}$$

and that it could also reduce $Sn^{2+}(aq)$ but with a lower thermodynamic drive, thus,

$$Fe(s) + Sn^{2+}(aq) \rightarrow Fe^{2+}(aq) + Sn(s); \; E^\circ(\text{net reaction}) = 0.30 \text{ V}$$

Finally, the E° values associated with the redox couple suggest that Sn could reduce Sn^{4+} to Sn^{2+}, according to the following redox reaction:

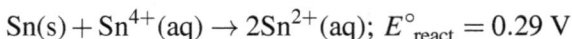

$$Sn(s) + Sn^{4+}(aq) \rightarrow 2Sn^{2+}(aq); \; E^\circ_{react} = 0.29 \text{ V}$$

4.2 Equilibrium constant and redox potential

Let us consider an equilibrium reaction:

$$aA + bB \rightleftharpoons cC + dD$$

Equilibrium constant,

$$K = [C]^c[D]^d / [A]^a[B]^b$$

For our purposes, activities are replaced by concentrations.

For a redox reaction: $Ox + ne^- \rightarrow Red$, E° (standard reduction potential)

In terms of Gibbs energy and equilibrium constant,

$$\Delta G = \Delta G^\circ + RT \ln K$$

$$-nFE = -nFE^\circ + RT \ln K \ (\text{since } \Delta G = -nFE)$$

$$E = E^\circ - (RT/nF) \ln K = E^\circ - (2.303 \, RT/F) \times 1/n \log\{[\text{Red}]/[\text{Ox}]\}$$

$$E = E^\circ - 0.059/n \log\{[\text{Red}]/[\text{Ox}]\} \tag{1}$$

[2.303 $RT/F = 2.303 \times 8.314 \text{J K}^{-1}\text{mol}^{-1} \times 298 \text{ K}/96 \ 500 \text{ J V}^{-1}$ mol^{-1} = 0.059 V; 1 F = 96 500 C mol^{-1} = 96 500 J V^{-1} mol^{-1}, since 1V = 1J/1C)]

This is the *Nernst equation* [German chemist W. Nernst (1864–1941); Nobel Prize in Chemistry: 1920]

A selected example on redox reaction:

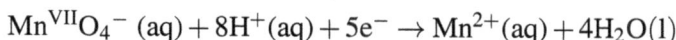

$$\text{Mn}^{\text{VII}}\text{O}_4^- \, (\text{aq}) + 8\text{H}^+(\text{aq}) + 5\text{e}^- \rightarrow \text{Mn}^{2+}(\text{aq}) + 4\text{H}_2\text{O}(\text{l})$$

$$(E^\circ = 1.5 \text{ V}; \, [\text{H}^+] = 1.0 \text{ M})$$

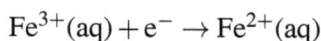

$$\text{Fe}^{3+}(\text{aq}) + \text{e}^- \rightarrow \text{Fe}^{2+}(\text{aq})$$

$$(E^\circ = 0.77 \text{ V}; [\text{H}^+] = 1.0 \text{ M})$$

Higher the value of E° of the redox couple, the oxidized form of the redox partners ($\text{Mn}^{\text{VII}}\text{O}_4^-/ \text{Mn}^{2+}$) is the stronger oxidizing agent. In the cited example, one equiv of MnO_4^- will oxidize 5 equiv of Fe^{2+} (the redox partners: $\text{Fe}^{3+}/\text{Fe}^{2+}$) [Note: E° (oxidized form/reduced form)].

Now,

$$\Delta G^\circ = -nFE^\circ \quad \text{and} \quad \Delta G^\circ = -RT \ln K$$

Hence,

$$-RT \ln K = -nFE^\circ$$

Then,

$$(RT/F) \ln K = nE^\circ; \, 2.303 \, (RT/F) \log K = nE^\circ$$

i.e.

$$0.059 \log K = nE^\circ$$

$$\log K = nE^\circ/0.059 \tag{2}$$

4.3 Stability field of water

Let us discuss how the redox potential for the oxidation and reduction of water changes as a function of pH.

Oxidation of water:

$$O_2(g) + 4H^+(aq) + 4e^- \rightarrow 2H_2O(l); \quad E^\circ = 1.23\,V \text{ vs NHE/SHE};$$

$$[H^+] = 1.0\,M \text{ (acidic medium)}$$

(normal hydrogen electrode/standard hydrogen electrode)

$$E = E^\circ - (0.059/4) \log\{[H_2O(l)]^2/[O_2(g)][H^+(aq)]^4\}$$

$$E = E^\circ - 0.059\,pH$$

(E = working potential; E° = *standard redox potential*)

[concentration (preciously it is activity) of pure solid and liquid is unity. Hence, for pure $H_2O(l)$ it is 1]
   At pH $= 7$, $[H^+] = 10^{-7}$ M (neutral medium)

$$E = 1.23 - 0.059\,pH = 1.23 - 0.059 \times 7 \simeq 0.82\,V$$

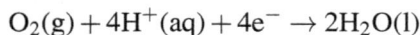

$$O_2(g) + 4H^+(aq) + 4e^- \rightarrow 2H_2O(l)$$

At pH $= 14$, $[H^+] = 10^{-14}$ M; i.e. $[OH^-] = 1.0$ M (basic medium)

$$E = 1.23 - 0.059\,pH = 1.23 - 0.059 \times 14 \simeq 0.40\,V$$

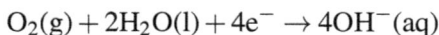

$$O_2(g) + 2H_2O(l) + 4e^- \rightarrow 4OH^-(aq)$$

Reduction of water:

$$2H^+(aq) + 2e^- \rightarrow H_2(g); \quad E^\circ = 0.0\,V \text{ (acidic medium)}$$

($H_2(g)$ pressure $= 1$ bar, $[H^+(aq)] = 1.0$ M, and $T = 298$ K; by definition)

$$E = E^\circ - (0.059/2) \log\{(p(H_2)/[H^+(aq)]^2\}; \quad E = E^\circ - 0.059\,pH$$
[$\log p = 0$ when $p = 1$ bar]

At pH $= 7$, $[H^+] = 10^{-7}$ M (neutral medium)

$$E = 0.0 - 0.059 \text{ pH} = 0.0 - 0.059 \times 7 \simeq -0.42 \text{ V}$$

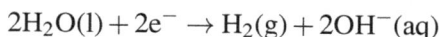

$$2H_2O(l) + 2e^- \rightarrow H_2(g) + 2OH^-(aq)$$

At pH $= 14$, $[H^+] = 10^{-14}$ M; $[OH^-] = 1.0$ M (basic medium)

$$E = 0.0 - 0.059 \text{ pH} = 0.0 - 0.059 \times 14 \simeq -0.84 \text{ V}$$

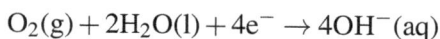

$$O_2(g) + 2H_2O(l) + 4e^- \rightarrow 4OH^-(aq)$$

The reduction potential becomes more negative (cathodic) by 59 mV or 0.059 V for each unit increase in pH.

Water is stable at pH 7 in the potential range –0.42 to 0.82 V vs NHE. This is called stability field of water.

4.4 Calculation of cell potential and equilibrium constant

Example 4.4 Given E° (acidic conditions) for O_2 (g)/H_2O_2 (aq) $= 0.70$ V and for $H_2O_2(aq)$/$H_2O(l) = 1.76$ V.

Estimate E° (acidic conditions) for $O_2(g)$/$H_2O(l)$ couple. Comment on the thermodynamic stability of $H_2O_2(aq)$ under acidic conditions.

Answer

$$O_2(g) + 2e^- + 2H^+(aq) \rightarrow H_2O_2(aq); \Delta G^\circ = -2F \times 0.70 \text{ kcal/mol}$$

$$H_2O_2(aq) + 2e^- + 2H^+(aq) \rightarrow 2H_2O(l); \Delta G^\circ = -2F \times 1.76$$

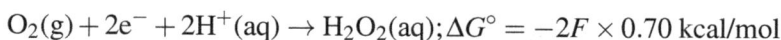

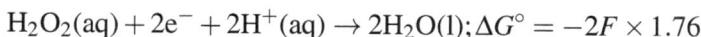

Adding two half-cell reactions,

$$O_2(g) + 4e^- + 4H^+(aq) \rightarrow 2H_2O(l)$$

$$\Delta G^\circ = -2F \times (0.70 + 1.76) = -2F \times 2.46 = -4F \times 1.23$$

Hence, E° (acidic conditions) $= 1.23$ V

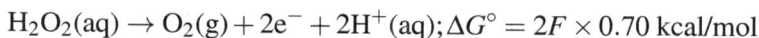

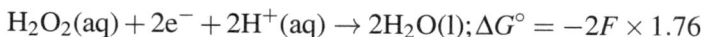

$$H_2O_2(aq) \rightarrow O_2(g) + 2e^- + 2H^+(aq); \Delta G^\circ = 2F \times 0.70 \text{ kcal/mol}$$

$$H_2O_2(aq) + 2e^- + 2H^+(aq) \rightarrow 2H_2O(l); \Delta G^\circ = -2F \times 1.76$$

Adding two half-cell reactions,

$$2H_2O_2(aq) \rightarrow 2H_2O(l) + O_2(g);$$

$$\Delta G° = -2F \times (-0.70 + 1.76) = -2F \times 1.06$$

Therefore, under acidic conditions H_2O_2 (aq) is expected to disproportionate to $H_2O(l)$ and $O_2(g)$.

Example 4.5
Given:

$$Cu^+(aq) + e^- \rightarrow Cu(s); \ E° = 0.52 \ V$$

$$Cu^{2+}(aq) + e^- \rightarrow Cu^+(aq); \ E° = 0.16 \ V$$

Justify why the reaction (disproportionation), $2Cu^+(aq) \rightarrow Cu^{2+}(aq) + Cu(s)$, is thermodynamically favorable.

Answer

$$Cu^+(aq) + e^- \rightarrow Cu(s); \ \Delta G° = -nFE° = -1F \times 0.52$$

$$Cu^{2+}(aq) + e^- \rightarrow Cu^+(aq); \ \Delta G° = -1F \times 0.16$$

For $Cu^+(aq) \rightarrow Cu^{2+}(aq) + e^-; \ \Delta G° = 1F \times 0.16$
    Adding two half-cell reactions,

$$2Cu^+(aq) \rightarrow Cu^{2+}(aq) + Cu(s)$$

$$\Delta G° = -1F \times (0.52 - 0.16) = -1F \times 0.36 \ kcal/mol$$

Hence, $E°$ (net potential) $= 0.36 \ V$.
    From eq 2, $\log K = 1 \times 0.36/0.059 \simeq 6.1; \ K \simeq (10)^{6.1} \ (K \simeq 1.26 \times 10^6)$
    Hence, the disproportionation of $Cu^+(aq)$ to $Cu^{2+}(aq)$ and $Cu(s)$ is justified.

Example 4.6
Given:

$$O_2(g) + 4H^+(aq) + 4e^- \rightarrow 2H_2O(l); \ E° = 0.82 \ V$$

$$NAD^+(aq) + H^+(aq) + 2e^- \rightarrow NADH(aq); \ E° = -0.32 \ V$$

Justify that the span of the respiratory chain is 1.14 V, which corresponds to ~ 53 kcal/mol.

Answer

For O_2 (g) $+ 4H^+$ (aq) $+ 4e^- \rightarrow 2H_2O(l)$; $\Delta G° = -4F \times 0.82$

    For $1/2O_2$ (g) $+ 2H^+$ (aq) $+ 2e^- \rightarrow H_2O(l)$; $\Delta G° = -2F \times 0.82$

    For NAD^+ (aq) $+ H^+$ (aq) $+ 2e^- \rightarrow NADH$ (aq); $\Delta G° = -2F \times -0.32$

    For NADH (aq) $\rightarrow NAD^+$ (aq) $+ H^+$ (aq) $+ 2e^-$; $\Delta G° = -2F \times 0.32$

    Adding two half-cell reactions, $1/2O_2(g) + H^+(aq) + NADH$ (aq) $\rightarrow$ $H_2O(l) + NAD^+(aq)$

$$\Delta G° = -2F \times 1.14 \text{ kcal mol}^{-1}; \ E° = 1.14 \text{ V}$$

$$= -2 \times 23.06 \text{ kcal mol}^{-1}\text{V}^{-1} \times 1.14 \text{ V}$$

$$= -52.6 \text{ kcal/mol} \simeq -53 \text{ kcal/mol}$$

$$(1F = 23.06 \text{ kcal mol}^{-1} \text{ V}^{-1})$$

Example 4.7 It is a known that Cu(II) ion can be estimated iodometrically by titrating the I_2 liberated from I^- ion:

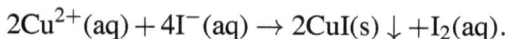

$$2Cu^{2+}(aq) + 4I^-(aq) \rightarrow 2CuI(s) \downarrow + I_2(aq).$$

A consideration of the $E°$ values:

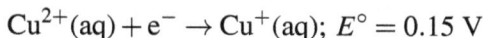

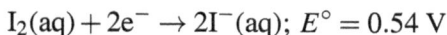

$$Cu^{2+}(aq) + e^- \rightarrow Cu^+(aq); \ E° = 0.15 \text{ V}$$

$$I_2(aq) + 2e^- \rightarrow 2I^-(aq); \ E° = 0.54 \text{ V}$$

The $E°$ values would indicate a reverse reaction: $2Cu^+(aq) + I_2(aq) \rightarrow 2Cu^{2+}(aq) + 2I^-(aq)$. Explain the anomaly.

Answer

This anomaly can be explained by considering the sparing solubility of CuI(s) (actually, $Cu_2I_2(s)$). Assuming the concentration of I^- (aq) as 0.1 M and taking the solubility product of CuI(s) as 10^{-12}, we have,

$$K_{sp} = [Cu^+][I^-]; [Cu^+] = 10^{-12}/10^{-1} = 10^{-11}M$$

The $E°$ of the $Cu^{2+}(aq)/Cu^+(aq)$ couple then becomes,

$$E(\text{apparent}) = 0.15 - 0.059/1 \log \{10^{-11}/[Cu^{2+}(aq)]\}$$

$$= 0.799 + 0.059 \log [Cu^{2+}(aq)]$$

Since 0.799 is larger than the $E°$ value of the $I_2(s)/I^-$ (aq) couple, it follows that the preferred reaction is

$$2Cu^{2+}(aq) + 4I^-(aq) \rightarrow 2CuI(s) \downarrow + I_2(aq)$$

Example 4.8   Given:

$$Mn^{2+}(aq)/Mn(s) \text{ (acidic conditions)}; E° = -1.19 \text{ V}$$

The solubility product (K_{sp}) of $Mn(OH)_2(s) = 2 \times 10^{-13}$

What is the $E°$ value (basic conditions) for $Mn(OH)_2(aq)/Mn(s)$ couple?

Answer

$Mn(OH)_2(s) \rightarrow Mn^{2+}(aq) + 2OH^-(aq)$

$\quad K_{sp} = [Mn^{2+}(aq)][OH^-(aq)]^2$

$\quad K_{sp} = [Mn^{2+}(aq)] = 2 \times 10^{-13}$ (since in alkaline medium, $[OH^-(aq)]$
$\simeq 1.0$ M)

$\quad$ Acidic conditions: Mn^{2+} (aq) $+ 2e^- \rightarrow Mn(s)$

E(basic conditions) $= E°$(acidic conditions) $- 0.059/2 \log\{1/[Mn^{2+}(aq)]\}$

$= -1.19 - 0.059/2 \log\{1/(2 \times 10^{-13})\} = -1.56 \text{ V}$

Therefore, $E°$ for $Mn(OH)_2(s) + 2e^- \rightarrow Mn(s) + 2OH^-$ (aq) redox process
$= -1.56$ V

Example 4.9

Given (under acidic conditions):

$$NO_3^-(aq)/N_2O_4(g); E° = 0.79 \text{ V}$$

$$N_2O_4(g)/HNO_2(aq); E° = 1.07 \text{ V}$$

$$HNO_2(aq)/NO(g); E° = 0.996 \text{ V}$$

Calculate the potential for the redox couple $NO_3^-(aq)/HNO_2(aq)$.

Also, calculate the potential for the redox couple $NO_3^-(aq)$, $NO(g)/HNO_2(aq)$.

Answer

$2NO_3^-$ (aq) $+ 2e^- + 4H^+$ (aq) $\rightarrow N_2O_4(g) + 2H_2O(l); \Delta G° = -2F \times 0.79$ kcal/mol

$\quad$ Hence, NO_3^- (aq) $+ e^- + 2H^+$ (aq) $\rightarrow 1/2N_2O_4(g) + H_2O(l); \Delta G° = -1F \times 0.79$

$N_2O_4(g) + 2e^- + 2H^+ (aq) \rightarrow 2HNO_2(aq); \Delta G° = -2F \times 1.07$

Hence, $1/2N_2O_4(g) + e^- + H^+ (aq) \rightarrow HNO_2(aq); \Delta G° = -1F \times 1.07$

Adding two half-cell reactions,

$$NO_3^- (aq) + 2e^- + 3H^+ (aq) \rightarrow HNO_2(aq) + H_2O(l);$$

$$\Delta G° = -1F \times (0.79 + 1.07)$$

$$= -1F \times 1.86$$

$$= -2F \times 0.93 \text{ kcal/mol}$$

Therefore, $E°$ for $NO_3^- (aq)/HNO_2(aq)$ couple $= 0.93$ V

$NO_3^- (aq) + 2e^- + 3H^+ (aq) \rightarrow HNO_2 (aq) + H_2O(l); \Delta G° = -2F \times 0.93$ kcal/mol

$HNO_2 (aq) + H^+ (aq) + e^- \rightarrow NO(g) + H_2O(l); \Delta G° = -1F \times 0.996$

Hence, $NO(g) + H_2O(l) \rightarrow HNO_2 (aq) + H^+ (aq) + e^-; \Delta G° = 1F \times 0.996$

Adding two half-cell reactions,

$$NO_3^- (aq) + NO(g) + 2H^+ (aq) + e^- \rightarrow 2HNO_2(aq)$$

$$\Delta G° = -1F \times (2 \times 0.93 - 0.996)$$

$$= -1F \times 0.864$$

Hence, under acidic conditions $E°$ for the redox couple $(NO_3^-(aq), NO(g)/HNO_2(aq)) = 0.864$ V

Example 4.10

Given:

$E°$ (basic conditions) for $NO_3^- (aq)/NO_2^- (aq) = 0.1$ V and for $NO_2^-(aq)/NO(g) = -0.46$ V.

Calculate $E°$ (basic conditions) for the $NO_3^-(aq)$, $NO(g)/NO_2^- (aq)$ couple.

Answer

$NO_3^- (aq) + H_2O(l) + 2e^- \rightarrow NO_2^- (aq) + 2OH^-(aq); \Delta G° = -2F \times 0.1$ kcal mol^{-1}

$2NO_2^- (aq) + 2H_2O(l) + 2e^- \rightarrow 2NO(g) + 4OH^-(aq); \Delta G° = -2F \times -0.46 = 2F \times 0.46$

Hence, $2NO(g) + 4OH^-(aq) \rightarrow 2NO_2^-(aq) + 2H_2O(l) + 2e^-$
$\Delta G^\circ = -2F \times 0.46$

Therefore, for the reaction

$$NO_3^-(aq) + 2NO(g) + 2OH^-(aq) \rightarrow 3NO_2^-(aq) + H_2O(l)$$

$$\Delta G^\circ = -2F \times (0.1 + 0.46)$$

$$= -2F \times 0.56 \text{ kcal mol}^{-1}$$

The E° (basic conditions) value for NO_3^- (aq), $NO(g)/NO_2^-$ (aq) $= 0.56$ V

Example 4.11

Given:

E° for ClO_4^- (aq)/ClO_3^- (aq) (acidic conditions) $= 1.19$ V

Calculate the E° value of this redox process in basic conditions.

Answer

$$ClO_4^-(aq) + 2H^+(aq) + 2e^- \rightarrow ClO_3^-(aq) + H_2O(l)$$

$$E = E^\circ \text{(acidic)} - 0.059/2 \log [ClO_3^-(aq)][H_2O(l)]/[ClO_4^-(aq)][H^+(aq)]^2$$

Under standard conditions, all concentrations are unity; $E = E^\circ$.

Under basic conditions, $[ClO_3^-(aq)]$, $[ClO_4^-(aq)]$, and $[OH^-(aq)]$ are unity but $[H^+(aq)]$ will be much less i.e. $[H^+] \simeq 1 \times 10^{-14}$ M, then,

$$E^\circ \text{(basic)} = 1.19 - 0.059/2 \log\{1/[1 \times 10^{-14}]^2\} = 0.36 \text{ V}$$

The redox reaction (basic condition): ClO_4^- (aq) $+ H_2O(l) + 2e^- \rightarrow ClO_3^-$ (aq) $+ 2OH^-$ (aq) Thus, the oxyacids are poorer oxidizing agents under basic conditions.

Example 4.12

Given (under basic conditions):

$$E^\circ \text{ for } Br_2(l)/Br^-(aq) = 1.07 \text{ V}$$

$$E^\circ \text{ for } BrO^-(aq)/Br_2(l) = 0.45 \text{ V}$$

$$E^\circ \text{ for } BrO^-(aq)/Br^-(aq) = 0.76 \text{ V}$$

$$E^\circ \text{ for } BrO_3^-(aq)/BrO^-(aq) = 0.54 \text{ V}$$

Estimate E° (basic conditions) for $Br_2(l)/Br^-(aq)$, $BrO^-(aq)$ redox couple.

Decide whether $Br_2(l)$ disproportionates to Br^- (aq) and BrO_3^- (aq).

Also decide whether BrO^-(aq) is expected to disproportionate to BrO_3^- (aq) and Br^- (aq).

Answer

$$Br_2(g) + 2e^- \rightarrow 2Br^-(aq); \Delta G^\circ = -2F \times 1.07 \text{ kcal/mol}$$

$$2BrO^-(aq) + 2H_2O(l) + 2e^- \rightarrow Br_2(l) + 4OH^-(aq); \Delta G^\circ = -2F \times 0.45$$

Hence, $Br_2(l) + 4OH^-(aq) \rightarrow 2BrO^-$ (aq) $+ 2H_2O(l) + 2e^-$; $\Delta G^\circ = 2F \times 0.45$

Adding two half-cell reactions,

$$2Br_2(l) + 4OH^-(aq) \rightarrow 2Br^-(aq) + 2BrO^-(aq) + 2H_2O(l)$$

$$\Delta G^\circ = -2F \times (1.07 - 0.45) = -2F \times 0.62$$

Hence, E° (basic conditions) for $Br_2(l)/ Br^-(aq)$, BrO^- (aq) redox couple $= 0.62$ V

Thus, under basic conditions the disproportionation reaction ($Br_2(l) \rightarrow Br^-$ (aq) $+ 2BrO^-$ (aq) must take place, as ΔG° is negative.

$$2BrO^-(aq) + 2H_2O(l) + 4e^- \rightarrow 2Br^-(aq) + 4OH^-(aq)$$

$$\Delta G^\circ = -4F \times 0.76 \text{ kcal/mol}$$

$$BrO_3^-(aq) + 2H_2O(l) + 4e^- \rightarrow BrO^-(aq) + 4OH^-(aq)$$

$$\Delta G^\circ = -4F \times 0.54$$

Therefore, for $BrO^-(aq) + 4OH^-(aq) \rightarrow BrO_3^-(aq) + 2H_2O(l) + 4e^-$ $\Delta G^\circ = 4F \times 0.54$ kcal/mol

Adding two half-cell reactions, $3BrO^-(aq) \rightarrow BrO_3^-(aq) + 2Br^-(aq)$ $\Delta G^\circ = -4F \times (0.76 - 0.54) = -4F \times 0.22$ kcal/mol

Therefore, under basic conditions $BrO^-(aq)$ is expected to disproportionate to $BrO_3^-(aq)$ and $Br^-(aq)$.

4.5 Inner-sphere electron transfer concepts through electrochemical studies

In 1969, a historical report from Creutz and Taube [Canadian-born American chemist Henry Taube (1915–2005): Nobel Prize in Chemistry in

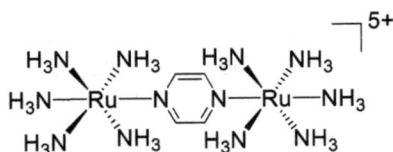

Fig. 4.1 The Creutz-Taube ion.

1983] described an intervalence charge transfer band in a symmetrical mixed-valence system, in which the initial and final electronic states were degenerate. This system is known as the Creutz-Taube ion [(NH$_3$)$_5$Ru(μ-pyrazine)Ru(NH$_3$)$_5$]$^{5+}$, where each Ru is considered to have an oxidation state of 2.5+ (Fig. 4.1).[3]

The Creutz-Taube ion and its analogs provide an excellent platform for studying inner-sphere electron transfer mechanisms. Comparing electrochemical data corresponding to the redox events of the mixed-valence system with 5+ charge to that of the reactant with 4+ charge and doubly oxidized 6+ species, we can obtain equilibrium constants that relate the degree of electronic communication (electronic coupling) between the redox centers.

$$[M^{3+} - M^{3+}] + [M^{2+} - M^{2+}] \overset{K_c}{\rightleftharpoons} 2[M^{3+} - M^{2+}]$$

We can use the Nernst equation to determine the comproportionation equilibrium constant, K_c.

Redox process 1:

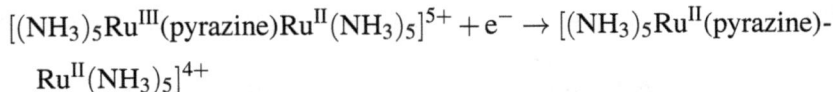

$$[(NH_3)_5Ru^{III}(pyrazine)Ru^{II}(NH_3)_5]^{5+} + e^- \rightarrow [(NH_3)_5Ru^{II}(pyrazine)\text{-}$$
$$Ru^{II}(NH_3)_5]^{4+}$$

$(E^{o\prime}$ (formal potential) $= 0.35\,V$ vs NHE)

$\Delta G^{o\prime}(1) = -1F \times 0.35$ kcal/mol

---

[3] (a) C. Creutz and H. Taube, *J. Am. Chem. Soc.* **1969**, *91*, 3988. (b) C. Creutz and H. Taube, *J. Am. Chem. Soc.* **1973**, *95*, 1086. (c) P. A. Lay, R. H. Magnuson, and H. Taube, *Inorg. Chem.* **1988**, *27*, 2364.

Hence,

$$[(NH_3)_5Ru^{II}(\text{pyrazine})Ru^{II}(NH_3)_5]^{4+} \rightarrow [(NH_3)_5Ru^{III}(\text{pyrazine})\text{-}$$
$$Ru^{II}(NH_3)_5]^{5+} + e^-$$

$$\Delta G^{\circ\prime}(1) = 1F \times 0.35 \text{ kcal/mol}$$

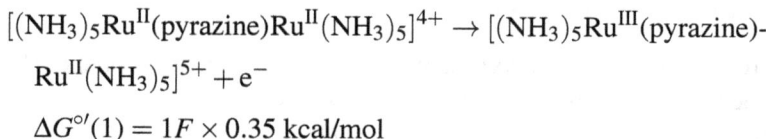

Redox process 2:

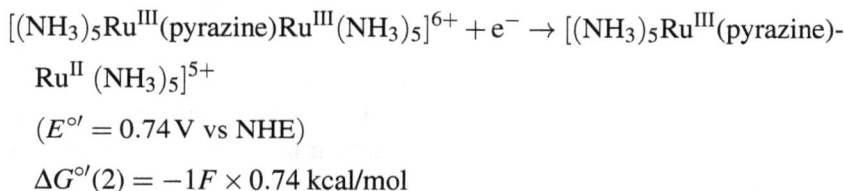

$$[(NH_3)_5Ru^{III}(\text{pyrazine})Ru^{III}(NH_3)_5]^{6+} + e^- \rightarrow [(NH_3)_5Ru^{III}(\text{pyrazine})\text{-}$$
$$Ru^{II}(NH_3)_5]^{5+}$$

$$(E^{\circ\prime} = 0.74\,V \text{ vs NHE})$$

$$\Delta G^{\circ\prime}(2) = -1F \times 0.74 \text{ kcal/mol}$$

Hence, for the reaction:

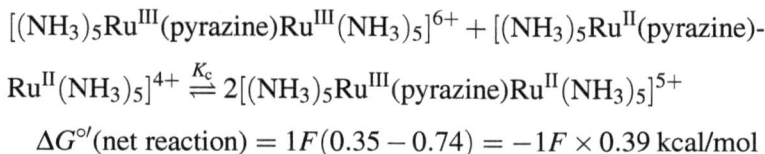

$$[(NH_3)_5Ru^{III}(\text{pyrazine})Ru^{III}(NH_3)_5]^{6+} + [(NH_3)_5Ru^{II}(\text{pyrazine})\text{-}$$
$$Ru^{II}(NH_3)_5]^{4+} \overset{K_c}{\rightleftharpoons} 2[(NH_3)_5Ru^{III}(\text{pyrazine})Ru^{II}(NH_3)_5]^{5+}$$

$$\Delta G^{\circ\prime}(\text{net reaction}) = 1F(0.35 - 0.74) = -1F \times 0.39 \text{ kcal/mol}$$

From eq 2, log $K_c = nE^{\circ}/\,0.059 = 0.39\,/\,0.059 = 6.61$; $K_c \simeq 4.1 \times (10)^6$

Once K_c is known, the Robin-Day classification[4] can aid us in determining the degree of electronic communication between systems:

$$\text{Class I}: K_c < 10^2$$

$$\text{Class II } 10^2 \leq K_c \leq 10^6$$

$$\text{Class III } K_c > 10^6$$

Thus, Creutz-Taube ion belongs to Class III highly delocalized mixed-valence system.

---

[4](a) M. B. Robin and P. Day, *Adv. Inorg. Chem. Radiochem.* **1967**, *10*, 247. (b) K. Ventura, M. B. Smith, J. R. Prat, L. E. Echegoyen, and D. Villagràin, *J. Chem. Educ.* **2017**, *94*, 526.

4.6 Proton-coupled electron transfer (PCET) and hydrogen atom transfer (HAT)

The formation of a phenoxyl radical (PhO$^\bullet$) from a neutral phenol (PhOH) can occur via a direct hydrogen atom transfer (HAT) or via a proton-coupled electron transfer (PCET) (Scheme 4.1) process. For the HAT and PCET mechanisms, homolytic O–H bond cleavage constitutes the rate-determining step of the reaction. In the HAT mechanism, the proton H$^+$ and the electron of the H$^\bullet$ radical both come from the same orbital. Conversely, proton and electron transfers are both rate-determining for the PCET process but occur from different orbitals in a concerted mechanism. Alternatively, the proton and electron transfers can be uncoupled (PT–ET) with either proton transfer (PT) or electron transfer (ET) being the rate-determining step (Scheme 4.1).

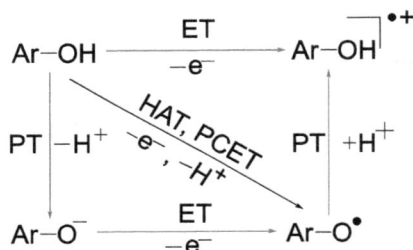

Scheme 4.1 Possible reaction pathways for the oxidation of phenols.

Proton-coupled electron transfer (PCET) is a fundamental and ubiquitous reaction process in many kinds of redox reactions.[5] In a PCET reaction, a proton is transferred to a proton-accepting site that should be basic and an electron is also transferred to an electron-accepting site that should have a hole. The two processes occur separately. The concept of PCET is described in Scheme 4.2.

---

[5](a) M. H. V. Huynh and T. J. Meyer, *Chem. Rev.* **2007**, *107*, 5004. (b) J. J. Warren, T. A. Tronic, and J. M. Mayer, *Chem. Rev.* **2010**, *110*, 6961. (c) D. R. Weinberg, C. J. Gagliardi, J. F. Hull, C. F. Murphy, C. A. Kent, B. C. Westlake, A. Paul, D. H. Ess, D. G. McCafferty, and T. J. Meyer, *Chem. Rev.* **2012**, *112*, 4016. (d) S. Kundu, E. Miceli, E. R. Farquhar, and K. Ray, *Dalton Trans.* **2014**, *43*, 4264.

The protonaton and redox reaction can be put into a simple thermodynamic cycle (Scheme 4.2):

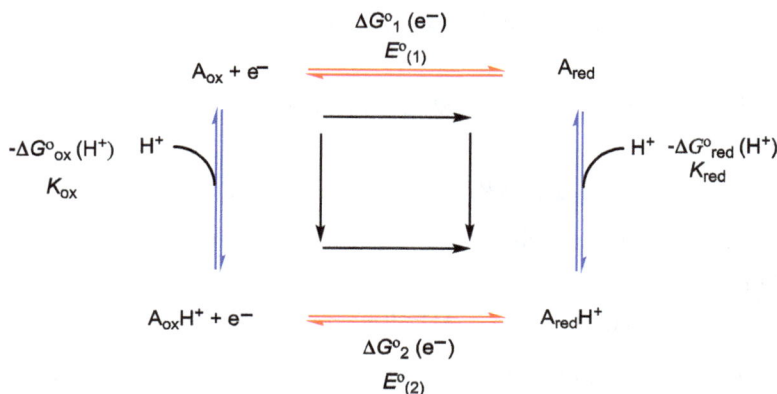

Scheme 4.2 The concept of PCET.

The deprotonation equilibria:

$$A_{Ox}H^+ \rightleftharpoons A_{Ox} + H^+ \quad \text{and} \quad A_{Red}H^+ \rightleftharpoons A_{Red} + H^+$$

Therefore,

$$K_{Ox} = [A_{Ox}][H^+]/[A_{Ox}H^+] \quad \text{and} \quad K_{Red} = [A_{Red}][H^+]/[A_{Red}H^+]$$

$$pK_{Ox} = -\log K_{Ox} \quad \text{and} \quad pK_{Red} = -\log K_{Red}$$

or,

$$[A_{Ox}H^+] = [A_{Ox}][H^+]/K_{Ox} \quad \text{and} \quad [A_{Red}H^+] = [A_{Red}][H^+]/K_{Red}$$

$$[A_{Ox}H^+]/[A_{Ox}] = [H^+]/K_{Ox} \quad \text{and} \quad [A_{Red}H^+]/[A_{Red}] = [H^+]/K_{Red}$$

The number of different species present in solution can be identified as A_{Ox}, $A_{Ox}H^+$, A_{Red}, and $A_{Red}H^+$. The protonation reactions are written in the direction of deprotonation, and the equilibrium constant is a proton dissociation equilibrium constant, used to define the pKs.

Now let us consider two reduction processes.

Reduction of the oxidized species, $A_{Ox} + e^- \rightarrow A_{Red}$

$$E = E_{(1)}^\circ - 2.303\, RT/nF \log\{([A_{Red}]/[A_{Ox}])\}$$

where $E_{(1)}^{\circ}$ is the thermodynamic redox potential for A_{Ox}/A_{Red} redox process; E is the experimentally determined solution potential.

$$= E_{(1)}^{\circ} - 0.059 \log\left([A_{Red}]/[A_{Ox}]\right)(n = 1, T = 298\,\mathrm{K})$$

and

Reduction of the protonated oxidized species, $A_{Ox}(H^+) + e^- \rightarrow A_{Red}(H^+)$

$$E = E_{(2)}^{\circ} - 2.303\,RT/nF\ \log\left\{[A_{Red}H^+]/[A_{Ox}H^+]\right\}$$

where $E_{(2)}^{\circ}$ is the thermodynamic redox potential for $A_{ox}(H^+)/A_{red}(H^+)$ redox process.

$$= E_{(2)}^{\circ} - 0.059 \log\left([A_{Red}H^+]/[A_{Ox}H^+]\right)(n = 1, T = 298\,\mathrm{K})$$

In Scheme 4.2 the minus signs, $-\Delta G_{Ox}^0(H^+)$ and $-\Delta G_{Red}^0(H^+)$, come from the fact that we defined these in the direction of deprotonation.

There are two different pathways to go from the oxidized species A_{Ox} to the protonated reduced species, $A_{Red}H^+$. Since, the transformed Gibbs reaction free-energy G is a state function, the ΔG° must be identical no matter which way we go. A_{Ox} can be either protonated first and then reduced, or reduced first and then protonated. The free-energy change will be the same.

Then,

$$-\Delta G_{Ox}^{\circ}(H^+) + \Delta G_{(2)}^{\circ}(e^-) = +\Delta G_{(1)}^{\circ}(e^-) - \Delta G_{Red}^{\circ}(H^+)$$

It follows that,

$$\Delta G_2^{\circ}(e^-) - \Delta G_1^{\circ}(e^-) = -\Delta G_{Red}^{\circ}(H^+) + \Delta G_{Ox}^{\circ}(H^+)$$

$$-FE_{(2)}^{\circ} + FE_{(1)}^{\circ} = RT \log K_{Red} - RT \log K_{Ox} = RT \log\left(K_{Red}/K_{Ox}\right)$$

or,

$$E_{(2)}^{\circ} - E_{(1)}^{\circ} = -RT/F \log\left(K_{Red}/K_{Ox}\right) = -0.059 \log\left(K_{Red}/K_{Ox}\right)$$

Hence,

$$E_{(2)}^{\circ} - E_{(1)}^{\circ} = 0.059 \log\left(K_{Ox}/K_{Red}\right) = 0.059(pK_{Red} - pK_{Ox}) \qquad (3)$$

If the difference in the E° values between the protonated and deprotonated forms is specified, this also defines the difference between the pK values of the reduced and oxidized forms.

Although we have defined $E^{\circ}_{(1)}$ and $E^{\circ}_{(2)}$, these cannot be directly measured. Instead, we measure the *apparent/formal* potential $E^{\circ\prime}$ $(E_{1/2})$ (see below).

Let us now consider that we are keeping the pH constant during the reaction. The value of the transformed thermodynamic parameter, $E^{\circ\prime}$ $(E_{1/2})$, is what we can actually measure, since we have no way to know whether the species is protonated or not. It is useful to see how $E^{\circ\prime}$ $(E_{1/2})$ is related to the non-transformed reduction potentials of the unprotonated and protonated species $E^{\circ}_{(1)}$ and $E^{\circ}_{(2)}$, respectively.

From Scheme 4.2, we have,

$$E = E^{\circ\prime}(E_{1/2}) - 0.059 \log\{[A_{Red} + A_{Red}H^+]/[A_{Ox} + A_{Ox}H^+]\}$$

Substituting from the proton-equilibria constants (deprotonation equilibria; see above),

$$E = E^{\circ\prime}(E_{1/2}) - 0.059 \log\{A_{Red}[1 + [H^+]/K_{Red}]/A_{Ox}[1 + [H^+]/K_{Ox}]\}$$

or,

$$E = E^{\circ\prime}(E_{1/2}) - 0.059 \log([A_{Red}]/[A_{Ox}])$$
$$- 0.059 \log\{[1 + [H^+]/K_{Red}]/[1 + [H^+]/K_{Ox}]\}$$

From the definition,

$$-0.059 \log\{[A_{Red}]/[A_{Ox}]\} = E - E^{\circ}_{(1)}$$

Therefore,

$$E = E^{\circ\prime}(E_{1/2}) + E - E^{\circ}_{(1)} - 0.059 \log\{[1 + [H^+]/K_{Red}]\}/$$
$$\{[1 + [H^+]/K_{Ox}]\}$$
$$E^{\circ\prime}(E_{1/2}) = E^{\circ}_{(1)} + 0.059 \log\{[1 + [H^+]/K_{Red}]/[1 + [H^+]/K_{Ox}]\} \quad (4)$$

As the proton concentration gets very low ($[H^+] \ll K_{Red}$ and $[H^+] \ll K_{Ox}$), eq 4 predicts that both $[1 + [H^+]/K_{Red}]$ and $[1 + [H^+]/K_{Ox}]$ will approach 1, so, the expression within the logarithm will also approach a value of 1. Since $\log 1 = 0$, at high pH (low $[H^+]$ concentration), $E^{\circ\prime}(E_{1/2}) \approx E^{\circ}_{(1)}$.

At high pH, $E^{\circ\prime}(E_{1/2}) \approx E^{\circ}_{(1)}$

At very acidic pH, the values of both $[H^+]/K_{Red}$ and $[H^+]/K_{Ox}$ become much larger than 1, so

Eq 4 becomes,

$$E^{o'}(E_{1/2}) \approx E_{(1)}^{\circ} + 0.059 \log \{[H^+]/K_{Red}\}/\{[H^+]/K_{Ox}]\}$$

$$E^{o'}(E_{1/2}) \approx E_{(1)}^{\circ} + 0.059 \log (K_{Ox}/K_{Red}) \tag{5}$$

Finally, we substitute from eq 3 to see that at low pH, we get,

$$E^{o'}(E_{1/2}) \approx E_{(1)}^{\circ} + 0.059 \log (K_{Ox}/K_{Red}) = E_{(1)}^{\circ} + (E_{(2)}^{\circ} - E_{(1)}^{\circ})$$

$$E^{o'}(E_{1/2}) \approx E_{(2)}^{\circ}, \text{ when } pH \ll pK_{Ox}, pK_{Red}$$

At very acidic pH, $E^{o'}(E_{1/2}) \approx E_{(2)}^{\circ}$

Thus, $E^{o'}(E_{1/2})$ has limiting values which corresponds to $E_{(2)}^{\circ}$ of the protonated form at low pH, and $E_{(1)}^{\circ}$ of the unprotonated form at high pH. When the pH is below the pK's of both the reduced and oxidized forms of A, then essentially A remains protonated whether it is oxidized or reduced, and there is no further dependence of $E^{o'}$ on pH. There is also no dependence of $E^{o'}$ on pH above the pK's of both the reduced and oxidized forms (pH > 8 in this problem, see below), and both the reduced and oxidized forms remain unprotonated. This effect is displayed in Fig. 4.2.

From eq 5,

$$E^{o'}(E_{1/2}) \approx 0.1 + 0.059 \log (10^2) \approx 0.1 + 0.118 \approx 0.22 \text{ V}$$

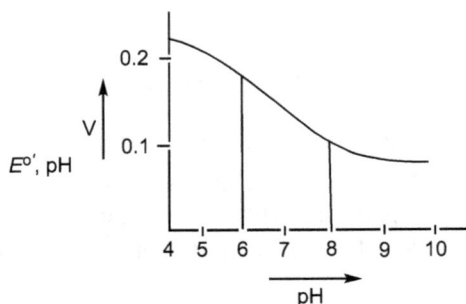

Fig. 4.2 Plot of eq 5, assuming a value of $E_{(1)}^{\circ} = 0.1$ V and $K_{Ox} = 10^{-6}$ (p$K_{Ox} = 6$) and $K_{Red} = 10^{-8}$ (p$K_{Red} = 8$).

(a) Effect of pH on the reduction potentials: quinone/hydroquinone redox couple

Let us consider the quinone (Q)/hydroquinone (H$_2$Q) redox couple. The half-call reaction for this couple (or for any redox couple involving protons) involves H$^+$ ions.[6]

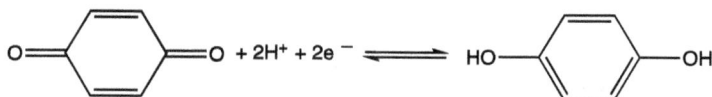

Changing the pH of the solution would alter the concentration of one of the species involved in the reaction and result in a shift in the redox potential.

The appropriate equation for the electrode potential of this half-cell reaction can be written as,

$$Ox + ne^- + mH^+ \rightleftharpoons Red$$

$$E = E^{o\prime}(H^+) - RT/nF \ \ln\{[Red]/[Ox][H^+]^m\}$$

$$= E^{o\prime}(H^+) - RT/nF \ \ln\{[Red]/[Ox]\} + RT/nF \ \ln[H^+]^m$$

$$= E^{o\prime}(H^+) - 0.059/n \log\{[Red]/[Ox]\} + 0.059(m/n) \log[H^+]$$

$$E = E^{o\prime}(H^+) - 0.059/n \log\{[Red]/[Ox]\} - 0.059(m/n)pH \qquad (6)$$

where $E^{o\prime}$ (H$^+$) is the formal electrode potential for the proton-coupled electron transfer (PCET) reaction.

In the redox process, where protons are not involved,

$$Ox + ne^- \rightarrow Red$$

$$E = E^{o\prime} - 0.059/n \log\{[Red]/[Ox]\}$$

where $E^{o\prime}$ is the formal electrode potential, without involvement of proton. When [Red] = [Ox], $E = E^{o\prime}$.

Substituting the value of $-0.059/n \log\{[Red]/Ox]\}$, eq 6 rearranges to,

$$E = E^{o\prime}(H^+) + E - E^{o\prime} - 0.059(m/n) \ pH$$

$$E^{o\prime} = E^{o\prime}(H^+) - 0.059(m/n) \ pH$$

$$E^{o\prime}(\simeq E_m \text{ or } E_{1/2}) = E^{o\prime}(H^+) - 0.059(m/n) \ pH \qquad (7)$$

---

[6](a) M. M. Walczak, D. A. Dryer, D. D. Jacobson, M. G. Foss, and N. T. Flynn, *J. Chem. Educ.* **1997**, *74*, 1195. (b) J. G. Mohanty and A. Chakravorty, *Inorg. Chem.* **1976**, *15*, 2912.

In eq 7, $E^{\circ\prime}$ is replaced by E_m, the measured redox potential of the redox process, when [Ox] = [Red]. Now, E is the measured potential i.e. $E = E^{\circ\prime} = E_m$.

In cyclic voltammetry (CV), E_m is approximated to half-wave potential, $E_{1/2}$. In the redox process, where protons are not involved ($m = 0$),

$$E = E^{\circ\prime}(\simeq E_m \text{ or } E_{1/2})$$

From eq 7 it is readily seen that,

$$m = -(n/0.059)(\Delta E_{1/2}/\Delta\text{pH}) \tag{8}$$

where $\Delta E_{1/2}$ is the shift of $E^{\circ\prime}$ due to the change in pH by ΔpH. The negative sign of eq 8 signifies that as pH increases, $E^{\circ\prime}$ for the couple decreases. Eq 8 provides a method for determination of m if n is known.

In CV, the potential applied to the working electrode is swept linearly between two switching potentials ca. 0.2 V past the anodic and cathodic peak potentials. Just prior to a cathodic sweep the electroactive species exists primarily as Q. As the electrode potential is swept cathodically the Q is reduced to H_2Q and the ratio [Q]/[H_2Q] approaches zero. Thus, a cycle between entirely oxidized and entirely reduced species occurs as the working electrode potential is swept cathodically between the switching potentials. For a more comprehensive understanding of the technique, the readers are referred to the articles on CV.[7]

The half-wave potential ($E_{1/2}$) is defined as the potential where the current is equal to one half of the limiting diffusion current. Usually, oxidant and reductant have similar diffusion rate. In that scenario, at the half-wave potential, the concentrations ratio of Ox and Red is unity, i.e. [Red]/[Ox] = 1.

The redox potential ($E^{\circ\prime}$) for Q/H_2Q redox process in 1 M acid is 0.50 V vs Ag |AgCl| saturated KCl. Under identical experimental conditions, a cyclic voltammogram (to determine $E^{\circ\prime}/E_{1/2}$) for a 1 mM solution of H_2Q in acetate/phosphate buffer of pH 1.6 was recorded to calculate the E ($E_{1/2}$) as 0.39 V. According to eq 7, $E(\simeq E^{\circ\prime}$ or $E_{1/2})$ should shift to more negative values as the solution pH increases. Cyclic voltammograms

---

[7](a) G. A. Mabbott, *J. Chem. Educ.* **1983**, *60*, 697. (b) J. J. Van Benschoten. J. Y. Lewis, W. R. Heineman, D. A. Roston, and P. T. Kissinger, *J. Chem. Educ.* **1983**, *60*, 772.

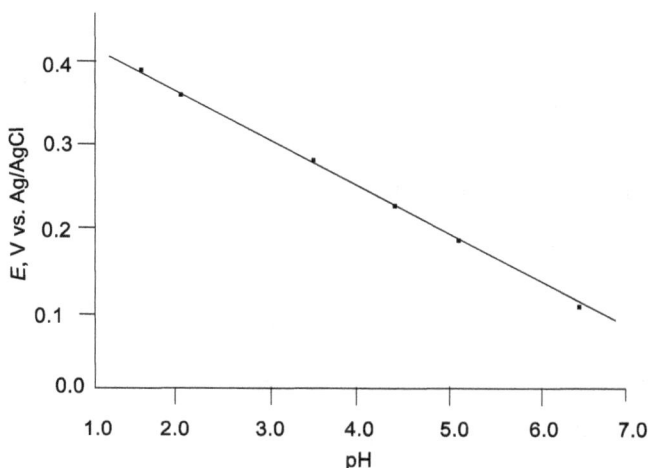

Fig. 4.3 Plot of $E^{\circ\prime}$ (H$^+$) ($E_{1/2}$ (H$^+$)) vs pH for 1 mM H$_2$Q in acetate/phosphate buffers of various pH.

were recorded for H$_2$Q solutions in the pH range 1.6 to 6.3. As predicted, the E values shift to more negative potentials as the solution pH increases. The pH dependence of potential can also be seen when E is plotted against pH (Fig. 4.3). The slope of the line is -0.057 (± 0.006) V/pH unit. From eq 7, the slope of the plot of E vs pH should be -0.059 V/pH unit. Using the slope obtained experimentally to calculate the number of electrons involved in the half-reaction, one finds for Q/H$_2$Q redox reaction to involve 2.07 ± 0.04 electrons. The value calculated is in reasonable agreement with that predicted for the reaction i.e. 2 electrons.

The Q/H$_2$Q redox couple is an excellent system for illustrating the Nernst equation. Changing the pH between 1 and 6 results in a total change in potential of 0.27 V. This large shift is easily discerned and follows the behavior expected from theory.

(b) Ru-bpy/py system

Clear evidence of PCET has been obtained for the presence of both III/II and IV/III redox couples in the '[Ru^{II}(bpy)$_2$(py)(OH$_2$)]$^{2+}$ system'.[8]

---

[8](a) B. A. Moyer and T. J. Meyer, *Inorg. Chem.* **1981**, *20*, 436. (b) R. A. Binstead and T. J. Meyer, *J. Am. Chem. Soc.* **1987**, *109*, 3287.

Consider the following redox reactions:

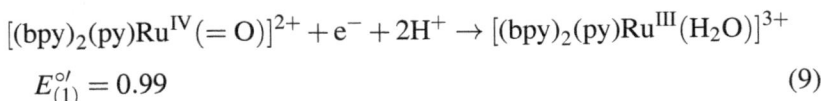

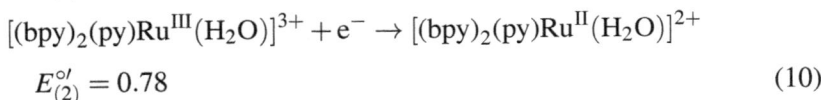

$$[(bpy)_2(py)Ru^{IV}(=O)]^{2+} + e^- + 2H^+ \rightarrow [(bpy)_2(py)Ru^{III}(H_2O)]^{3+}$$

$$E^{o\prime}_{(1)} = 0.99 \tag{9}$$

$$[(bpy)_2(py)Ru^{III}(H_2O)]^{3+} + e^- \rightarrow [(bpy)_2(py)Ru^{II}(H_2O)]^{2+}$$

$$E^{o\prime}_{(2)} = 0.78 \tag{10}$$

Potentials are in V vs SCE, $[H^+] = 1.0$ M.

Consider the following deprotonation reactions:

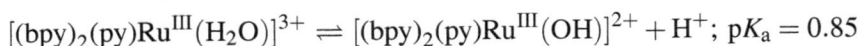

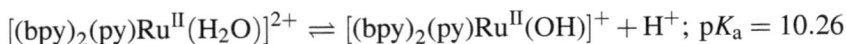

$$[(bpy)_2(py)Ru^{III}(H_2O)]^{3+} \rightleftharpoons [(bpy)_2(py)Ru^{III}(OH)]^{2+} + H^+; \ pK_a = 0.85$$

$$[(bpy)_2(py)Ru^{II}(H_2O)]^{2+} \rightleftharpoons [(bpy)_2(py)Ru^{II}(OH)]^+ + H^+; \ pK_a = 10.26$$

For pH < 1, Ru(III) species is present predominantly as the aqua complex, $[Ru^{III}(bpy)_2(py)(H_2O)]^{3+}$ ($pK_a = 0.85$), so that the Ru(IV)/(III) couple assumes a two-proton/one-electron pH dependence (-118 mV/pH), while the Ru(III)/(II) couple becomes pH independent. For pH > 10, $[Ru^{II}(bpy)_2(py)(H_2O)]^{2+}$ ($pK_a = 10.26$) is deprotonated to form the hydroxo complex. Electrochemical measurements for this pH range have shown that the Ru(III)/(II) couple becomes independent of pH, while the Ru(IV)/(III) couple continues to decrease with increasing pH.

The pH dependence of these redox couples can be calculated from eq 9 and eq 10 with use of the formal reduction potentials, $E^{o\prime}_{IV/III} = 0.99$ V and $E^{o\prime}_{III/II} = 0.78$ V vs SCE (determined experimentally), where $E_{IV/III}$ and $E_{III/II}$ are the calculated potentials and K_a^{III} and K_a^{II} are the acid dissociation constants for the Ru(III) and Ru(II) aqua complexes, respectively. When the above pK_a values are used, the redox potentials at pH 7 are calculated to be $E_{1/2} = 0.526$ V for Ru(IV)/(III) and $E_{1/2} = +0.416$ V for Ru(III)/(II).

$$E_{IV/III} = E^{o\prime}_{IV/III} - 0.059[pH - \log(K_a^{III} + [H^+])] \tag{11}$$

$$E_{III/II} = E^{o\prime}_{III/II} - 0.059[\log(K_a^{III} + [H^+]) - \log(K_a^{II} + [H^+])] \tag{12}$$

In the interval $2 < pH < 9$, where $[(bpy)_2(py)Ru^{III}(H_2O)]^{3+}$ is completely dissociated into the hydroxy complex, eq 9 and eq 10 no longer apply, and the Ru(IV)/Ru(III) and Ru(III)/Ru(I1) couples become

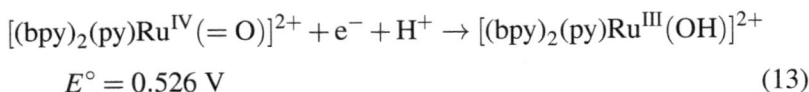

$$[(bpy)_2(py)Ru^{IV}(=O)]^{2+} + e^- + H^+ \rightarrow [(bpy)_2(py)Ru^{III}(OH)]^{2+}$$

$$E^\circ = 0.526 \text{ V} \tag{13}$$

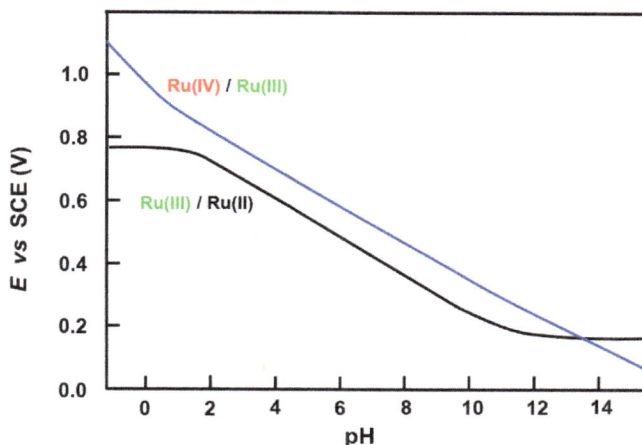

Fig. 4.4 Calculated pH-dependence of the redox potentials (V vs SCE) for the couples $[(bpy)_2(py)Ru^{IV}(=O)]^{2+}/[(bpy)_2(py)Ru^{III}(OH)]^{2+}$, and $[(bpy)_2(py)Ru^{III}(OH)]^{2+}/[(bpy)_2(py)Ru^{II}(H_2O)]^{2+}$ at $T = 298$ K.

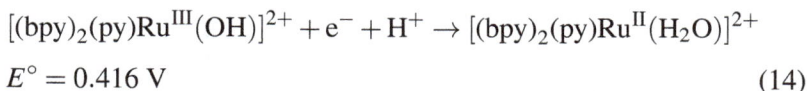

$$[(bpy)_2(py)Ru^{III}(OH)]^{2+} + e^- + H^+ \rightarrow [(bpy)_2(py)Ru^{II}(H_2O)]^{2+}$$

$$E^\circ = 0.416 \text{ V} \tag{14}$$

The effect of pH on the reduction potential of the IV/III and III/II couples is shown in Fig. 4.4.

Let us derive eq 11 and eq 12 and justify the answer.

Let us consider a thermodynamic cycle (Scheme 4.3).

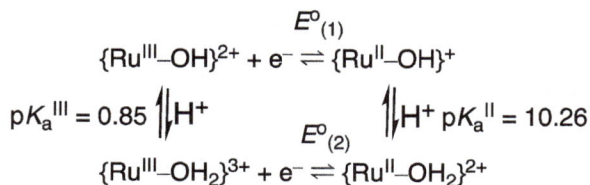

$$E^\circ_{(1)}$$
$$\{Ru^{III}\text{-}OH\}^{2+} + e^- \rightleftharpoons \{Ru^{II}\text{-}OH\}^+$$

$$pK_a^{III} = 0.85 \; \big\updownarrow H^+ \qquad E^\circ_{(2)} \qquad \big\updownarrow H^+ \; pK_a^{II} = 10.26$$

$$\{Ru^{III}\text{-}OH_2\}^{3+} + e^- \rightleftharpoons \{Ru^{II}\text{-}OH_2\}^{2+}$$

Scheme 4.3 The thermodynamic cycle.

Considering Scheme 4.2 and Scheme 4.3, from eq 3 and eq 4, we have,

$$E^{\circ\prime}(E_{1/2}) = E^\circ_{(2)} + 0.059 \log\{(K_{Red} + [H^+])/(K_{OX} + [H^+])\}$$

$$E^\circ_{(2)} = E^{\circ\prime}(E_{1/2}) - 0.059 \log\{(K_{Red} + [H^+])/(K_{OX} + [H^+])\}$$

$$E_{III/II}(\text{Calcd}) = E_{III/II}^{o\prime}(E_{1/2} \text{ of } \{Ru^{III}-OH_2\}^{3+}/$$

$$\{Ru^{II}-OH_2\}^{2+} \text{ redox couple})$$

$$-0.059\,[\log\,(K_a^{III} + [H^+]) - \log\,(K_a^{II} + [H^+])]$$

$$[E_{III/II}(\text{Calcd at pH 7}) = 0.78 - 0.059[\log\{(10)^{-0.85} + (10)^{-7}\}$$

$$- \log\{(10)^{-10.26} + (10)^{-7}\}] = 0.416 \text{ V.}$$

For the redox process,

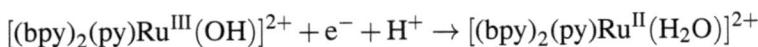

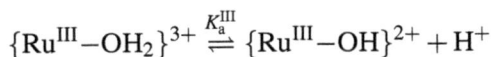

$$[(bpy)_2(py)Ru^{III}(OH)]^{2+} + e^- + H^+ \rightarrow [(bpy)_2(py)Ru^{II}(H_2O)]^{2+}$$

$$\{Ru^{III}-OH_2\}^{3+} \overset{K_a^{III}}{\rightleftharpoons} \{Ru^{III}-OH\}^{2+} + H^+$$

$$K_a^{III} = [\{Ru^{III}-OH\}^{2+}][H^+]/[\{Ru^{III}-OH_2\}^{3+}]$$

$$K_a^{III} + [H^+] = [H^+]([\{Ru^{III}-OH\}^{2+}] + [\{Ru^{III}-OH_2\}^{3+}]/$$

$$[\{Ru^{III}-OH_2\}^{3+}])$$

Since, total Ru^{III} species $[\{Ru^{III}-OH\}^{2+}] + [\{Ru^{III}-OH_2\}^{3+}] = [\{Ru^{IV}=O\}^{2+}]$

$$1/(K_a^{III} + [H^+]) = [\{Ru^{III}-OH_2\}^{3+}]/[\{Ru^{IV}=O\}^{2+}] \times 1/[H^+] \quad (15)$$

Let us consider the following two redox reactions:

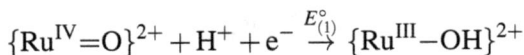

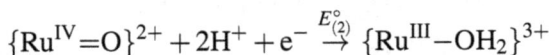

$$\{Ru^{IV}=O\}^{2+} + H^+ + e^- \overset{E_{(1)}^{\circ}}{\rightarrow} \{Ru^{III}-OH\}^{2+}$$

$$\{Ru^{IV}=O\}^{2+} + 2H^+ + e^- \overset{E_{(2)}^{\circ}}{\rightarrow} \{Ru^{III}-OH_2\}^{3+}$$

For $2H^+/1e^-$ process, $[(bpy)_2(py)Ru^{IV}=O]^{2+} + 2H^+ + e^- \rightarrow [(bpy)_2(py)Ru^{III}(OH_2)]^{3+}$

$$E = E_{(1)}^{o\prime}(E_{1/2}) - 0.059 \log([\{Ru^{III}-OH_2\}^{3+}]/[\{Ru^{IV}=O\}^{2+}][H^+])$$

$$+0.059 \log[H^+]$$

$$= E^{o\prime}_{(1)}(E_{1/2}) - 0.059 \log \{1/(K^{III}_a + [H^+])\} - 0.059 \text{ pH (From eq 15)}$$

$$= E^{o\prime}_{(1)}(E_{1/2}) + 0.059 \log (K^{III}_a + [H^+]) - 0.059 \text{ pH}$$

Therefore, $E_{IV/III}(\text{Calcd}) = E^{o\prime}_{IV/III}$ ($E_{1/2}$ of $\{Ru^{IV}{=}O\}^{2+}/\{Ru^{III}{-}OH_2\}^{3+}$ redox couple) $- 0.059 [\text{pH} - \log (K^{III}_a + [H^+])]$

$[E_{IV/III}(\text{Calcd at pH 7}) = 0.99 - 0.059 [7 - \log \{(10)^{-0.85} + (10)^{-7}\}] = 0.526$ V, for the redox process: $[(bpy)_2(py)Ru^{IV}(=O)]^{2+} + e^- + H^+ \rightarrow [(bpy)_2(py)Ru^{III}(OH)]^{2+}]$

(c) Ferrocenecarboxylic acid

Ferrocenecarboxylic acid (Fc–COOH) displays cyclic voltammetric Fe^{III}/Fe^{II} responses in its protonated and deprotonated forms at 0.53 and 0.34 V, respectively (see Fig. 4.5). The pK_a of Fc^{II}–COOH is 7.79. Using appropriate equations, calculate the pK_a of Fc^{III}–COOH form.[9]

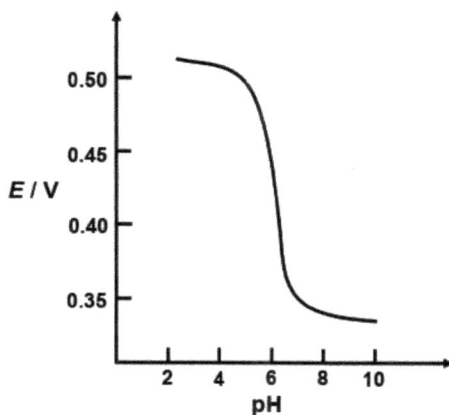

Fig. 4.5 pH-dependence of the redox potentials (V vs SCE) for the ferrocenecarboxylic acid.

[9]G. De Santis, L. Fabbrizzi, M. Licchelli, and P. Pallavicini, *Inorg. Chim. Acta* **1994**, *225*, 239.

Let us consider a thermodynamic cycle (Scheme 4.4).

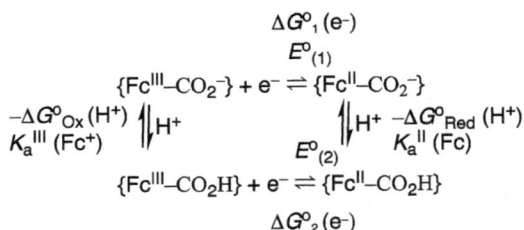

Scheme 4.4

Acid dissociation reactions:

$$Fc^{III}-CO_2H \rightleftharpoons Fc^{III}-CO_2^- + H^+$$

$$K_a^{III}(Fc^+) = [Fc^{III}-CO_2^-][H^+]/[Fc^{III}-CO_2H]$$

Therefore,

$$[Fc^{III}-CO_2H]/[Fc^{III}-CO_2^-] = [H^+]/K_a^{III}(Fc^+)$$

Similarly,

$$Fc^{II}-CO_2H \rightleftharpoons Fc^{II}-CO_2^- + H^+$$

$$K_a^{III}(Fc) = [Fc^{II}-CO_2^-][H^+]/[Fc^{II}-CO_2H]$$

Therefore,

$$[Fc^{II}-CO_2H]/[Fc^{II}-CO_2^-] = [H^+]/K_a^{II}(Fc)$$

Redox reactions:

$$Fc^{III}-CO_2^- + e^- \rightarrow Fc^{II}-CO_2^-$$

$$E = E_{(1)}^\circ - 0.059 \log\{[Fc^{II}-CO_2^-]/[Fc^{III}-CO_2^-]\}$$

Similarly,

$$Fc^{III}-CO_2H + e^- \rightarrow Fc^{II}-CO_2H$$

$$E = E_{(2)}^\circ - 0.059 \log\{[Fc^{II}-CO_2H]/[Fc^{III}-CO_2H]\}$$

In these equations, E is the solution potential and $E_{(1)}^\circ$ and $E_{(2)}^\circ$ are standard redox potentials for deprotonated and protonated forms, respectively. Remember that we are keeping the pH constant during the reaction.

From the thermodynamic cycle (Scheme 4.4),

$$E = E° - 0.059 \log\left[\{[Fc^{II}-CO_2^-]+[Fc^{II}-CO_2H]\}/\{[Fc^{III}-CO_2^-]\right.$$
$$\left.+[Fc^{III}-CO_2H]\}\right]$$
$$= E° - 0.059 \log[Fc^{II}-CO_2^-]\{1+[Fc^{II}-CO_2H]/[Fc^{II}-CO_2^-]\}/$$
$$[Fc^{III}-CO_2^-]\{1+[Fc^{III}-CO_2H]/[Fc^{III}-CO_2^-]\}$$
$$= E° - 0.059 \log\{[Fc^{II}-CO_2^-]/[Fc^{III}-CO_2^-]\}$$
$$- 0.059 \log\{1+[Fc^{II}-CO_2H]/[Fc^{II}-CO_2^-]\}/$$
$$\{1+[Fc^{III}-CO_2H]/[Fc^{III}-CO_2^-]\}$$
$$= E° + E - E°_{(1)} - 0.059 \log\{1+[Fc^{II}-CO_2H]/[Fc^{II}-CO_2^-]\}/$$
$$\{1+[Fc^{III}-CO_2H]/[Fc^{III}-CO_2^-]\}$$

(from definition; see above)

$$E° = E°_{(1)} + 0.059 \log\{(1+[H^+]/K_a^{II}(Fc))/(1+[H^+]/K_a^{III}(Fc^+))\}$$
$$= E°_{(1)} + 0.059 \log\{(K_a^{II}(Fc)+[H^+])\}/\{(K_a^{III}(Fc^+)+[H^+])\}$$
$$+ 0.059 \log\{K_a^{III}(Fc^+)/K_a^{II}(Fc)\}$$
$$= E°_{(1)} + 0.059 \log[\{K_a^{II}(Fc)+[H^+])\}/\{K_a^{III}(Fc^+)+[H^+]\} + E°_{(2)} - E°_{(1)}$$

(from eq 3)

$$E° = E°_{(2)} + 0.059 \log[\{K_a^{II}(Fc)+[H^+]\}/\{K_a^{III}(Fc^+)+[H^+]\}]$$
$$= E°_{(2)} - 0.059 \log[\{K_a^{III}(Fc^+)+[H^+]\}/\{K_a^{II}(Fc)+[H^+]\}] \qquad (16)$$

It should be noted also that for $[H^+] \ll K_a(Fc) < K_a(Fc^+)$ i.e. at pH ≥ 9, eq 16 becomes

$$E(=E°) = E°_{(2)}(E_{1/2}) - 0.059 \log[\{K_a^{III}(Fc^+)\}/\{(K_a^{II}(Fc)\}]$$
$$= 0.53 - 0.059 \log\{K_a^{III}(Fc^+)+0.059 \log\{K_a^{II}(Fc)\}$$
$$E° = 0.53 + 0.059 \times 4.57 - 0.059 \times 7.79$$
$$= 0.34 = E°'(E_{1/2}) \text{ (see below)}.$$

It is obvious from Fig. 4.5.

Rewriting eq 3,

$$E^\circ_{(2)} - E^\circ_{(1)} = 0.059(pK_{Red} - pK_{Ox}) = 0.059(K_{Ox} - K_{Red})$$

where, $E^\circ_{(2)}$ ($\simeq E^{o\prime}_{(2)}$) = formal potential for protonated form and $E^\circ_{(1)}$ ($\simeq E^{o\prime}_{(1)}$) = formal potential for deprotonated form.

Therefore,

$$0.53 - 0.34 = 0.059(7.79 - pK_{Ox})$$

$$0.19 = 0.059(7.79 - pK_{Ox})$$

$$(7.79 - pK_{Ox}) = 3.22;$$

Hence,

$$pK_{Ox} = 7.79 - 3.22 = 4.57$$

Example 4.12 Consider a proton-coupled electron transfer reaction: $A + H^+ + e^- \rightarrow AH^\bullet$ and the acid-base equilibrium: $AH^\bullet \rightleftharpoons A^{\bullet-} + H^+$ (Equilibrium constant, K_a).

Write appropriate Nernst equation, involving K_a and $[H^+]$.

Answer

$$A + H^+ + e^- \rightarrow AH^\bullet$$

$$E = E^\circ - 0.059 \log\{[AH^\bullet]/[A][H^+]\}$$

$$AH^\bullet \rightleftharpoons A^{\bullet-} + H^+$$

$$K_a = [A^{\bullet-}][H^+]/[AH^\bullet]$$

or,

$$K_a + H^+ = [H^+]([A^{\bullet-}] + [AH^\bullet])/[AH^\bullet]$$

$$= [H^+]([A]/[AH^\bullet])(\text{since } [A] = [A^{\bullet-}] + [AH^\bullet])$$

or,

$$1/\{K_a + [H^+]\} = 1/\{[H^+] \times [AH^\bullet]/[A]\}$$

Therefore,

$$E = E^\circ - 0.059 \log\{[AH^\bullet]/[A][H^+]\} = E^\circ - 0.059 \log(1/\{K_a + [H^+]\})$$

(d) Thermodynamic justification for hydrogen atom abstraction

Consider the thermodynamic data:[10]

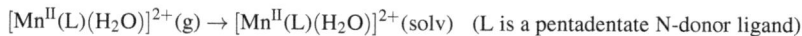

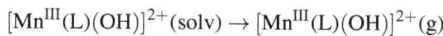

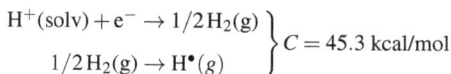

$[Mn^{II}(L)(H_2O)]^{2+}(g) \rightarrow [Mn^{II}(L)(H_2O)]^{2+}(solv)$ (L is a pentadentate N-donor ligand)

$[Mn^{II}(L)(H_2O)]^{2+}(solv) \rightarrow [Mn^{II}(L)(OH)]^{+}(solv) + H^{+}(soln.)$ $pK_a = 13$

$[Mn^{II}(L)(OH)]^{+}(soln.) \rightarrow [Mn^{III}(L)(OH)]^{2+}(soln.) + e^{-}$ $E_{1/2} = 0.81$ V vs NHE; MeCN

$[Mn^{III}(L)(OH)]^{2+}(solv) \rightarrow [Mn^{III}(L)(OH)]^{2+}(g)$

$\left. \begin{array}{l} H^{+}(solv) + e^{-} \rightarrow 1/2\,H_2(g) \\ 1/2\,H_2(g) \rightarrow H^{\bullet}(g) \end{array} \right\} C = 45.3$ kcal/mol

The last two equations describe the free energy of formation of a hydrogen atom and its solvation in MeCN; these values, plus the entropic contribution at 298 K, are conveniently combined into a single solvent-dependent constant C.[11] The constant C accounts for the thermodynamic properties of the H atom in solution and is dependent on solvent and the reference electrode used to measure the redox potentials.

Assuming, solvation free energy for two processes

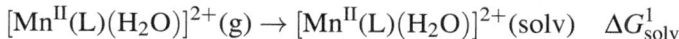

$$[Mn^{II}(L)(H_2O)]^{2+}(g) \rightarrow [Mn^{II}(L)(H_2O)]^{2+}(solv) \quad \Delta G^1_{solv}$$

$$[Mn^{III}(L)(OH)]^{2+}(solv) \rightarrow [Mn^{III}(L)(OH)]^{2+}(g) \quad -\Delta G^2_{solv}$$

are equal.

Species followed by the notation (solv) are in MeCN solution, while those followed by the notation (g) are gaseous.

Let us calculate bond dissociation energy (kcal/mol) in the gas phase for the H atom transfer (HAT) reaction $[Mn^{II}(L)(H_2O)]^{2+}(g) \rightarrow [Mn^{III}(L)(OH)]^{2+}(g) + H^{\bullet}(g)$.[10]

$$[Mn^{II}(L)(H_2O)]^{2+}(g) \rightarrow [Mn^{II}(L)(H_2O)]^{2+}(solv.) \quad \Delta G^1_{solv}$$

$$[Mn^{III}(L)(OH)]^{2+}(solv.) \rightarrow [Mn^{III}(L)(OH)]^{2+}(g) \quad -\Delta G^2_{solv}$$

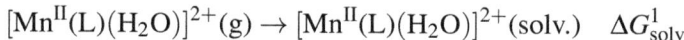

Since $\Delta G^{\circ} = -RT \ln K$

---

[10]C. R. Goldsmith, A. P. Cole, and T. D. P. Stack, *J. Am. Chem. Soc.* **2005**, *127*, 9904.
[11](a) F. G. Bordwell, J. P. Cheng, and J. A. Harrelson, Jr. *J. Am. Chem. Soc.* **1988**, *110*, 1229. (b) K. Wang and J. M. Mayer, *J. Am. Chem. Soc.* **1997**, *119*, 1470.

For the deprotonation reaction,

$$\Delta G° = -RT \ln K_a = 1.37 \times pK_a \text{ kcal/mol}$$
$$= 1.37 \times 13$$
$$= 17.8 \text{ kcal/mol}$$

For the redox reaction,

$$\Delta G° = nFE° = 1F \times 0.81 \times 23.06 \text{ (since 1 V} = 23.06 \text{ kcal mol}^{-1} \text{ V}^{-1})$$
$$= 18.7 \text{ kcal/mol}$$

Note: $\Delta G° = nFE°$, as the reaction is written as oxidation.
Hence,

$$\text{O–H bond dissociation energy (BDE)} = (17.8 + 18.7 + 45.3) \text{ kcal/mol}$$
$$= 81.8 \text{ kcal/mol}$$

The energy calculation for the overall reaction is an enthalpic, gas-phase BDE. This analysis provides an estimate of O–H bond strength of the H_2O ligand in $[Mn^{II}(L)(H_2O)](CF_3SO_3)_2$ of 82 (± 2) kcal/mol.[10]

Further reading

R. L. Dutta, *Inorganic Chemistry: Part-I Principles*, 6th edition, The New Book Stall, Kolkata (2009).

C. E. Housecroft and A. G. Sharpe, *Inorganic Chemistry*, 2nd edition, Pearson Education Limited (2005).

D. F. Shriver and P. W. Atkins, *Inorganic Chemistry*, 3rd edition, Oxford University Press (1999).

Exercises

4.1 Copper metal exposed to open air develops a green coated surface. Estimate the $E°$ value for oxidation of copper by atmospheric dioxygen in a damp environment.
Given (under acidic conditions):

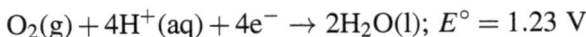

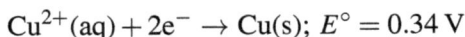

$$O_2(g) + 4H^+(aq) + 4e^- \rightarrow 2H_2O(l); \ E° = 1.23 \text{ V}$$
$$Cu^{2+}(aq) + 2e^- \rightarrow Cu(s); \ E° = 0.34 \text{ V}$$

4.2 Given (under acidic conditions):

$$Fe^{2+}(aq) + 2e^- \rightarrow Fe(s); \; E^\circ = -0.44 \text{ V}$$

$$Fe^{3+}(aq) + e^- \rightarrow Fe^{2+}(aq); E^\circ = 0.77 \text{ V}$$

Will Fe^{2+} disproportionate?

4.3 Given:

$$Cu^{2+}(aq)/Cu(s); \; E^\circ = 0.34 \text{ V}$$

$$Cu(OH)_2(s)/Cu(s), OH^-(aq); E^\circ = -0.22 \text{ V}$$

Find out the solubility product (K_{sp}) of $Cu(OH)_2(s)$.

4.4 Given:[12]

$$[M^{II}(L)(OH)]^{2-} \longrightarrow [M^{III}(L)(OH)]^- + e^-$$

$$[M^{III}(L)(OH)]^- \longrightarrow [M^{III}(L)(O)]^{2-} + H^+ \qquad pK_a = 28.3 \text{ (Mn)}; 25.0 \text{ (Fe)}$$

$$[M^{III}(L)(O)]^{2-} \longrightarrow [M^{IV}(L)(O)]^- + e^- \; E_{1/2} \text{ (DMSO)} = -0.076 \text{ V vs } Fc^+/Fc \text{ (Mn)}$$
$$E_{pa} \text{ (DMSO)} = 0.34 \quad \text{V vs } Fc^+/Fc \text{ (Fe)};$$
$$(E_{pa} = \text{anodic peak potential})$$

(L^{3-} is a tripodal N_4 ligand)

$C = 73.3 \text{ kcal mol}^{-1}$ (the value determined previously for DMSO and the Fc^+/Fc reference electrode)

Estimate BDE_{O-H} for the following:

$$[M^{III}(L)(OH)]^- \longrightarrow [M^{III}(L)(O)]^{2-}$$

$$[M^{III}(L)(O)]^{2-} \longrightarrow [M^{IV}(L)(O)]^-$$

[12]R. Gupta and A. S. Borovik, *J. Am. Chem. Soc.* **2003**, *125*, 13234.

Chapter 5

Spectroscopic Terms and Spin-Orbit Coupling

In Chapter 2 we have dealt with the symmetry properties of atomic orbitals. In this chapter the electronic energy states of isolated atoms or ions that are characterized by term symbols will be discussed. The term symbols correspond closely in symmetry properties to atomic orbitals.

The accurate description of atomic structure relies on the coupling of angular momenta in the valence electrons of the atom. Such coupling is normally described in two limiting representations: L-S and j-j coupling schemes, with which we derive the electronic states of a given electronic configuration.

Term symbols for electronic configurations are useful not only to explain the spectroscopic properties but also in understanding electronic and magnetic properties of inorganic molecules in general and transition metal complexes in particular (see Chapter 6). An understanding of the concepts of *energy terms*, *energy levels*, and *microstates* is of paramount importance for the study of the crystal field theory of bonding in transition metal complexes (see Chapter 6).

For a multielectron system, there is a correlation in the nature of movement of electrons about the nucleus. For a single electron, the expressions for orbital angular momentum and spin angular momentum, respectively, are,

$$\mu_l = \{l(l+1)\}^{1/2}\mu_B \quad \text{and} \quad \mu_s = g\{s(s+1)\}^{1/2}\mu_B$$

where l and s are single electron quantum numbers, g is the g-factor, and μ_B is the Bohr magneton (see below).

For multielectron systems,

$$\mu_L = \{L(L+1)\}^{1/2}\mu_B \quad \text{and} \quad \mu_S = g\{S(S+1)\}^{1/2}\mu_B$$

where L and S are the multielectron quantum numbers.

5.1 *L-S* and *j-j* coupling

In the *L-S* coupling scheme, appropriate for lighter atoms such as first-row transition elements, the electrostatic energy terms for the valence electrons are much larger than the spin-orbit terms so that the total orbital angular momenta L and total spin angular momenta S are good quantum numbers. In this scheme, the valence electrons' individual orbital angular momenta l's couple to yield the total orbital angular momenta $L(= \Sigma l)$, which is a constant of the motion. The resultant L has the allowed values given by the sequence: $l_1 + l_2, l_1 + l_2 - 1, \ldots, |l_1 - l_2|$. Similarly, individual electrons' spin angular momenta s's couple to yield the total spin angular momenta $S(= \Sigma s)$, which is also a constant of the motion. The resultant S has the allowed values given by the sequence: $s_1 + s_2, s_1 + s_2 - 1, \ldots, |s_1 - s_2|$. The M_L and M_S values for multielectron systems correspond to Σm_l and Σm_s, respectively. They have the allowed values given by the sequence: $L, L-1, \ldots, -L+1, -L$ and $S, S-1, \ldots, -S+1, -S$, respectively. The total number of M_L and M_S values are $2L+1$ and $2S+1$, respectively.

5.2 Term symbols

The determination of R-S term symbols[1] for different electron configurations is presented in this chapter in a simplified manner. In general, several R-S terms will result from any one electron configuration. For each term, the spin-orbit interaction is treated as a small perturbation yielding a representation in which the total electronic angular momentum J of the atom is the vector sum of the orbital (L) and spin (S) angular momenta. The result is that each term is split into levels that consist of states with the same value of J that are $(2J+1)$-fold degenerate corresponding to the possible values of M_J. The M_J values have the allowed values given by

---

[1] (a) M. German, *J. Chem. Educ.* **1973**, *50*, 189. (b) K. E. Hyde, *J. Chem. Educ.* **1975**, *52*, 87. (c) J. Vicente, *J. Chem. Educ.* **1983**, *60*, 560. (d) P. Coppo, *J. Chem. Educ.* **2016**, *93*, 1085.

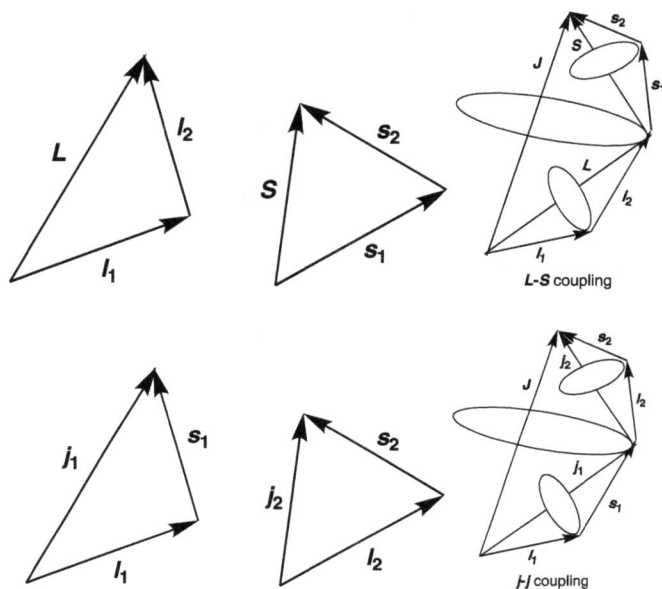

Fig. 5.1   (a) *L-S* and (b) *j-j* coupling schemes.

the sequence: $J, J-1, \ldots, -J+1, -J$ and the total number of M_J values is $2J+1$.

Remembering the assumption of the Russell-Saunders (American astronomer H. N. Russell (1877–1957) and American physicist F. A. Saunders (1875–1963)), R-S or *L-S* coupling scheme, proposed in 1925 that the electronic repulsion is much larger than the spin-orbit interaction. The *L-S* coupling scheme involves the orbital angular momentum quantum number L and the spin angular momentum quantum number S. They define an electronic term conventionally designated by a term symbol written in the form $^{2S+1}L_J$. The addition of L and S gives the total angular momentum quantum number J, which has the allowed values given by the sequence: $L+S, L+S-1, \ldots, |L-S|$ (Fig. 5.1(a)).

For a given L, a spin multiplicity of $(2S+1)$ is written as a superscript preceding the code symbol S, P, D, F, G, H, I ... for $L = 0, 1, 2, 3, 4, 5, 6 \ldots$

The term symbol is represented as $^{2S+1}L_J$ (^{2S+1}L, excluding J). As for example, the term symbol ^3P indicates that there are two unpaired electrons ($2 \times 1 + 1 = 3$) in a state with maximum $L = 1$.

In the j-j coupling scheme, more appropriate coupling scheme for heavier elements such as the actinides, in contrast to the L-S scheme, the spin-orbit interaction is much larger than the electronic repulsion. Each electron's l and s couple to give j, the electron's total angular momentum i.e. $j_1 = l_1 + s_1, l_1 + s_1 - 1, \ldots, |l_1 - s_1|$. As with L-S coupling, several terms result from the different ways in which each electron can couple its angular momenta. For each term the electrostatic interaction is treated as a small perturbation yielding a representation in which the total electronic angular momentum J of the atom is the vector sum of each electron's angular momentum $J = \Sigma j$ (Fig. 5.1(b)). As like L and S, J takes up values of $j_1 + j_2, j_1 + j_2 - 1, \ldots, |j_1 - j_2|$. In this scheme the electrons appear to move independently of one another and in these circumstances the individual values j, l, and s are good quantum numbers. The result is that each term is split into levels, which again consist of states with the same value of J that are $(2J + 1)$-fold degenerate.

It should be remembered that the term symbol S for $L = 0$ and the total spin angular momentum quantum number S are different. Spin-orbit interaction between the total orbital and spin angular momenta gives rise to different energies for each allowed value of J. For the lighter elements, these energy separations are relatively small.

When atomic states are accurately represented by R-S coupling, the energy ordering of different terms arising from a given electron configuration follows Hund's rules (German physicist F. H. Hund (1896–1997) formulated around 1927).

Hund's Rule:

1. State with the largest value of spin multiplicity $2S + 1$ is most stable and stability decreases with decreasing the value of $(2S + 1)$ i.e. states with higher spin multiplicities have lower energies.
   Electrons are negatively charged and, as a result, they repel each other. Electrons tend to minimize repulsion by occupying their own orbitals, rather than sharing an orbital with another electron. Electrons always enter an empty orbital before they pair up.
2. For states with same values of $(2S + 1)$, the state with the largest value of L is the most stable i.e. states with larger L values have lower energies.

In order to minimize interelectronic repulsion, electrons tend to move in the same direction, as long as possible by Hund's rule and Pauli exclusion principle, such that the resultant L value is maximized.

3. For a subshell that is less than half-filled, state with smallest $(2J+1)$ is most stable and for subshells that are more than half-filled, state with largest value of $(2J+1)$ is most stable.

Lower-case letters define single electron quantum numbers and upper-case letters to define multiple electron terms.

A state represented by a particular combination of L and S is called a *term*. Microstates, of the same energy (degenerate), are grouped together into *terms*.

The ground-state energy term of an electron configuration is represented as $^{2S+1}L_J$.

A combination of S, L, and J is called a *level*.

Now let us discuss how to find out R-S terms for a given electron configuration.

Following Hund's rule and considering L and S values, the ground-state electron configuration for p^3, d^4, and d^7 are determined as follows.

p^3 configuration:

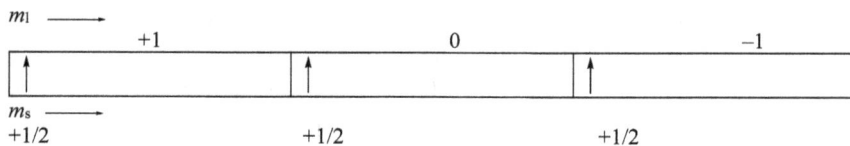

$M_S = \Sigma m_s = +3/2$ corresponds to $S = 3/2$ and $M_L = \Sigma m_l = 0$ corresponds to $L = 0$

Hence, the ground-state term symbol for p^3 is ^4S.

d^4 configuration:

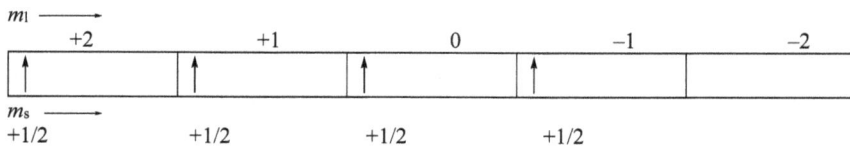

$M_S = \Sigma m_s = +2$ corresponds to $S = 2$ and $M_L = \Sigma m_l = +2$ corresponds to $L = 2$.

Hence, the ground-state term symbol for d^4 is ^5D.

d^7 configuration:

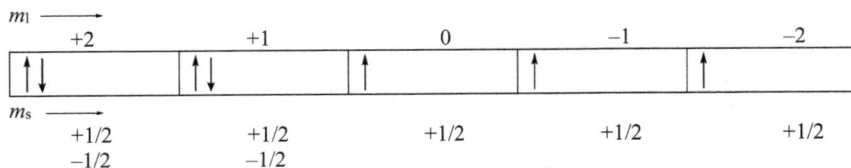

$M_S = \Sigma m_s = +3/2$ corresponds to $S = 3/2$ and $M_L = \Sigma m_l = +3$ corresponds to $L = 3$.

Hence, the ground-state term symbol for d^4 is ^4F.

A spectral term corresponds to a set of microstates.

A microstate is described by the full electron configuration showing the population of each orbital. We can find the spectral terms for any electron configuration by writing all possible microstates.

Example 5.1 Find out the R-S terms for p^2 configuration.

Answer Each arrangement of an electron configuration in terms of m_l and m_s values is called a microstate. The number (N) of microstates is predictable by using the simple formula (see below), where x is twice the number of orbitals in the subshell (three p orbitals can accommodate a total of six electrons, five d orbitals can accommodate a total of ten electrons etc.), and n is the number of electrons:

$$N = x!/\{n!(x-n)!\}.$$

For $p^2 : N = 6!/(2!4!) = 15$

Approach 1: General

For two p-electrons with each $l = 1$ and m_l values $+1$, 0, -1, the $L(= \Sigma l)$ values $(l_1 + l_2)$, $(l_1 + l_2 - 1)$,... through $(l_1 - l_2)$ i.e. $+L$ to $-L$ leads to $M_L(= \Sigma m_l)$ values $+2$, $+1$, 0, -1, -2. For two electrons, each m_s

value $+1/2$ or $-1/2$, the $S\,(=\Sigma s)/M_S(=\Sigma m_s)$ values $|s_1+s_2|$ to $|s_1-s_2|$ leads to M_s values $+1, 0, -1$.

$$
\begin{array}{cccccc}
a & b & c & d & e & f \\[4pt]
\binom{m_l}{m_s} \quad \binom{+1}{+1/2} & \binom{0}{+1/2} & \binom{-1}{+1/2} & \binom{+1}{-1/2} & \binom{0}{-1/2} & \binom{-1}{-1/2}
\end{array}
$$

Case 1.

	ab	ac	bc	bd	cd	bf	de	df	ef
$M_L=\Sigma m_l$	+1	0	−1	+1	0	−1	+1	0	−1
$M_S=\Sigma m_s$	+1	+1	+1	0	0	0	−1	−1	−1

$(+1^+, 0^+),\ (+1^+, -1^+),\ (0^+, -1^+),\ (0^+, +1^-),\ (-1^+, +1^-),\ (0^+, -1^-),\ (+1^-, 0^-),\ (+1^-, -1^-),\ (0^-, -1^-)$
$\quad\ \ \text{ab}\qquad\quad\ \text{ac}\qquad\quad\ \text{bc}\qquad\quad\ \ \text{bd}\qquad\quad\ \ \text{cd}\qquad\qquad\ \text{bf}\qquad\quad\ \ \text{de}\qquad\quad\ \ \text{df}\qquad\quad\ \ \text{ef}$

$+1^+$ represents an electron whose m_l and m_s values are $+1$ and $+1/2$, respectively.

$m_l \rightarrow$

+1	0	−1	M_L	M_S
↑	↑		+1	+1
↑		↑	0	+1
	↑	↑	−1	+1
↓	↑		+1	0
↓		↑	0	0
	↑	↓	−1	0
↓	↓		+1	−1
↓		↓	0	−1
	↓	↓	−1	−1

$M_L == +1, 0, -1$ and $M_S = +1, 0, -1$ correspond to $L=1$ and $S=1$, respectively.

Term symbol: ^3P.

The degeneracy of a spectral term is the product of the spin-multiplicity $(2S+1)$ and orbital degeneracy $(2L+1)$ of the term.

Hence, the number of microstates associated with ^3P term $= (2S+1)(2L+1) = 3 \times 3 = 9$.

Case 2.

	ad	ae	af	ce	cf	ad	ae	af	ce	cf
$M_L = \Sigma m_l$	+2	+1	0	-1	-2	$(+1^+, +1^-)$,	$(+1^+, 0^-)$,	$(+1^+, -1^-)$,	$(-1^+, 0^-)$,	$(-1^+, -1^-)$
$M_S = \Sigma m_s$	0	0	0	0	0					

$m_l \rightarrow$

+1	0	-1	M_L	M_S
↑↓			+2	0
↑	↓		+1	0
↑		↓	0	0
	↓	↑	-1	0
		↑↓	-2	0

$M_L == +2, +1, 0, -1, -2$ and $M_S = 0$ correspond to $L = 2$ and $S = 0$, respectively. Hence, the term symbol: ^1D. Number of microstates associated with ^1D term $= (2S+1)(2L+1) = 1 \times 5 = 5$.

Case 3.

	be	be
$M_L = \Sigma m_l$	0	$(0^+, 0^-)$
$M_S = \Sigma m_s$	0	

$m_l \rightarrow$

+1	0	-1	M_L	M_S
	↑↓		0	0

$M_L = 0$ and $M_S = 0$ corresponds to $L = 0$ and $S = 0$, respectively. This leads to the term symbol ^1S.

Number of microstate associated with ^1S term $= (2S+1)(2L+1) = 1 \times 1 = 1$.

The term symbols to represent p^2 electron configuration of 15 microstates: ^3P (9), ^1D (5), ^1S (1). Each term ^3P, ^1D, ^1S defines a state (group of microstates of the same energy).

The R-S terms for p^2 valence electron configuration are ^3P, ^1D, ^1S and the ground-state energy term is ^3P. This implies that the term symbols ^1D and ^1S are higher in energy. The energy-level ordering is $^3P < {}^1D < {}^1S$.

Approach 2: General and applying the 'branching rule'

Case 1. Maximum spin multiplicity (see Example 5.1). The R-S term is ^3P $(L = 1, S = 1)$.

Case 2. Let us consider first p^1 configuration. The term symbol is ^2P ($L = 1$ and $S = 1/2$).

Now to this electron configuration, we add one p-electron with term symbol ^2P. We get p^2 configuration.

$$p^1 \longrightarrow {}^2P\ (L = 1)$$

$$\Big\downarrow +1p\ (l = 1) \qquad \Big\downarrow + {}^2P\ (L = 1)$$

$$p^2 \qquad\qquad L = 2, 1, 0$$
$$\text{D, P, S}$$

Remembering that now spin-multiplicity $(2S + 1)$ should decrease by one unit (2 to 1) and $L = 2, 1, 0$ $(L = 1 + 1 = 2$ to $1 - 1 = 0)$, we have ^1D, ^1P, ^1S terms.

The value of Ms for $S = 1$ are ± 1 and 0 (Case 1; ^3P: $L = 1, S = 1$), so D, P, S terms with $Ms = 0$ will appear from other configurations as ^1D, ^1P, and ^1S. We must subtract one P term of lower spin multiplicity i.e. ^1P ($L = 1$, $S = 0$), as one P term of higher spin multiplicity ^3P is already present. Now from the singlet terms, eliminating ^1P term, we have the terms ^1D and ^1S.

Term symbols for p^2 electron configuration to represent 15 microstates: ^3P (9), ^1D (5), ^1S (1).

Example 5.2 Showing all the microstates, find out the R-S terms for d^2 configuration.

Answer For d^2 : $N = 10!/(2!8!) = 45$

For two d-electrons with each $l = 2$, the (m_l, m_s) values:

	a	b	c	d	e	f	g	h	i	j
$\binom{m_l}{m_s}$	$\binom{+2}{+1/2}$	$\binom{+1}{+1/2}$	$\binom{0}{+1/2}$	$\binom{-1}{+1/2}$	$\binom{-2}{+1/2}$	$\binom{+2}{-1/2}$	$\binom{+1}{-1/2}$	$\binom{0}{-1/2}$	$\binom{-1}{-1/2}$	$\binom{-2}{-1/2}$

$\longrightarrow$

Case 1.

$m_l \rightarrow$

+1	0	-1	M_L	M_S
↑	↑		+1	+1
↑		↑	0	+1
	↑	↑	-1	+1
↓	↑		+1	0
↓		↑	0	0
	↑	↓	-1	0
↓	↓		+1	-1
↓		↓	0	-1
	↓	↓	-1	-1

	ab	ac	ad	ae	cd	ce	de
$M_L = \Sigma m_l$	+3	+2	+1	0	-1	-2	-3
$M_S = \Sigma m_s$	+1	+1	+1	+1	+1	+1	+1

$$(+2^+, +1^+), (+2^+, 0^+), (+2^+, -1^+), (+2^+, -2^+), (0^+, -1^+), (0^+, -2^+), (-1^+, -2^+)$$
$$\text{ab} \qquad \text{ac} \qquad \text{ad} \qquad \text{ae} \qquad \text{cd} \qquad \text{ce} \qquad \text{de}$$

	bc	bd	be	
$M_L = \Sigma m_l$	+1	0	-1	$(+1^+, 0^+), (+1^+, -1^+), (+1^+, -2^+)$
$M_S = \Sigma m_s$	+1	+1	+1	

bc   bd   be

$m_l \rightarrow$

+2	+1	0	-1	-2	M_L	M_S
↑	↑				+3	+1
↑		↑			+2	+1
↑			↑		+1	+1
↑				↑	0	+1
		↑	↑		-1	+1
		↑		↑	-2	+1
			↑	↑	-3	+1
	↑	↑			+1	+1
	↑		↑		0	+1
	↑			↑	-1	+1

Seven microstates correspond to $M_L = +3, +2, +1, 0, -1, -2, -3$ and $M_S = +1$

Three microstates correspond to $M_L = +1, 0, -1$ and $M_S = +1$

	ag	ah	ai	aj	ci	cj	dj
$M_L = \Sigma m_l$	+3	+2	+1	0	−1	−2	−3
$M_S = \Sigma m_s$	0	0	0	0	0	0	0

$(+2^+, +1^-), (+2^+, 0^-), (+2^+, -1^-), (+2^+, -2^-), (0^+, -1^-), (0^+, -2^-), (-1^+, -2^-)$
$\quad$ ag $\qquad$ ah $\qquad\qquad$ ai $\qquad\qquad$ aj $\qquad\qquad$ ci $\qquad\qquad$ cj $\qquad\quad$ dj

	bh	bi	ci
$M_L = \Sigma m_l$	+1	0	−1
$M_S = \Sigma m_s$	0	0	0

$\quad$ bh $\qquad$ bi $\qquad$ ci
$(+1^+, 0^-), (+1^+, -1^-), (0^+, -1^-)$

$m_l \rightarrow$

+2	+1	0	−1	−2	M_L	M_S
↑	↓				+3	0
↑		↓			+2	0
↑			↓		+1	0
↑				↓	0	0
		↑	↓		−1	0
		↑		↓	−2	0
			↑	↓	−3	0
	↑	↓			+1	0
	↑		↓		0	0
		↑	↓		−1	0

These microstates correspond to $M_L = +3, +2, +1, 0, -1, -2, -3$ and $M_S = 0$ and $M_L = +1, 0, -1$ and $M_S = 0$

	fg	fh	fi	fj	hj	hj	ij
$M_L = \Sigma m_l$	+3	+2	+1	0	−1	−2	−3
$M_S = \Sigma m_s$	−1	−1	−1	−1	−1	−1	−1

$(+2^-, +1^-), (+2^-, 0^-), (+2^-, -1^-), (+2^-, -2^-), (0^-, -1^-), (0^-, -2^-), (-1^-, -2^-)$
$\quad$ fg $\qquad$ fh $\qquad\qquad$ fi $\qquad\qquad$ fj $\qquad\qquad$ hi $\qquad\qquad$ hj $\qquad\quad$ ij

	gh	gi	gj
$M_L = \Sigma m_l$	+1	0	−1
$M_S = \Sigma m_s$	−1	−1	−1

$(+1^-, 0^-), (+1^-, -1^-), (+1^-, -2^-)$

$m_l \rightarrow$

+2	+1	0	−1	−2	M_L	M_S
↓	↓				+3	−1
↓		↓			+2	−1
↓			↓		+1	−1
↓				↓	0	−1
		↓	↓		−1	−1
		↓		↓	−2	−1
			↓	↓	−3	−1
	↓	↓			+1	−1
	↓		↓		0	−1
	↓			↓	−1	−1

These 30 microstates correspond to $M_L = +3, +2, +1, 0, -1, -2, -3$
and $M_S = +1, 0, -1$ and $M_L = +1, 0, -1$ and $M_S = +1, 0, -1$

$M_L = +3$ to -3 leads to $L = 3$. Hence, F term

$M_S = +1, 0, -1$ leads to $S = 1, \therefore 2S + 1 = 3$

Term symbol: ^3F

Number of microstates associated with ^3F term $= 3 \times 7 = 21$.

$M_L = +1$ to -1 leads to $L = 1$. Hence P term

$M_S = +1, 0, -1$ leads to $S = 1, \therefore 2S + 1 = 3$

Term symbol: ^3P

Number of microstates associated with ^3F term $= 3 \times 3 = 9$.

Case 2.

	af	bf	bg	df	ch	eg	di	ei	ej
$M_L = \Sigma m_l$	+4	+3	+2	+1	0	-1	-2	-3	-4
$M_S = \Sigma m_s$	0	0	0	0	0	0	0	0	0

$(+2^+, +2^-), (+1^+, +2^-), (+1^+, +1^-), (-1^+, +2^-), (0^+, 0^-), (-2^+, +1^-), (-1^+, -1^-), (-2^+, -1^-),$
       af   bf   bg   df   ch   eg   di   ei

$(-2^+, -2^-)$
       ej

$m_l \rightarrow$

+2	+1	0	-1	-2	M_L	M_S
↑↓					+4	0
↑	↓				+3	0
↑		↓			+2	0
↑			↓		+1	0
↑				↓	0	0
	↑			↓	-1	0
		↑		↓	-2	0
			↑	↓	-3	0
				↑↓	-4	0

$M_L = +4, +3, +2, +1, 0, -1, -2, -3, -4$ leads to $L = 4$. Hence G term

$M_S = 0$ leads to $S = 0, \therefore 2S + 1 = 1$

Term symbol: ^1G

Number of microstates associated with ^1G term $= 1 \times 9 = 9$.

	cf	cg	dg	dh	eh		cf	cg	dg	dh	eh
$M_L = \Sigma m_l$	+2	+1	0	−1	−2		$(0^+, +2^-),$	$(0^+, +1^-),$	$(-1^+, +1^-),$	$(-1^+, 0^-),$	$(-2^+, 0^-)$
$M_S = \Sigma m_s$	0	0	0	0	0						
$m_l \to$											

+2	+1	0	−1	−2	M_L	M_S
↓						
		↑			+2	0
	↓	↑			+1	0
	↓		↑		0	0
		↓	↑		−1	0
		↓		↑	−2	0

$M_L = +2, +1, 0, -1, -2$ correspond to $L = 2$. Hence D term

$M_S = 0$ corresponds to $S = 0, \therefore 2S+1 = 1$

Term symbol: ^1D

Number of microstates associated with ^1D term $= 1 \times 5 = 5$.

Case 3.

	c	c
$M_L = \Sigma m_l$	0	$(0^+, 0^-)$
$M_S = \Sigma m_s$	0	
$m_l \to$		

+2	+1	0	−1	−2	M_L	M_S
		↑↓			0	0

$M_L = 0$ leads to $L = 0$. Hence S term

$M_S = 0$ leads to $S = 0, \therefore 2S+1 = 1$

Term symbol: ^1S

Number of microstate associated with ^1S term $= 1 \times 1 = 1$.

Term symbols for d^2 electron configuration to represent 45 microstates:
^3F (21), ^3P (9), ^1G (9), ^1D (5), ^1S (1).

Example 5.3 Applying general and the 'branching rule', find out the R-S terms for d^2 configuration.

Answer Case 1.

The R-S terms are ^3F (21 microstates) and ^3P (9 microstates) (see Example 5.2).

Case 2. Applying the 'branching rule',

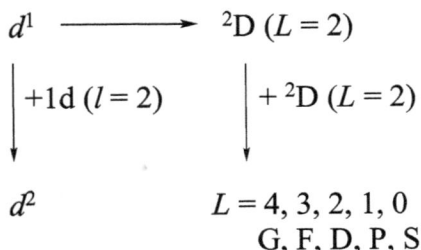

$$d^1 \longrightarrow {}^2D \ (L=2)$$

$$\Big| +1d \ (l=2) \qquad \Big| + {}^2D \ (L=2)$$

$$d^2 \qquad\qquad L=4, 3, 2, 1, 0$$
$$\text{G, F, D, P, S}$$

Remembering that now spin-multiplicity should decrease by one unit (2 to 1) and $L = 4, 3, 2, 1, 0$ ($h = 2+2 = 4$ to $2-2 = 0$), we have ${}^1G$, ${}^1F$, ${}^1D$, ${}^1P$, ${}^1S$.

Now from the singlet terms, eliminating one ${}^1F$ and one ${}^1P$ terms (see above for the p^2 case) we have the terms ${}^1G$, ${}^1D$, ${}^1S$.

Term symbols for d^2 electron configuration to represent 45 microstates: ${}^3F$ (21), ${}^3P$ (9), ${}^1G$ (9), ${}^1D$ (5), ${}^1S$ (1).

Example 5.4 Showing all the microstates, find out the R-S terms for d^3 configuration.

Answer For $d^3 : N = 10!/(3!7!) = 120$

For three d-electrons with each $l = 2$, the (m_l, m_s) combinations:

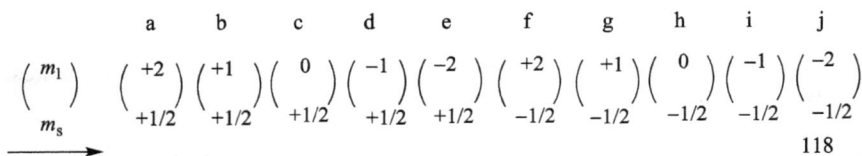

$$
\begin{array}{cccccccccc}
\text{a} & \text{b} & \text{c} & \text{d} & \text{e} & \text{f} & \text{g} & \text{h} & \text{i} & \text{j} \\
\end{array}
$$

$$
\binom{m_l}{m_s} \longrightarrow
\binom{+2}{+1/2}
\binom{+1}{+1/2}
\binom{0}{+1/2}
\binom{-1}{+1/2}
\binom{-2}{+1/2}
\binom{+2}{-1/2}
\binom{+1}{-1/2}
\binom{0}{-1/2}
\binom{-1}{-1/2}
\binom{-2}{-1/2}
$$

Case 1. Maximum spin multiplicity

$m_l \rightarrow$

+2	+1	0	−1	−2	M_L	M_S
↑	↑	↑			+3	+3/2
↑	↑		↑		+2	+3/2
↑	↑			↑	+1	+3/2
↑		↑		↑	0	+3/2
↑			↑	↑	−1	+3/2
	↑		↑	↑	−2	+3/2
		↑	↑	↑	−3	+3/2
↑	↑			↑	+1	+3/2
	↑	↑	↑		0	+3/2
	↑	↑		↑	−1	+3/2

These 10 microstates (all spins up) correspond to $M_L = +3, +2, +1, 0, -1, -2, -3$ and $M_S = +3/2$ also $M_L = +1, 0, -1$ and $M_S = +3/2$

Similarly, there are 10 microstates with all three spins down.

Then, $M_L = +3, +2, +1, 0, -1, -2, -3$ and $M_S = -3/2$ and $M_L = +1, 0, -1$ and $M_S = -3/2$

$m_l \rightarrow$

+2	+1	0	−1	−2	M_L	M_S
↑	↑	↓			+3	+1/2
↑	↑		↓		+2	+1/2
↑	↑			↓	+1	+1/2
↑		↓		↑	0	+1/2
↓			↑	↑	−1	+1/2
	↓		↑	↑	−2	+1/2
		↓	↑	↑	−3	+1/2
↑	↓			↑	+1	+1/2
↑		↓		↑	0	+1/2
↑			↓	↑	−1	+1/2

These 10 microstates (two spins up and one spin down) give rise to $M_L = +3, +2, +1, 0, -1, -2, -3$ and $M_S = +1/2$ also $M_L = +1, 0, -1$ and $M_S = +1/2$

Similarly, there are 10 microstates with one spin up and two spins down.

Then, $M_L = +3, +2, +1, 0, -1, -2, -3$ and $M_S = -1/2$ and $M_L = +1$, $0, -1$ and $M_S = -1/2$

These considerations give rise to:

$M_L = +3, +2, +1, 0, -1, -2, -3$ and $M_S = +3/2, +1/2, -1/2, -3/2$ corresponds to $L = 3$ and $S = 3/2$. This leads to the term symbol ^4F.

Number of microstates associated with ^4F term $= 4 \times 7 = 28$

$M_L = +1, 0, -1$ and $M_S = +3/2, +1/2, -1/2, -3/2$ corresponds to $L = 1$ and $S = 3/2$. This leads to the term symbol ^4P.

Number of microstates associated with ^4P term $= 4 \times 3 = 12$

The values of Ms for $S = 3/2$ are $\pm 3/2$ and $\pm 1/2$, so F and P terms with $Ms = 1/2$ will appear from other configurations as ^2F and ^2P, and we must subtract these terms below.

Case 2. Applying the 'branching rule',

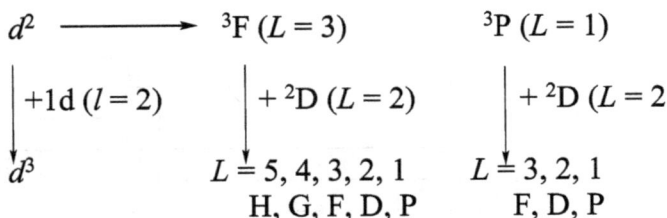

$$d^2 \longrightarrow \quad ^3F\ (L = 3) \qquad\qquad ^3P\ (L = 1)$$

$$\left| +1d\ (l = 2) \right. \qquad\quad \left| + ^2D\ (L = 2) \right. \qquad\quad \left| + ^2D\ (L = 2 \right.$$

$$d^3 \qquad\qquad\qquad L = 5, 4, 3, 2, 1 \qquad L = 3, 2, 1$$
$$\qquad\qquad\qquad\qquad H, G, F, D, P \qquad\quad F, D, P$$

For R-S terms ^3F and ^3P, see Example 5.2.

The spin-multiplicity must decrease by one unit i.e. $2S + 1$ changes from 3 to 2. Subtracting one ^2F and one ^2P terms, we have, $^2H, ^2G, ^2F$, $^2D\ (2), ^2P$

Term symbols to represent d^3 electron configuration for 120 microstates:

$^4F\ (28),\ ^4P\ (12),\ ^2H\ (22),\ ^2G\ (18),\ ^2F\ (14),\ ^2D\ (10),\ ^2D\ (10),\ ^2P\ (6)$

Exercise 5.5 Find out the R-S terms for d^4 configuration.

Answer For $d^4 : N = 10!/(4!6!) = 210$

Case 1. Maximum spin multiplicity

Let us consider the following ten microstates:

$$M_L = \Sigma m_l \quad +2 \ +1 \ 0 \ -1 \ -2 \quad \text{correspond to } L = 2$$
$$M_S = \Sigma m_s \quad 2 \quad 2 \quad 2 \quad 2 \quad 2 \quad \text{correspond to } S = 2$$

This leads to the term symbol ^5D. Number of microstates associated with ^5D term $= 5 \times 5 = 25$.

Case 2. For maximum spin multiplicity of three spins the term symbols are ^4F and ^4P (see Example 5.4).

Applying the 'branching rule',

d^3	^4F ($L = 3$)	^4P ($L = 1$)
$\Big\downarrow$ +1d ($l = 2$)	$\Big\downarrow$ + ^2D ($L = 2$)	$\Big\downarrow$ + ^2D ($L = 2$)
d^4	$L = 5, 4, 3, 2, 1$	$L = 3, 2, 1$
	H, G, F, D, P	F, D, P

The spin-multiplicity must decrease by one unit i.e. $2S + 1$ changes from 4 to 3. Subtracting one ^3D (as we already have ^5D), we have, ^3H, ^3G, $^3F(2)$, ^3D, $^3P(2)$

Case 3. For maximum spin multiplicity with two spins, the term symbols are ^3F and ^3P (see Example 5.3).

Applying the 'branching rule',

d^2	^3F ($L = 3$)		^3P ($L = 1$)	
$\Big\downarrow$ ^3F, ^3P	$\Big\downarrow$ + ^3F ($L = 3$)	$\Big\downarrow$ + ^3P ($L = 1$)	$\Big\downarrow$ + ^3F ($L = 3$)	$\Big\downarrow$ + ^3P ($L = 1$)
d^4	$L = 6, 5, 4, 3, 2, 1, 0$	$L = 4, 3, 2$	$L = 4, 3, 2$	$L = 2, 1, 0$
	I, H, G, F, D, P, S	G, F, D	G, F, D	D, P,

The spin-multiplicity must decrease by two unit i.e. $2S + 1$ changes from 3 to 1. Subtracting one ^1H (as we already have ^3H), one ^1G (as we already have ^3G), two ^1F (as we already have two ^3F), and two ^1D (as we already have one ^5D, one ^3D), we have, ^1I, $^1G(2)$, ^1F, $^1D(2)$, ^1P, $^1S(2)$

Table 5.1 Russell-Saunders term symbols for s, p, d electron configuration[a]

Configuration(s)	Russell-Saunders Term(s)
s^1	^2S
s^2	^1S
p^1 also p^5	^2P
p^2 also p^4	^3F, ^1D, ^1S
p^3	^4S, ^2D, ^2P
p^6	^1S
d^1 and d^9	^2D
d^2 and d^8	^3F, ^3P, ^1G, ^1D, ^1S
d^3 and d^7	^4F, ^4P, ^2H, ^2G, ^2F, $^2D(2)$, ^2P
d^4 and d^6	^5D, ^3H, ^3G, $^3F(2)$, ^3D, $^3P(2)$, ^1I, $^1G(2)$, ^1F, $^1D(2)$, $^1S(2)$
d^5	^6S, ^4G, ^4F, ^4D, ^4P, ^2I, ^2H, $^2G(2)$, $^2F(2)$, $^2D(3)$, ^2P, ^2S
d^{10}	^1S

[a]When multiple terms are present, the ground state term is identified by bold.

Term symbols to represent d^4 electron configuration for 210 microstates:

^5D (25), ^3H (33), ^3G (27), ^3F (21), ^3F (21), ^3D (15), ^3P (9), ^3P (9), ^1I (13), ^1G (9), ^1G (9), ^1F (7), ^1D (5), ^1D (5), ^1S (1), ^1S (1)

Considering the concept of *hole formalism*, which states that for many electronic properties one may consider systems with electron (e) or $(n-e)$, the number of unoccupied sites or "holes", to be equivalent. For a set of p orbitals, $n = 6$ since there are 2 positions in each orbital. Thus,

p^1 (1 electron) and p^5 $(6 - 1 = 5$ holes) are equivalent. Similarly, p^2 and p^4 are equivalent.

Similarly, for a set of d orbitals, $n = 10$ since there are 2 positions in each orbital.

d^1 (1 electron) and d^9 $(10 - 1 = 9$ holes), d^2 and d^8, d^3 and d^7, d^4 and d^6 are equivalent.

The various R-S terms for the p^n and d^n electron configurations are given in Table 5.1.

Example 5.6 Present flow charts of electron-electron interactions, showing the creation of atomic terms and spin-orbit states for (a) p^2 and (b) d^2 electron configuration.

Answer

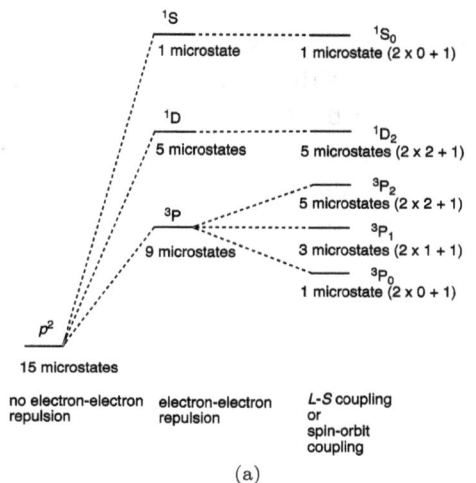

^{1}S
1 microstate .... ____ $^1S_0$
                        1 microstate (2 x 0 + 1)

^{1}D
5 microstates .... ____ $^1D_2$
                         5 microstates (2 x 2 + 1)

                        ____ $^3P_2$
                        5 microstates (2 x 2 + 1)
^{3}P
9 microstates ◄ .... ____ $^3P_1$
                        3 microstates (2 x 1 + 1)
                        ____ $^3P_0$
                        1 microstate (2 x 0 + 1)

p^2

15 microstates

no electron-electron   electron-electron   *L-S* coupling
repulsion   repulsion   or
                                               spin-orbit
                                               coupling

(a)

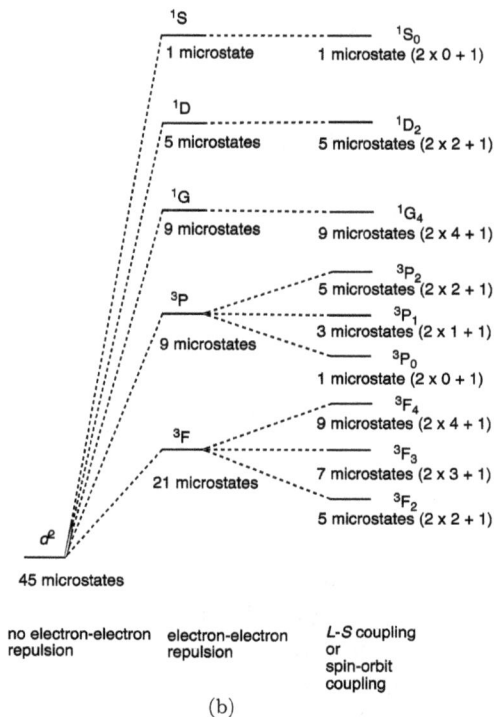

^{1}S
1 microstate .... ____ $^1S_0$
                        1 microstate (2 x 0 + 1)

^{1}D
5 microstates .... ____ $^1D_2$
                         5 microstates (2 x 2 + 1)

^{1}G
9 microstates .... ____ $^1G_4$
                         9 microstates (2 x 4 + 1)

                        ____ $^3P_2$
                        5 microstates (2 x 2 + 1)
^{3}P
9 microstates ◄ .... ____ $^3P_1$
                        3 microstates (2 x 1 + 1)
                        ____ $^3P_0$
                        1 microstate (2 x 0 + 1)

                        ____ $^3F_4$
                        9 microstates (2 x 4 + 1)
^{3}F
21 microstates ◄ .... ____ $^3F_3$
                         7 microstates (2 x 3 + 1)
                        ____ $^3F_2$
                        5 microstates (2 x 2 + 1)

d^2

45 microstates

no electron-electron   electron-electron   *L-S* coupling
repulsion   repulsion   or
                                               spin-orbit
                                               coupling

(b)

Example 5.7 Write out the electronic state (L, S, J) for carbon atom.

Answer Spin-orbit interaction between L and S causes relatively small intervals of each term for different possible values of J. For example, the lowest-energy term for the C atom (electronic configuration: $1s^2\ 2s^2\ 2p^2$) can have three possible J values, with the energy-level ordering: $^3P_0 < {}^3P_1 < {}^3P_2$. It is clear that with this L-S scheme, each electronic state is defined by quantum numbers L, S, and J.

5.3 Spin-orbit interaction

A relatively simple and quantitative description of the spin-orbit interaction for an electron bound to an atom, using some semiclassical electrodynamics and non-relativistic quantum mechanics, up to first order in perturbation theory, is discussed here.

An electron moving in an electrical field experiences an effective magnetic field, which acts on the electron spin magnetic moment. The interaction of the electron magnetic moment with the effective magnetic field is called the spin-orbit interaction.

The magnetic fields created by l and s are not isolated from one another; they interact through spin-orbit coupling (*Russell-Saunders coupling*).

Let us consider an electron of mass m is moving in an orbit about a nucleus of charge Ze. The electron will experience an electric field created by the nucleus.

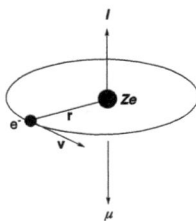

The magnitude of this electric field, $E = Ze/r^2 = (Ze/r^2)r/r = (Ze/r^3)$ r, where r is the radius vector. In addition to the electric field the electron experiences a magnetic field. The magnitude of the magnetic field, $H = (E \times v)/c$, where $E \times v$ indicates a cross-product, i.e. $E \times v \sin\theta$ (θ is the angle between E and v).

Then, $H = (E \times v)/c = (Ze/r^3)r \times v/c = (Ze/cr^3)r \times v$

The orbital angular momentum vector l of the electron $= mr \times v$.

Therefore, $H = (Ze/mcr^3)\, l$. The electron is moving under the influence of a magnetic field H, created by the nucleus. But the spinning electron has its own magnetic dipole of moment m. This magnetic dipole will interact with magnetic field H and according to classical electtromagnetic theory, the energy (ΔW) of a magnetic dipole of moment μ in a magnetic field H is $\Delta W = -\mu \bullet H = -\mu H \cos\theta$, where θ is the angle between the dipolar axis μ and the field H, s is the spin angular momentum, μ_B is the Bohr magneton, and $g = 2$ is the electron spin g-factor.

The magnetic moment (μ) expression has a negative sign, as the magnetic moment is antiparallel to the angular momentum (see above).[2]

$$\mu = -(g\mu_B s)/(h/2\pi) = -\{2(eh/4\pi mc)s\}/h/2\pi = (-e/mc)s$$

[the magnitude of the orbital and spin angular momentum vectors are expressed (quantum mechanically) in terms of the quantum numbers l and s, respectively, as $l = \{l(l+1)\}^{1/2}h/2\pi$ and $s = \{s(s+1)\}^{1/2}h/2\pi$; the magnitude of spin-only magnetic moment $\mu_S = g\{S(S+1)\}^{1/2}\mu_B = 2\{ns(ns+1)\}^{1/2} = 2\{n \times 1/2(n \times 1/2 + 1)\}^{1/2} = \{n(n+2)\}^{1/2}$ (n is the number of unpaired electron and since $s = 1/2$)].

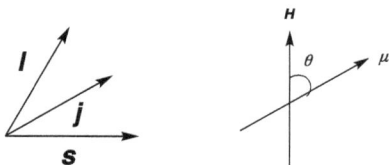

$$\Delta W = -(-e/mc)s(Ze/mcr^3)l\cos(ls) = (Ze^2/m^2c^2r^3)ls\cos(l,s)$$

$$= (Ze^2/m^2c^2r^3)(l \cdot h/2\pi)(s \cdot h/2\pi) \times \cos(l,s)$$

(since both l and s are expressed in terms of $h/2\pi$)

$$= (Ze^2h^2/4\pi^2m^2c^2r^3)ls\cos(l,s)$$

Now we should consider the fact that both the electron and the nucleus are moving with very different velocities. When one looks at the nucleus,

---

[2] R. L. Carlin, *J. Chem. Educ.* **1966**, *43*, 521.

the coordinate system on the electron seems to rotate by 180°, when the electron has completed one full turn around the nucleus. This means that to the observer the electron appears to be spinning with only half its rate. It is due to this relativistic effect that ΔW gets reduced by half.

Thus, $\Delta W = 1/2(Ze^2h^2/4\pi^2m^2c^2r^3)ls\cos(l,s) = (Ze^2h^2/8\pi^2m^2c^2r^3)ls$ $\cos(l,s)$

The factor (1/2) is the Thomas correction (British physicist and applied mathematician L. H. Thomas: 1903–1992).

Considering the spin-orbit interaction in a single electron,

$$\text{spin-orbit interaction energy, } \Delta W \propto ls\cos(l,s).$$

The strength of spin-orbit interaction is manifested by the spin-orbit coupling constant ξ. Thus, ξ is a proportionality constant and it is a positive quantity. Its magnitude indicates the effectiveness with which the l and s vectors couple to produce j. A knowledge of ξ is essential to get an idea about the energy separation between the successive j levels (see above).

$$\xi = (Ze^2h^2/8\pi^2m^2c^2r^3)$$

For a multielectron system the spin-orbit coupling interaction is shared between the total number of electrons. This in turn lowers the spin-orbit effect due to a single electron. The expression of ξ becomes, where Z_{eff} covers screening constant of the electrons due to intervening electron shells,

$$\xi = (Z_{\text{eff}}e^2h^2/8\pi^2m^2c^2r^3)$$

In this situation ξ is replaced by λ. Thus, λ is the spin-orbit coupling constant of the electronic configuration of a multielectron system. The relation between ξ and λ is,

$$\lambda = \pm\xi/(2S) = \pm\xi/(2\times ns) = \pm\xi/(2\times n\times 1/2) = \pm\xi/n$$

where n is the number of unpaired electron. In this relation, λ will be '+' for half-filled or less than half-field shells and will be '–' for more than half-filled shells.

Some generalizations:

1. The higher the oxidation state of the metal ion, the higher is ξ or λ.
   Example: the values of ξ are 135, 170, and $210\,cm^{-1}$ for V^+, V^{2+}, and V^{3+}, respectively.
   It is because Z_{eff} increases but r decreases.
2. The value of ξ increases very substantially as we move from first-transition series to the second, and to the third.
   The values of ξ (in cm^{-1}) are 515 and 1220 for Co^{2+} and Rh^{2+}, respectively, and 630 and 1460 for Ni^{2+} and Pd^{2+}, respectively.
   It is because r will increase to some extent, which is likely to lower ξ. But this is very effectively countered by a great increase in Z_{eff}.
3. For invariant oxidation state and the transition series, ξ increases with increase in atomic number.
   The values of ξ (in cm^{-1}) increase in the series Ti^{2+} (123), V^{2+} (170), Cr^{2+} (230), Mn^{2+} (300), Fe^{2+} (400), Co^{2+} (515), Ni^{2+} (630), and Cu^{2+} (830).
   It is because roughly there is a trend of decreasing size, i.e. atomic number increases which leads to an increase in Z_{eff}.

Further reading

C. E. Housecroft and A. G. Sharpe, *Inorganic Chemistry*, 2nd ed., Person Education Ltd. (2005).

D. F. Shriver and P. W. Atkins, *Inorganic Chemistry*, 3rd ed., Oxford University Press (1999).

R. L. Dutta and A. Shyamal, *Elements of Magnetochemistry*, 2nd ed., Affiliated East-West Press Pvt. Ltd. (1993).

B. E. Douglas and C. A. Hollingsworth, *Symmetry in Bonding and Spectra – An Introduction*, Academic Press, Inc. (1985).

B. E. Douglas, D. H. McDaniel, and J. J. Alexander, *Concepts and Models of Inorganic Chemistry*, 2nd edition, John Wiley & Sons, Inc. (1983).

B. N. Figgis, *Introduction to Ligand Fields*, Wiley Eastern Limited (1976).

Exercises

5.1 Find out the R-S terms for p^3 configuration.
5.2 Find out the R-S terms for d^5 configuration.

Chapter 6

Chemistry of *d*-Block Elements

Since the discovery of the theory of coordination compounds by Swiss chemist Alfred Werner (1866–1919; Nobel Prize in Chemistry: 1913), transition metal complexes have played a very important role in many sub-disciplines of chemistry, including inorganic and bioinorganic chemistry, and organometallic chemistry, and polymer and materials science. The metal-ligand coordination unit has also gained importance for the construction of supramolecular assemblies and solid-state structures.

Coordination chemistry deals with the study of a broad class of diversified compounds in which a central metal atom or ion, acting as a Lewis acid is surrounded by molecules or anions, acting as Lewis bases, known as ligands. It focuses on the chemistry of metal ions and especially transition metal ions. The principles of coordination chemistry are multidisciplinary in character, and hence they played a pivotal role in the development of transition metal chemistry and bioinorganic chemistry in one hand, and organometallic chemistry and supramolecular chemistry in the other. These varied fields are connected by a common thread of metal-ligand coordination units. Depending on the electronic requirement for certain geometries, metal ions have preferences for certain ligands with specific donor sites (hard and soft) and ligand geometry.

General descriptions of *d*-block elements, types of ligands, the nomenclature of coordination complexes, the chirality in coordination complexes, and polymetallic complexes (coordination clusters) are outside the scope of this book, as many excellent textbooks and reference material are available on these topics. However, as the properties of *d*-block metal complexes are due to their variable valence, color, structure, and magnetic properties, necessary discussions on these aspects will be the central theme of this chapter. In this chapter the structures of complexes in terms of bonding theories, magnetic and electronic spectroscopic properties are discussed in reasonable detail.

Bonding theories

The understanding of the magnetic and electronic spectral properties of transition metal complexes has always attracted the attention of inorganic chemists. To account for these properties, three theories have been developed: (i) valence bond, (ii) crystal field, and (iii) molecular orbital.

6.1 Valence bond theory

The idea that atoms form covalent bonds by sharing pairs of electrons was first proposed by American chemist G. N. Lewis (1875–1946) in 1916. In 1927, German physicists Walter Heitler (1904–1981) and Fritz London (1900–1954) showed how the sharing of pairs of electrons holds a covalent molecule together. In 1931, American chemist, biochemist, chemical engineer, peace activist Linus Pauling (1901–1994; Nobel Prizes in Chemistry (1954) and in Peace (1962)) integrated Lewis' proposal and the Heitler-London theory to give rise to two additional key concepts in valence bond (VB) theory: resonance and hybridization. According to Pauling, a covalent bond is formed between two atoms by the overlap of their half-filled valence orbitals, each of which contains one unpaired electron. Valence bond structures and Lewis structures are similar, except where a single Lewis structure provides inadequate description, several valence bond structures can be used to explain experimental results. It is in this aspect of VB theory that we see the concept of resonance. The orbital hybridization means that atomic orbitals mix to form hybrid orbitals (directed orbitals), such as, sp, sp^2, sp^3, dsp^2, sp^3d, dsp^3, sp^3d^2, and d^2sp^3 orbitals (see Chapter 1).

Given the number of unpaired electron(s) known in a metal complex, VB theory can be applied to predict the geometry of the complex. The ligands are assigned to empty metal orbitals (in fact hybridized orbitals of metal) in the order $3d$, $4s$, $4p$, $4d$, so on and so forth. We must remember that the empty hybridized orbitals of metal, available for the overlap with suitable ligand orbitals, steadily increase in energy.

The different steps involved in VB approach are:

(i) Find out the oxidation state of M^{n+} ion i.e. the number of d-electron.

(ii) Identify the metal orbitals that are available i.e. not occupied by metal ion electrons, and hence available for bonding with the ligands.

(iii) Hybridization of metal orbitals is considered for the formation of hybrid orbitals, which will be so disposed as to satisfy the actual geometry (tetrahedral, square planar, octahedral etc.).

(iv) The set of hybrid orbitals for occupation by electrons, donated by the ligands, should be reserved.

(v) The nonbonding orbitals i.e. those outside the hybridization are then utilized to accommodate metal d electrons, maintaining maximum number of unpaired electrons.

Selected examples of coordination complexes in varied geometries, their magnetic properties (the number of unpaired electron is mentioned), and hybridization-type are considered.

$[Fe^{II}Cl_4]^{2-}$ (paramagnetic: four unpaired electrons, sp^3), $[Fe^{III}Cl_4]^-$ (paramagnetic: five unpaired electrons, sp^3), $[Ni^{II}(CN)_4]^{2-}$ (diamagnetic, dsp^2), $[Co^{II}(H_2O)_6]^{2+}$ (paramagnetic: three unpaired electrons, sp^3d^2), $[Co^{III}(NH_3)_6]^{3+}$ (diamagnetic, d^2sp^3), $[Ni^{II}(H_2O)_6]^{2+}$ (paramagnetic: two unpaired electrons, sp^3d^2), $[Fe^{II}(CN)_6]^{4-}$ (diamagnetic, d^2sp^3), $[Fe^{III}(CN)_6]^{3-}$ (paramagnetic: one unpaired electron, d^2sp^3) (cf. magnetic properties).

Selected examples are presented in Fig. 6.1.

Given the fact that $[Ni^{II}Cl_4]^{2-}$, $[Ni^{II}(CN)_4]^{2-}$, and $[Ni^{II}(NH_3)_6]^{2+}$ has two, zero (diamagnetic), and two unpaired electrons, the sp^3, dsp^2, and sp^3d^2 hybridizations correctly predict their tetrahedral, square planar, and octahedral structure, respectively. As $[Co^{III}(H_2O)_6]^{3+}$ has no unpaired electron (diamagnetic) the d^2sp^3 hybridization correctly predicts its geometry.

Shortcomings of VB theory:

i) Structure (metal site geometry) cannot always be distinguished by magnetic measurements alone. For example, sp^3 and sp^3d^2 hybridization predict the same number of nonbonding orbitals and hence to accommodate the same number of unpaired electron(s).

(a) $[Ni^{II}Cl_4]^{2-}$ (paramagnetic with two unpaired electrons):

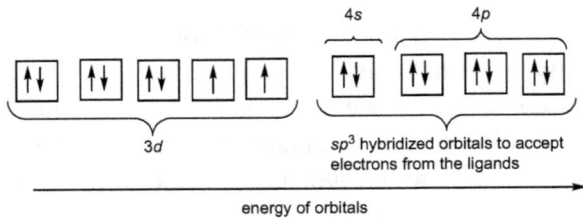

(b) $[Ni^{II}(CN)_4]^{2-}$ (diamagnetic):

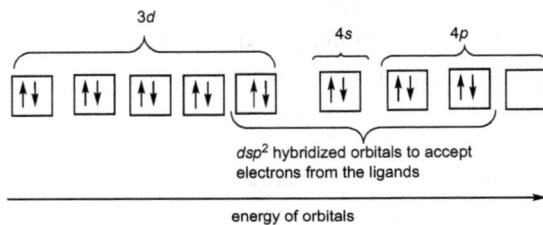

(c) $[Ni^{II}(NH_3)_6]^{2+}$ (paramagnetic with two unpaired electrons):

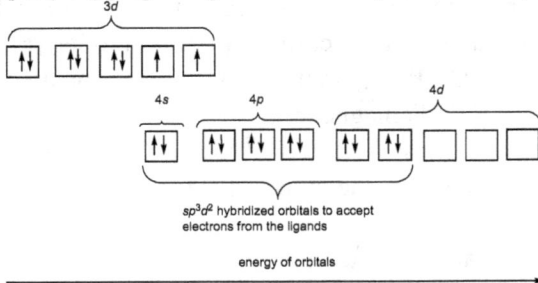

(d) $[Co^{III}(H_2O)_6]^{3+}$ (diamagnetic):

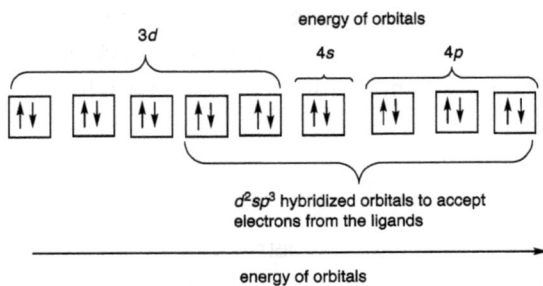

Fig. 6.1 Conventional orbital box-diagram (VB model) for (a) $[Ni^{II}Cl_4]^{2-}$, (b) $[Ni^{II}(CN)_4]^{2-}$, (c) $[Ni^{II}(NH_3)_6]^{2+}$, and (d) $[Co^{III}(H_2O)_6]^{3+}$.

ii) The success in predicting the structure is countered by the failure of the VB theory in the matter of understanding the electronic transitions responsible for spectral bands exhibited by transition metal complexes.

iii) In transition metal complexes, spectral and magnetic properties, both being dependent on the arrangement of the d-electrons, must be interrelated. VB theory provides no indication of how they may be correlated.

6.2 Crystal field theory

Crystal field (CF) theory is a model for the bonding interaction between transition metal ions and ligands, which was proposed by German-American theoretical physicist Hans A. Bethe (1906–2005; Nobel Prize in Physics in 1967) in 1929 for crystalline solids. CF theory was developed to describe important properties of complexes (oxidation state, structure, magnetic, absorption spectral, etc.). Subsequent modifications were proposed by American physicist and mathematician J. H. Van Vleck (1899–1980; Nobel Prize in Physics in 1977) in 1935 to allow for some covalency in the interactions (see below); British chemist L. E. Orgel (1927–2007) and British chemist, mathematician, and biophysicist J. S. Griffith (1928–1972) further popularized CF theory.

CF theory provides a way of calculating, by simple electrostatic (ion-ion or ion-dipole) considerations, how the relative energies of the d-orbitals of the metal ion orbitals will be affected by the static electrostatic field created by a set of surrounding ligands (ions or dipoles). Here a point charge (or a point charge-point dipole) model is considered. The well-known diagram presented in Fig. 6.2(a) is the mean energy of the five d-orbitals shown to be raised by the presence of the ligands. Here only the energy of the d-orbitals is considered. If the total energy of the system is considered, it may be found that the energy gained from the attraction between the positive charge on the metal ion and the ligands is larger than the destabilization energies of the d and other (s, p) electrons of the metal ion. Consequently, there may be a net stabilization energy for the complex (Fig. 6.2(b)). It is obvious that the relative energies of the electrons in the split d-orbitals is but one of many interactions contributing to the overall energy of the formation of a coordination complex.

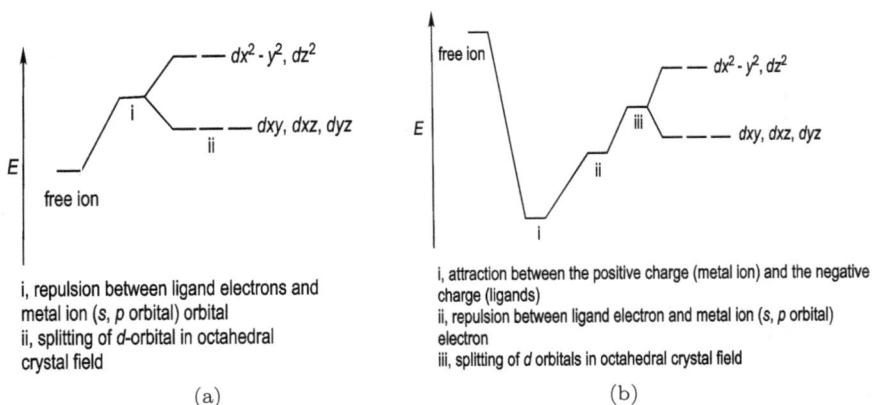

i, repulsion between ligand electrons and metal ion (s, p orbital) orbital
ii, splitting of d-orbital in octahedral crystal field

(a)

i, attraction between the positive charge (metal ion) and the negative charge (ligands)
ii, repulsion between ligand electron and metal ion (s, p orbital) electron
iii, splitting of d orbitals in octahedral crystal field

(b)

Fig. 6.2 (a) The energy profile of only the five d-orbitals in an octahedral crystal field. (b) The total energy profile of a coordination complex on the crystal field model.

(a) Crystal field splitting diagrams for octahedral and tetrahedral geometry

Crystal field theory describes the breaking of degeneracy of d-orbitals, due to a static electric field (or crystal/ligand field) produced by a surrounding charge distribution. *The overlap between metal orbitals and ligand orbitals is totally ignored.* In order to understand clearly the CF interactions in transition metal complexes, it is necessary to look at the geometrical or spatial disposition (shapes) of the five d-orbitals (Fig. 6.3). In addition, it should be realized that the dz^2 orbital can be regarded as a linear combination of two hypothetical equivalent orbitals, $dz^2 - x^2$ and $dz^2 - y^2$ (Fig. 6.4), each of which is shaped like the other four d-orbitals.

According to CF theory, the interaction between a transition metal ion and ligands arises from the attraction between the positively charged metal ion and the negative charge on the ligand. Let us suppose that the metal ion is placed in an octahedral array of six ligands, as shown in Fig. 6.5. It is easy to visualize that the $dx^2 - y^2$ and $dz^2 (dz^2 - x^2 + dz^2 - y^2)$ orbitals are all oriented in one way, while the dxy, dxz, and dyz orbitals are all oriented in another way. Specifically, all those in the first set have each of their lobes going toward a ligand, while each one in the other set has each lobe going between ligands. Thus, as a ligand approaches the metal ion, the electrons in the d-orbitals and those in the ligand repel each other due to repulsion between like charges. Due to this situation, the electrons from the ligand

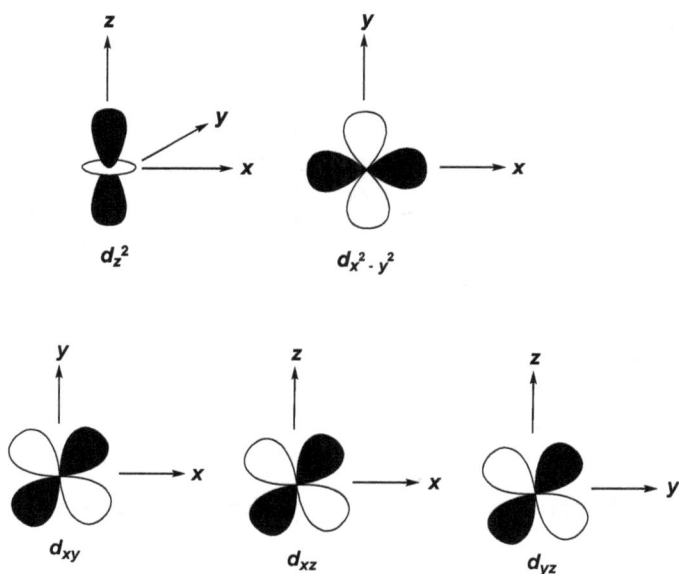

Fig. 6.3 The set of five d-orbitals, showing the sign of the wave functions in each lobe (shaded lobes represent '+' and unshaded lobes represent '−').

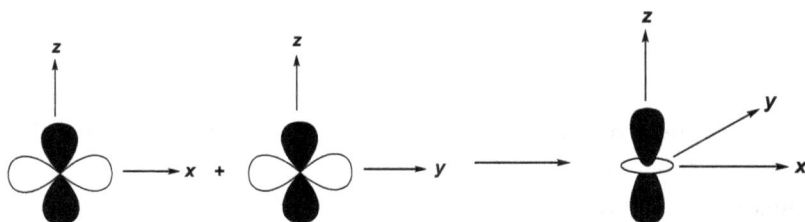

Fig. 6.4 Drawings showing how the dz^2 orbital consists of a $dz^2 - x^2$ and a $dz^2 - y^2$ orbital in equal proportions.

will be closer to some of the *d*-orbitals and farther away from others, caus-ing a loss of degeneracy. The idea is that electrons will stay away from the point charges (or point dipoles) as much as possible. Thus, we get the result that the *dxy*, *dxz*, and *dyz* orbitals (t_{2g} set; cf. Chapter 2) are equiva-lent to one another, whereas the $dx^2 - y^2$ and dz^2 (e_g set; cf. Chapter 2) are different from the first three, but equivalent to one another. This leads to the standard *d*-orbital splitting diagram for an octahedral complex with the

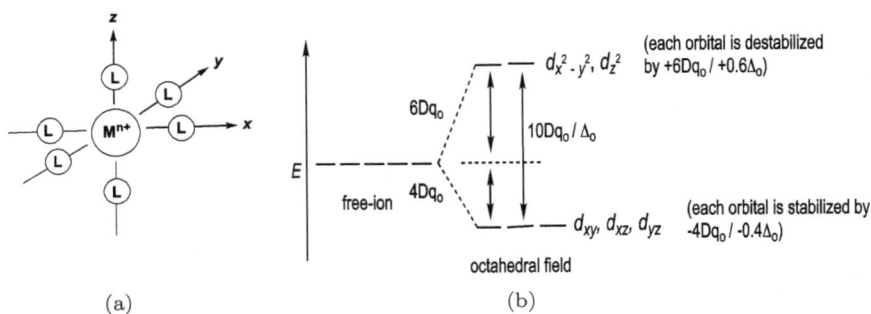

Fig. 6.5 (a) The arrangement of the six ligands in an octahedral complex. (b) d-orbital splitting diagram for octahedral crystal field.

$dx^2 - y^2$ and d_z^2 more unstable than the dxy, dxz, and dyz orbitals (Fig. 6.5). The $dx^2 - y^2$ and dz^2 atomic orbitals are destabilized by $-6Dq_0$ while the dxy, dyz, and dxz atomic orbitals are stabilized by $+4Dq_0$.

The separation between the two set of orbitals is represented as $10Dq_0$ or Δ_0 (the subscript o denotes octahedral field). This pattern of splitting, in which the algebraic sum of all energy shifts of all orbitals is zero, is said to preserve the *center of gravity or barycenter* of the set of levels.

Similarly, for a tetrahedral set of four ligands, arranged as shown in Fig. 6.6 at alternate vertices of a cube, it is obvious that the dxy, dxz, and dyz orbitals are in one kind of relationship to the ligands (with their lobes pointing to cube edges) while the $dx^2 - y^2$ and $dz^2 (dz^2 - x^2 + dz^2 - y^2)$ orbitals are in another relationship (with their lobes point to the centers of cube faces). Hence, we note that dxy, dxz, and dyz (t_2 set; cf. Chapter 2) form one equivalent set while $dx^2 - y^2$ and dz^2 (e set; cf. Chapter 2) form a second equivalent set. Thus, with respect to their energy in a spherical field (the barycentre, a kind of 'centre of gravity'), the $dx^2 - y^2$ and dz^2 atomic orbitals are stabilized by $-6Dq_t$ (the subscript t denotes tetrahedral field) while the dxy, dxz, and dyz atomic orbitals are destabilized by $+4Dq_t$.

From Fig. 6.6 it is evident that the lobes of $dx^2 - y^2$ and dz^2 orbitals and the ligands are half the face-diagonal away ($\sqrt{2}\,a/2 = a/\sqrt{2}$) and the lobes of dxy, dxz, and dyz orbitals and the ligands are half the side away ($a/2$). So, the repulsions between dxy, dxz, and dyz orbitals and the ligands are larger than that between $dx^2 - y^2$ and dz^2 orbitals and the ligands.

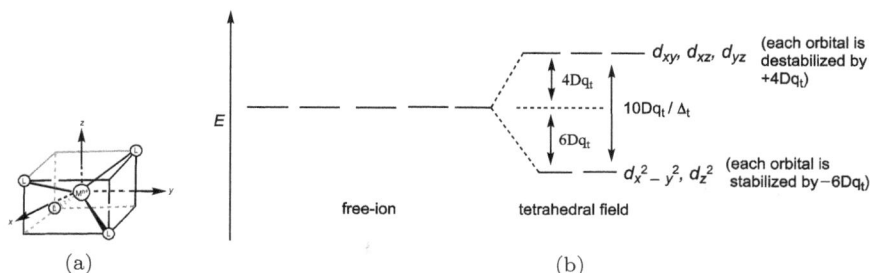

Fig. 6.6 (a) The arrangement of the four ligands in a tetrahedral complex. (b) d-orbital splitting diagram for tetrahedral crystal field. The sign convention used here for CFSE follows the thermodynamic convention.

(b) Strength of the ligands

We consider first the octahedral geometry. The magnitude of Δ_o is assessed by the strength of the crystal field or ligand field created by the ligand(s). The two extreme situations are called *weak field* and *strong field* ligands. The energy gap Δ_o can be measured easily by recording the UV-vis spectrum of the complex (see below). The following three factors decide the magnitude of Δ_o:

(i) The nature of the ligands (spectrochemical series, see below)
(ii) The value of Δ_o increases with increasing charge on the metal ion or in other words the oxidation state of the metal ion. Δ_o values for complexes of the first-transition series are 7500–$12\,500\,\text{cm}^{-1}$ for M^{2+} ions and $14\,000$–$25\,000\,\text{cm}^{-1}$ for M^{3+} ions.
(iii) The value of Δ_o varies significantly depending on whether the metal ion is a first-, second- or third-row transition element. The Δ_o values for complexes of the $[Co^{III}(NH_3)_6]^{3+}$, $[Rh^{III}(NH_3)_6]^{3+}$, and $[Ir^{III}(NH_3)_6]^{3+}$ are $\sim 23\,000$, $\sim 34\,000$, and $\sim 41\,000\,\text{cm}^{-1}$, respectively. It is evident that complexes of metal ions in the same group and with the same oxidation and spin state the Δ_o values increase significantly.

(c) Spectrochemical series

The extent of crystal field splitting of ligand field splitting is tuned by the nature of ligand(s). Ligands which cause only a small splitting are termed *weak field* ligands and which cause a large splitting are called *strong field* ligands. The common ligands can be arranged in increasing order of crystal field strength. The order remains practically invariant for different metal ions.

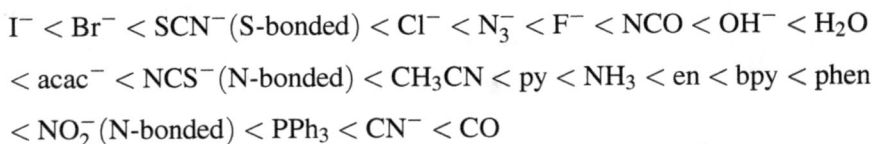

$$I^- < Br^- < SCN^- \text{(S-bonded)} < Cl^- < N_3^- < F^- < NCO < OH^- < H_2O$$

$$< acac^- < NCS^- \text{(N-bonded)} < CH_3CN < py < NH_3 < en < bpy < phen$$

$$< NO_2^- \text{(N-bonded)} < PPh_3 < CN^- < CO$$

($acac^-$ = acetylacetonate ion, py = pyridine, en = 1,2-diaminoethane, bpy = 2,2′-bipyridine, phen = 1,10-phenanthroline; py is a monodentate and $acac^-$, en, bpy, and phen are bidentate ligands)

This series is called the *spectrochemical series*. It is an experimentally determined series. In an octahedral complex, the stronger is the crystal field caused by the ligands the greater is the splitting (Δ_o). The original order of the spectrochemical series was first proposed by Japanese chemist R. Tsuchida in 1938, based on the results of absorption spectra of Co(III) complexes from spectral shifts for complexes such as $[Co^{III}(NH_3)_5X]^{n+}$ with $X = I^-$, Br^-, Cl^-, H_2O, and NH_3. The observable parameter was the colors of the complexes, which range from deep purple for $X = I^-$ through pink for Cl^- to yellowish orange with NH_3. It is understandable that to establish a series like this, the gross geometry and the oxidation and spin state of the metal ion should be invariant and the only variable is the ligands.

(d) Octahedral (high spin and low spin) and tetrahedral complexes, and CFSE

Let us consider the consequences of filling up the two set of d-orbitals in octahedral field (Fig. 6.5). The splitting of the energy levels causes no change in the energy of the system, if all five d-orbitals are evenly occupied. Hence, the net change is zero for high-spin d^5 ($3 \times -4Dq_o + 2 \times + 6Dq_o$) or d^{10} ($6 \times -4Dq_o + 4 \times + 6Dq_o$) ions. Barring these two, for all other electron distributions, crystal field splitting lowers the total energy of the system. The situation is quite straightforward for $d^1 - d^3$ systems. When we reach the d^4 configuration, there are two possible choices for the fourth electron: it can occupy either one of the empty e_g orbitals or one of the singly occupied t_{2g} orbitals. Recall that placing an electron in an already occupied orbital results in electrostatic repulsion that increase the energy of the system. This increase in energy is called the spin-pairing energy (P). If Δ_o is less than P, then the lowest-energy arrangement has the fourth electron in one of the empty e_g orbitals. Because this arrangement results in four unpaired electrons, it is called a high-spin (hs) configuration, and a complex with this electron configuration, such as the $[Cr^{II}(H_2O)_6]^{2+}$ ion (d^4 system), is called a hs complex. Conversely, if Δ_o is greater than P, then the lowest-energy arrangement has the fourth electron in one of the occupied t_{2g} orbitals. Because this arrangement results in only two unpaired electrons, it is called a low-spin (ls) configuration, and a complex with this electron configuration, such as the $[Mn^{III}(CN)_6]^{3-}$ ion (d^4 system), is called a ls complex. As for d^4, metal ions with the d^5, d^6 or d^7 electron configurations can be either hs or ls, depending on the magnitude of Δ_o.

The energy gained (decrease in energy) by a d^n ion due to preferential occupation of electrons in the split orbitals (say, t_{2g} set and e_g set in octahedral geometry), caused by the crystal field created by the ligand, is called the *crystal field stabilization energy* (CFSE) or *ligand field stabilization energy* (LFSE).

Even though H_2O is a weak field ligand, the complex $[Co^{III}(H_2O)_6]^{3+}$ is ls because in the ls state the CFSE is $-24Dq_o$ (in the hs state the CFSE is $-4Dq_o$). As the F^- ion is a weak field ligand ($\Delta_o < P$), the complexes $[Co^{III}F_6]^{3-}$ and $[Co^{III}(H_2O)_3F_3]$ are hs.

As for a tetrahedral complex the magnitude of the crystal field splitting (Δ_t) is about half (actually it is $\sim$4/9) of the splitting in octahedral field (Δ_o), all tetrahedral complexes are high-spin. Table 6.1 summarizes the values of CFSE for d^1 to d^9 systems in both octahedral and tetrahedral crystal field.

Table 6.1 CFSE in octahedral (hs and ls) (Dq_o/Δ_o) and tetrahedral (Dq_t/Δ_t) crystal fields

	Octahedral	Octahedral	Tetrahedral
d^n ion	spin state	spin state	spin state
	hs	ls	hs
d^1	$-4Dq_o$ (or $-0.4\Delta_o$)	—	$-6Dq_t$(or $-0.6\Delta_t$)
d^2	-8	—	-12
d^3	-12	—	-8
d^4	-6	-16	-4
d^5	0	-20	0
d^6	-4	-24	-6
d^7	-8	-18	-12
d^8	-12	—	-8
d^9	-6	—	-4

For d^4–d^7 systems the possibility exists of observing either hs (maximum number of unpaired electrons) or ls (maximum number of paired up electrons) complexes, depending on the strength of ligands and the magnitude of P.

While weak field ligands support the formation of hs complexes $(\Delta_o < P)$, strong field ligands support the formation of ls complexes $(\Delta_o > P)$.

All octahedral cobalt(III) (d^6 system) complexes are ls, including $[Co^{III}(H_2O)_6]^{3+}$ and excepting $[Co^{III}F_6]^{3-}$ and $[Co^{III}(H_2O)_3F_3]$.

The *spin-pairing energy* is defined as the energy necessary to force two electrons to stay in the same orbital with one spin up and the other down (Pauli's exclusion principle).

All tetrahedral complexes are high-spin.

The CFSE is expressed as a multiple of the crystal field splitting parameter Δ_o/Δ_t. For paired electrons in a single orbital, the term P represents the spin-pairing energy.

Example 6.1 For which d^n configuration in octahedral or tetrahedral complexes could maximum values of CFSE be achieved?

Answer According to Fig. 6.5 and Table 6.1, it is seen that for d^3 and d^8 and ls d^6 configuration in O_h complexes achieve maximum values of CFSE. For T_d complexes (Fig. 6.6 and Table 6.1) maximum values of CFSE are achieved for d^2 and d^7 configuration.

Example 6.2 An aqueous solution of $FeSO_4 \cdot 7H_2O$ is paramagnetic but when excess CN^- ion is added to it, the resulting species is diamagnetic. Explain.

Answer An aqueous solution of $FeSO_4 \cdot 7H_2O$ consists of $[Fe^{II}(H_2O)_6](SO_4) \cdot H_2O$. H_2O is a weak field ligand and hence the six-coordinate Fe(II) complex must be high-spin. According to Fig. 6.5 the complex is paramagnetic with four unpaired electrons ($t_{2g}^4 \, e_g^2$).

On addition of CN^- ion, the complex $[Fe^{II}(H_2O)_6]^{2+}$ changes to $[Fe^{II}(CN)_6]^{4-}$ species. In this complex CN^- is a strong field ligand and now the octahedral complex must be low-spin ($t_{2g}^6 \, e_g^0$) and hence diamagnetic.

Example 6.3 Arrange the following complexes in increasing order of their CFSE:

$$[CoCl_4]^{2-}, \ [NiCl_4]^{2-}, \ [CuCl_4]^{2-}.$$

Answer As Cl^- is a weak field ligand the geometry of the tetrachloro metal(II) complexes $[Co^{II}Cl_4]^{2-}$ ($e^4 \, t_2^3$), $[Ni^{II}Cl_4]^{2-}$ ($e^4 \, t_2^4$), $[Cu^{II}Cl_4]^{2-}$ ($e^4 \, t_2^5$) are tetrahedral. According to Fig. 6.6 and Table 6.1, the CFSE values are $-12Dq_t$ (or $-1.2\Delta_t$), $-8Dq_t$ (or $-0.8\Delta_t$), and $-4Dq_t$ (or $-0.4\Delta_t$), respectively. The increasing order of their CFSE values is $[CoCl_4]^{2-} > [NiCl_4]^{2-} > [Cu^{II}Cl_4]^{2-}$.

Example 6.4 Which of the following pairs of complexes will have larger Δ_o / Δ_t values. Give reasons.

(a) $[Cr(H_2O)_6]^{2+}$, $[Cr(CN)_6]^{4-}$; (b) $[Co(H_2O)_6]^{2+}$, $[Co(H_2O)_6]^{3+}$; (c) $[Co^{II}Cl_4]^{2-}$, $[Co^{III}F_6]^{3-}$.

Answer The complexes $[Cr^{II}(H_2O)_6]^{2+}$, $[Cr^{II}(CN)_6]^{4-}$; $[Co^{II}(H_2O)_6]^{2+}$, $[Co^{III}(H_2O)_6]^{3+}$; $[Co^{II}Cl_4]^{2-}$, $[Co^{III}F_6]^{3-}$. H_2O, Cl^-, and F^- are all weak field ligands and hence the complexes will be high-spin octahedral or tetrahedral. However, the complex $[Co^{III}(H_2O)_6]^{3+}$ is low spin (see above). CN^- is a strong field ligand and hence $[Cr^{II}(CN)_6]^{4-}$ will be low-spin octahedral complex. Therefore,

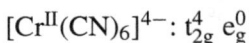

$[Cr^{II}(H_2O)_6]^{2+}: t_{2g}^3\, e_g^1$

$[Cr^{II}(CN)_6]^{4-}: t_{2g}^4\, e_g^0$

(a) Δ_o of $[Cr(CN)_6]^{4-}$ will be larger than Δ_o of $[Cr(H_2O)_6]^{2+}$, the former complex is a low-spin complex.

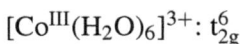

$[Co^{II}(H_2O)_6]^{2+}: t_{2g}^5\, e_g^2$

$[Co^{III}(H_2O)_6]^{3+}: t_{2g}^6$

(b) Δ_o of $[Co(H_2O)_6]^{3+}$ will be larger than Δ_o of $[Co(H_2O)_6]^{2+}$, as one complex is of Co(III) and the other is of Co(II). The reason is enhanced electrostatic energy in the case of $[Co^{III}(H_2O)_6]^{3+}$.

$[Co^{II}Cl_4]^{2-}: e^4\, t_2^3$

$[Co^{III}F_6]^{3-}: t_{2g}^4\, e_g^2$

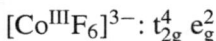

(c) Δ_o of $[CoF_6]^{3-}$ will be larger than Δ_t of $[CoCl_4]^{2-}$, since $\Delta_t \simeq 4/9\,\Delta_o$.

The total electron pairing energy (P) has two components P_c and P_e. P_c is a destabilizing energy due to coulombic repulsion associated with placing two electrons into the same orbital and P_e is a stabilizing energy for electron-exchange associated with two degenerate electrons having parallel spin.

> CFSE, including P, is defined as the energy of the electron configuration under the application of crystal field minus the energy of the electronic configuration in the isotropic field (free ion).

Example 6.5 Analyze the various components of P with $[Fe(H_2O)_6]^{2+}$ as an example.

Answer $[Fe(H_2O)_6]^{2+}$ is a grossly octahedral Fe(II) complex (d^6 system), as H_2O is a weak field ligand.

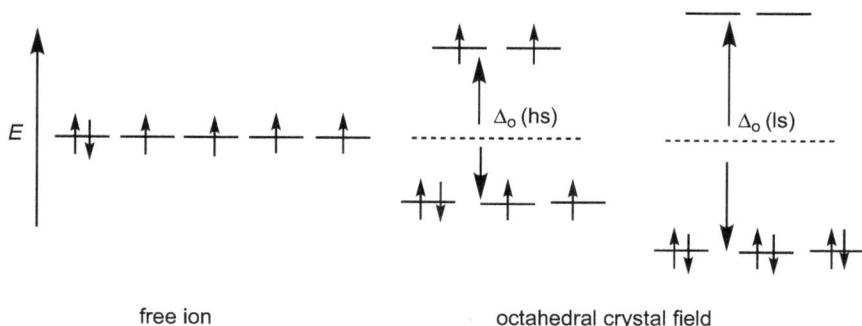

free ion   octahedral crystal field

The d-orbital splitting for a d^6 ion [Fe^{II}(H$_2$O)$_6$]$^{2+}$ in both (a) hs and (b) ls state.

For a given M^{n+} ion, P is constant. It does not vary with crystal field of the ligand and oxidation state of the M^{n+} ion.

Energy of the isotropic field (free ion), $E_{iso} = 6 \times 0 + 1P = P$

For hs case, the energy of the octahedral crystal field $(E_{CF}) = 4 \times -0.4\Delta_o + 2 \times +0.6\Delta_o + 1P = -0.4\Delta_o + P$

Therefore, CFSE for hs case, including $P = \{-0.4\Delta_o + P\} - P = -0.4\Delta_o$

Similarly, for ls case, the energy of the octahedral crystal field $(E_{CF}) = 6 \times -0.4\Delta_o + 3P = -2.4\Delta_o + 3P$

Therefore, CFSE for ls case, including $P = \{-2.4\Delta_o + 3P)\} - P = -2.4\Delta_o + 2P$

For *ls* d^4, d^5, and d^7 ions the CFSE's, including P, are

d^4	d^5	d^7
$\{-1.6\Delta_o + P)\} - 0$	$\{-2.0\Delta_o + 2P)\} - 0$	$\{-1.8\Delta_o + 3P)\} - 2P$
$= -1.6\Delta_o + P$	$= -2.0\Delta_o + 2P$	$= -1.8\Delta_o + P$

Example 6.6 Decide whether [Fe(H$_2$O)$_6$]$^{2+}$ is a hs or a ls complex. Given: $\Delta_o = 9350$ cm^{-1}; $P_c = 19\,600$ cm^{-1}; $P_e = -2000$ cm^{-1}.

Answer For hs case, the CFSE (without P) $= -0.4\Delta_o = -0.4 \times 9350$ cm$^{-1} = -3740$ cm^{-1}

For ls case (hypothetical), the CFSE (without P) $= -2.4\Delta_o = -2.4 \times 9350$ cm$^{-1} = -22\,440$ cm^{-1}

Total energy (E) for hs case $= 1P_c + 4P_e + \text{CFSE} = \{19\ 600 + 4 \times (-2000) + (-3740)\}\ \text{cm}^{-1} = 7860\ \text{cm}^{-1}$

Total energy (E) for ls case (hypothetical) $= 3P_c + 6P_e + \text{CFSE} = \{3 \times 19\ 600 + 6 \times (-2000) + (-22\ 440)\}\ \text{cm}^{-1} = 24\ 360\ \text{cm}^{-1}$

Therefore, $[\text{Fe}^{II}(\text{H}_2\text{O})_6]^{2+}$ is a hs complex.

As for H_2O $\Delta_o < P$, hexa-aqua complexes of $3d$ metal ions are always hs.

(e) Consequences of d-orbital splitting

i) Ionic radii

Let us consider the structures of first-transition series M^{2+} ions in their oxides of composition $M^{II}O$. In these ionic solids of face-centered cubic close packing of O^{2-} ions, each M^{II} ion is octahedrally surrounded by six O^{2-} ions. The O^{2-} ion is a weak field ligand and hence the resulting oxides are uniformly high-spin. From structural studies the M^{II}–O bond distances could be determined. Now if O^{2-} radius (~ 1.4 Å) is subtracted from M^{II}–O distances one gets the ionic radii of M^{II} ions in $M^{II}O$ oxides.

In the absence of crystal field stabilization effects, the ionic radii would be expected to decrease steadily from d^0 Ca^{II} to d^5 Mn^{II} to d^{10} Zn^{II}, CFSE $= 0$ for each of these ions, because of the increasing nuclear charge, as electrons are added to the same orbital. A plot of ionic radii of M^{II} ions against atomic number is displayed in Fig. 6.7(a). A line can be drawn through the points corresponding to Ca^{II}, Mn^{II}, and Zn^{II} ions (Fig. 6.7(a)). The change in size is not regular. In fact, we see two minima, one at V^{II} (d^3 system) and the other at Ni^{II} (d^8 system), and a maximum at Mn^{II}(d^5 system). The higher the CFSE, the smaller is the ionic radius (Fig. 6.7(b)), related to the hypothetical spherically symmetric ion.

The electrons in the t_{2g} orbital set does not face O^{2-} ligands directly (orbital lobes are directed between the ligands) and hence less repulsion is felt but the attractive influence of the increasing number of protons in the nucleus (nuclear charge) leads to a contraction in size. But when electrons are added to the e_g orbital set, they face the O^{2-} ligands directly (head on repulsion) and the repulsion is large. Increase in the nuclear charge cannot compensate the repulsive effect. As a result, M^{II}–O bond length increases, i.e., the ionic radii of M^{II} ions increase.

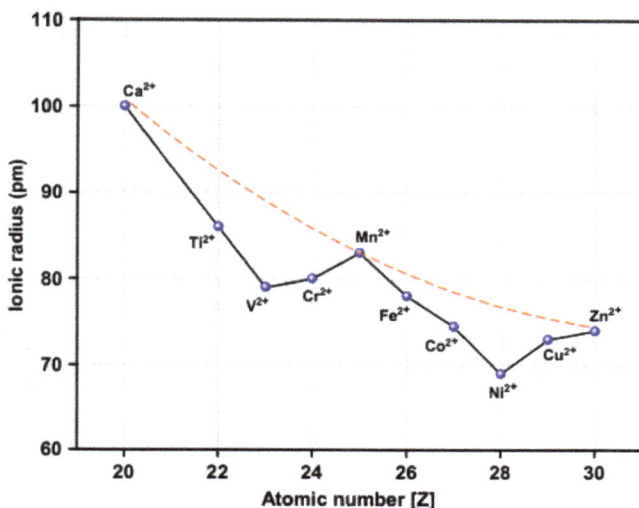

(a)

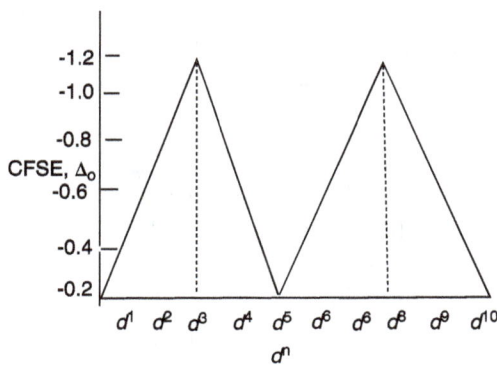

(b)

Fig. 6.7 (a) Plot of ionic radii against atomic number. (b) Plot of CFSE against atomic number. M^{II} ionic radii, 6-coordination (pm): Ca 100, Ti 86, V 79, Cr 80, Mn 83, Fe 78, Co 74.5, Ni 69, Cu 73, Zn 74.

Example 6.7 In the structure of bivalent metal oxides, each metal ion is surrounded octahedrally by six oxide ions. The Mn–O bond length in MnO is longer than Cr–O bond length in CrO. Explain.

Answer As O^{2-} ion is a weak field ligand, the complexes are hs. While in $Cr^{II}O$ the three d-electrons are in the t_{2g} level (t_{2g}^3), the five d-electrons in

$Mn^{II}O$ are symmetrically distributed over five d-orbitals with $t_{2g}^3 \, e_g^2$ electronic distribution. As in the case of $Mn^{II}O$, two electrons occupy two e_g levels, which face the ligand orbitals directly along the axes, the repulsion between oxide electrons and Mn(II) electrons is enhanced. As a consequence, the Mn–O bond length in $Mn^{II}O$ is longer than Cr–O bond length in $Cr^{II}O$.

ii) Lattice enthalpy

Lattice enthalpy (LE) is defined as the energy released (heat liberated), when a mole of M^{n+} ions and a mole of n X^- ions are brought together from infinity to the distance they occupy in the crystal lattice of $M^{II}X_n$ (s).

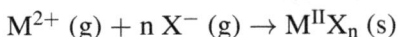

$$M^{2+} \text{ (g)} + n \, X^- \text{ (g)} \rightarrow M^{II}X_n \text{ (s)}$$

Let us consider the LE of first-transition series bivalent metal fluorides of composition $M^{II}F_2$. In these ionic solids of face-centered cubic close packing of F^- ions, each M^{II} ion is octahedrally surrounded by six F^- ions. The F^- ion is a weak field ligand and hence the resulting fluorides are uniformly hs.

In the absence of CF stabilization effects, the ionic radii would be expected to decrease steadily with increasing atomic number. Thus, with decrease in ionic radii the LE should gradually increase. Plot of LE against atomic number (Fig. 6.8) passes through two maxima and a minimum. Given the fact that plot of ionic radii against atomic number passes through two minima and a maximum (Fig. 6.7), the observed result is understandable. It is reasonable to assume that the difference between actual LE and hypothetical LE (in the absence of crystal field effects) is a measure of CFSE.

The LE for d^5 $Mn^{II}F_2$ and d^{10} $Zn^{II}F_2$, each with CFSE = 0, are 663 kcal/mol and 714 kcal/mol, respectively (1 kcal = 4.18 J). We can now calculate hypothetical LE of $Ni^{II}F_2$, without any CFSE.

Expected LE of $Ni^{II}F_2$ = LE for $Zn^{II}F_2$ − 2/5 (LE for $Zn^{II}F_2$ − LE for $Mn^{II}F_2$) (remembering the series Mn, Fe, Co, Ni, Cu, Zn) = 714 − 2/5(714 − 663) ≃ 714 − 2/5 × 50 = 694 kcal/mol.

Actual LE for $Ni^{II}F_2$ is 729 kcal/mol. The difference is (729 − 694) kcal/mol = 35 kcal/mol.

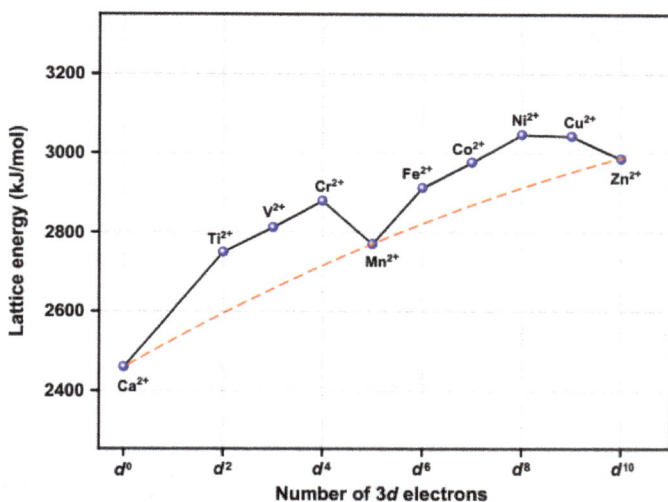

Fig. 6.8 Plot of lattice enthalpy for bivalent metal fluorides $M^{II}F_2$ against atomic number. The LE values (kJ/mol): -2459 (CaF_2), -2749 (TiF_2), -2812 (VF_2), -2879 (CrF_2), -2770 (MnF_2), -2912 (FeF_2), -2976 (CoF_2), -3046 (NiF_2), -3042 (CuF_2), -2985 (ZnF_2).

Moreover, CFSE of $Ni^{II}F_2$ (in octahedral crystal field) $= -1.2\Delta_o$

Hence, $1.2\Delta_o = 35$ kcal/mol and it leads to $\Delta_o \simeq 29.2$ kcal/mol $\simeq$ $10\,200$ cm^{-1} (1 kcal/mol $= 350$ cm^{-1}). The computed value is in conformity with experimentally determined Δ_o value for $Ni^{II}F_2$.

Example 6.8 The lattice enthalpy (LE) values (in kJ/mol) of the first-transition series bivalent metal oxides $M^{II}O$ are 3460 (CaO), 3880 (TiO), 3910 (VO), 3810 (MnO), 3920 (FeO), 3990 (CoO), 4070 (NiO). Justify the trend.

Answer In these ionic solids of face-centered cubic close packing of O^{2-} ions, each M^{II} ion is octahedrally surrounded by six O^{2-} ions. The O^{2-} ion is a weak field ligand and hence the resulting oxides are hs.

While in $Ca^{II}O$ the number of d electron is zero. In TiO, VO, MnO, FeO, CoO, and NiO the number of d electron is two, three, five, six, seven, and eight, respectively. As the number of d electron increases the ionic size decreases up to d^3, as the electrons are added to the t_{2g} set. As a consequence, the LE increases. Further increase to five the size increases and

LE decreases, as the electrons are added to the e_g set. For d^6, d^7, and d^8 the electrons are added to the t_{2g} set and the size decreases, and in turn LE increases (cf. ionic radii and Example 6.7). Thus the observed trend is justified.

iii) Hydration enthalpy

The hydration enthalpy is the amount of energy (heat) released, when a mole of M^{2+} ion dissolves in excess water forming an infinite dilute solution of the hydrated species $[M^{II}(H_2O)_6]^{2+}(aq)$.

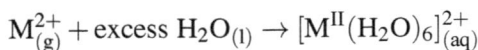

$$M^{2+}_{(g)} + \text{excess } H_2O_{(l)} \rightarrow [M^{II}(H_2O)_6]^{2+}_{(aq)}$$

Let us now discuss the plot of hydration energies of the M^{2+} ions of the first-transition series against atomic number (Fig. 6.9). The shape of this plot resembles the plot of LE of $M^{II}F_2$ against atomic number (Fig. 6.8). The d^0 Ca^{2+}, d^5 Mn^{2+}, and d^{10} Zn^{2+} ions have CFSE $= 0$. As in the LE against atomic number plot (Fig. 6.8), joining the points corresponding to these ions, a line is obtained. The concept of CFSE helps to explain the double-humped hydration enthalpies of $3d$-metal M^{2+} ions (Fig. 6.9).

iv) Spinel structures

The simple d-block mixed oxides have structures related to spinel, $MgAl_2O_4$, with the general formula $A^{II}B^{III}_2O_4$. In fact, two types of spinel structures (*normal* and *inverse*) are observed.

In the face-centered cubic close packing of O^{2-} ions two kinds of sites are generated: tetrahedral and octahedral. In the normal spinel structure with the general formula $A^{II}[B^{III}_2O_4]$, bivalent A^{II} ions occupy tetrahedral sites and trivalent B^{III} ions occupy octahedral sites. In inverse spinel structure with the general formula $B^{III}[A^{II}B^{III}O_4]$, bivalent A^{II} ions move to the tetrahedral sites and half of trivalent B^{III} ions occupy tetrahedral sites. The other half of B^{III} ions occupy octahedral sites.

Let us consider a few mixed oxides. Consideration of CFSE (O^{2-} is a weak field ligand), based on site preferences of the ions, helps to decide the spinel structure, whether it will be normal or inverse.

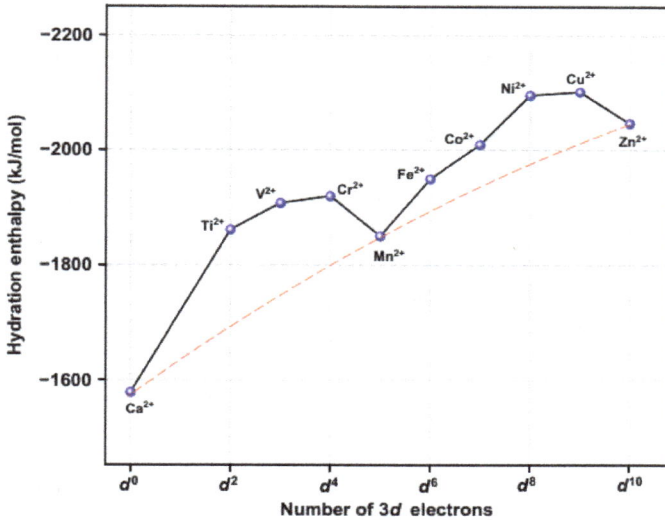

Fig. 6.9 Plot of hydration enthalpy against atomic number: Ca^{2+} -1579, Ti^{2+} -1862, V^{2+} -1895, Cr^{2+} -1908, Mn^{2+} -1851, Fe^{2+} -1950, Co^{2+} -2010, Ni^{2+} -2096, Cu^{2+} -2099, Zn^{2+} -2047 kJ/mol.

Example 6.9 Arrive at the spinel structure of $NiAl_2O_4$ whether it is normal or inverse.

Answer Normal spinel: $Ni^{II}[Al_2^{III}O_4]$
CFSE $= -8Dq_t$ (Ni^{II} in tetrahedral site) $+ 0$ (Al^{III} in octahedral site) $=$ $-8Dq_t \simeq -4Dq_o$ (for the purpose of comparison: either Dq_t or Dq_o).
Inverse spinel: $Al^{III}[Ni^{II}Al^{III}O_4]$
CFSE $= 0$ (Al^{III} in tetrahedral site) $- 12Dq_o$ (Ni^{II} in octahedral site) $+ 0$ (Al^{III} in octahedral site) $= -12Dq_o$

Hence, the structure of $NiAl_2O_4$ is inverse spinel.

Example 6.10 Arrive at the spinel structure of Fe_3O_4 whether it is normal or inverse.

Answer Normal spinel: $Fe^{II}[Fe_2^{III}O_4]$
CFSE $= -6Dq_t$ (Fe^{II} in tetrahedral site) $+ 0$ (Fe^{III} in octahedral site) $=$ $-6Dq_t \simeq -3Dq_o$

Inverse spinel: $Fe^{III}[Fe^{II}Fe^{III}O_4]$

CFSE $= 0 + -4Dq_o$ (Fe^{II} in octahedral site) $+ 0$ (Fe^{III} in octahedral site)

$= -4Dq_o$

Hence, the structure of Fe_3O_4 is inverse spinel.

Example 6.11 Arrive at the spinel structure of Co_3O_4 whether it is normal or inverse.

Answer Normal spinel: $Co^{II}[Co_2^{III}O_4]$

CFSE $= -12Dq_t$ (Co^{II} in tetrahedral site) $+ 2 \times -24Dq_o$ (Co^{III} in octahedral site)

$= -12Dq_t - 48Dq_o \simeq -6Dq_o - 48Dq_o = -54Dq_o$

Inverse spinel: $Co^{III}[Co^{II}Co^{III}O_4]$

CFSE $= -6Dq_t + -8Dq_o$ (Co^{II} in octahedral site) $+ -24Dq_o$ (Co^{III} in octahedral site)

$\simeq -3Dq_o - 8Dq_o - 24Dq_o = -35Dq_o$

Hence, the structure of Co_3O_4 is normal spinel.

v) The Irving-Williams series (Stability constants)

Fig. 6.10 displays the plot of log K (K is formation constant) for ethylenedi-iminodiacetate complexes of the M^{2+} ions of the $3d$-series against atomic number.[1] This kind of variation was first summarized by British chemists H. M. N. H. Irving (1905–1993) and R. J. P. Williams (1926–2015) in 1953 for the order of formation (stability) constants. The order (see below) is insensitive to the choice of ligands.

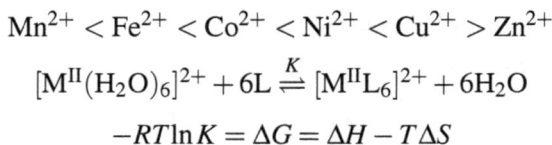

$$Mn^{2+} < Fe^{2+} < Co^{2+} < Ni^{2+} < Cu^{2+} > Zn^{2+}$$

$$[M^{II}(H_2O)_6]^{2+} + 6L \overset{K}{\rightleftharpoons} [M^{II}L_6]^{2+} + 6H_2O$$

$$-RT\ln K = \Delta G = \Delta H - T\Delta S$$

Since monodentate H_2O is replaced by monodentate L, there is no change in the randomness of the solution i.e. $T\Delta S$ term is not expected to impart much influence on the free energy (ΔG) value. If H_2O is replaced by

---

[1] P. R. Varadwaj, A. Varadwaj, and B.-Y. Jin, *Phys. Chem. Chem. Phys.* **2015**, *17*, 805.

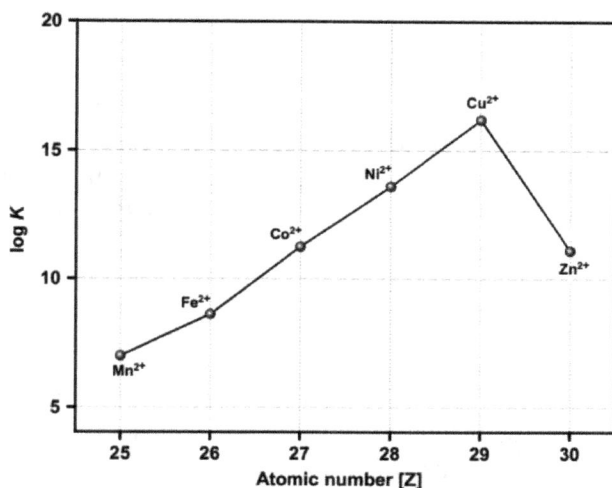

Fig. 6.10 Plot of $\log K$ values (Mn^{2+} 7.0, Fe^{2+} 8.63, Co^{2+} 11.25, Ni^{2+} 13.6, Cu^{2+} 16.2, Zn^{2+} 11.1) against atomic number.

a bidentate ligand such as en = 1,2-diaminoethane then $T\Delta S$ term would have increased. So, in the abovementioned reaction, the value of K will depend primarily on the variation of the enthalpy term ΔH.

Let us describe the formation enthalpy for $[M^{II}(H_2O)_6]^{2+}$ (ΔH_H, subscript 'H' signifies hydration) is due to the reaction,

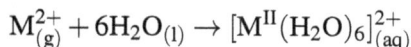

$$M^{2+}_{(g)} + 6H_2O_{(l)} \rightarrow [M^{II}(H_2O)_6]^{2+}_{(aq)}$$

Similarly, the formation enthalpy for $[M^{II}(L)_6]^{2+}$ (ΔH_L, subscript 'L' signifies L coordination) is due to the reaction,

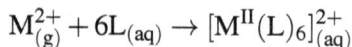

$$M^{2+}_{(g)} + 6L_{(aq)} \rightarrow [M^{II}(L)_6]^{2+}_{(aq)}$$

The difference $\Delta H_L^{corr} - \Delta H_H^{corr}$, is expected to be a measure of ΔH for the complexation reaction. Now the individual ΔH_L and ΔH_H terms carry a contribution due to CF stabilization effect of a hexa-aqua complex and a hexa-L complex. When both these terms are corrected for their respective CF contribution, we are left with enthalpy terms which do not carry CF stabilization effect. The difference $\Delta H_L^{corr} - \Delta H_H^{corr}$ is equal to ΔH term without CF effect. When such ΔH terms are plotted against atomic number of Mn to Zn all the values are on the smoothly increasing curve from

Mn^{2+} to Zn^{2+}. However, there is one important exception: the stability of Cu^{II} complexes is greater than that of Ni^{II} (CFSE $= -1.2\Delta_o$) despite the fact that for Cu^{II} the CFSE $= -0.6\Delta_o$. This anomaly is a consequence of the stabilizing influence of the Jahn-Teller effect, which results in strong binding of the four ligands in the plane of the tetragonally distorted Cu(II) complex and that stabilization enhances the value of K. The axial positions are correspondingly weakly bound.

(f) Tetragonal distortion and Jahn-Teller distortion

In 1937, British scientist of German descent H. Jahn (1907–1979) and Hungarian-American theoretical physicist E. Teller (1908–2003) postulated a theorem stating that nonlinear molecules/ions with a degenerate electronic ground state undergo a geometrical distortion that removes this degeneracy to lower the overall energy of the system. The consequence is the stabilization of the molecule/ion. The Jahn-Teller effect is manifested mainly by various hs and ls octahedral complexes.

For a hs $[Mn^{III}(H_2O)_6]^{3+}$ (d^4 system), there is one vacancy in the e_g orbital set, either in the $dx^2 - y^2$ or in the dz^2 orbital. The two possible ways to fill these orbitals, $(dx^2 - y^2)^1(dz^2)^0$ or $(dx^2 - y^2)^0(dz^2)^1$, are of equal energy (Fig. 6.11). Similarly, for $[Cu^{II}(H_2O)_6]^{2+}$ (d^9 system), the two $(dx^2 - y^2)^2(dz^2)^1$ or $(dx^2 - y^2)^1(dz^2)^2$ occupations of the e_g orbital set are of equal energy.

According to Jahn-Teller theorem, in both situations the octahedron distorts in such a way that the energy of the two electronic configurations are not equal (Fig. 6.12). The change in energy of the d-orbitals associated with distortions of both kinds, elongation (Z-out) and compression (Z-in)

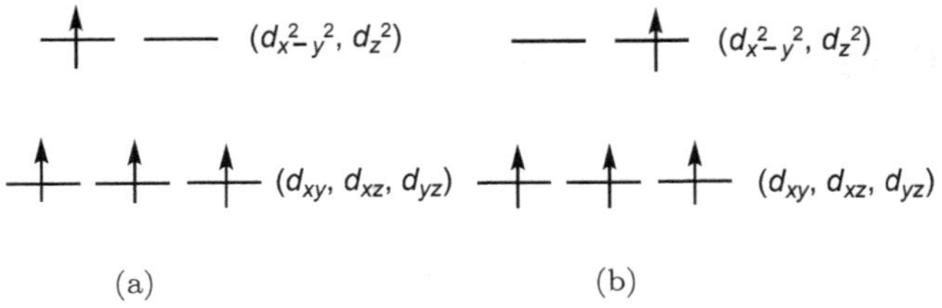

(a)   (b)

Fig. 6.11 A d^4 ion in hs form, differing in e_g set electron distribution.

——— ——— $(d_{x^2-y^2}, d_{z^2})$ <table><tr><td>⇅ δ'</td></tr><tr><td>—</td></tr></table> $d_{x^2-y^2}$ (+1/2 δ') <table><tr><td>⇅ δ'</td></tr><tr><td>—</td></tr></table> d_{z^2} (+1/2 δ')

d_{z^2} (-1/2 δ') $d_{x^2-y^2}$ (-1/2 δ')

↑ ——— ——— (d_{xy}, d_{xz}, d_{yz}) ↑⇅ δ d_{xy} (+2/3 δ) — — d_{xz}, d_{yz} (+1/3 δ)

↑— d_{xz}, d_{yz} (-1/3 δ) ↑⇅ δ d_{xy} (-2/3 δ)

(a) (b) (c)

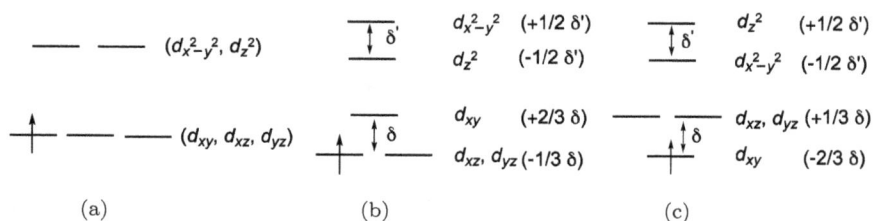

Fig. 6.12 The orbital splitting of a d^1 ion in three crystal fields (a) a perfect (hypothetical) octahedral, (b) tetragonal elongation (Z-out), and (c) tetragonal compression (Z-in).

(Fig. 6.12), of M–L bonds along z direction in octahedral geometry can be explained.

Let us consider a d^1 system $[Ti^{III}(H_2O)_6]^{3+}$.

For an ideal situation (perfect octahedral) all M–L bonds are equal (Fig. 6.12(a)). Now let us consider a situation where two ligands along the z axis are pulled out (z-out) (along z direction the M–L bond lengths increase). This situation is called tetragonal elongation. As a consequence the d-orbitals with z-component such as dxz, dyz from the lower set of orbitals and dz^2 from the upper set of orbitals are stabilized (less electrostatic repulsion, as the M–L bond distances along z direction are longer than M–L bond distances in the x, y directions). Let us label the two energy splitting parameters as δ (lower splitting) and δ' (upper splitting) (Fig. 6.12(b)). The opposite situation is called the tetragonal compression along z-axis (z-in) (Fig. 6.12(c)). Simple arithmetic consideration leads to stabilization of energy of $(-1/3)\delta$ for elongation (Fig. 6.12(b)) and $(-2/3)\delta$ for compression (Fig. 6.12(c)). From this simple analysis of the energy of d-orbitals, it is evident that for a d^1 ion z-in (compression) situation is better stabilized than z-out (elongation).

The Jahn-Teller distortions are possible whenever there is a partially filled degenerate set of energy levels. Thus, the J-T distortion can, of course, occur in tetrahedral complexes if either the e or t_2 levels are only partially filled. However, in this case the ligands are not pointing towards the orbitals directly and hence there is less stabilization to be gained upon distortion. Consequently, tetra-coordinate Cu(II) complexes often show pseudotetrahedral structure, where the most common type of distortions are tetragonal D_{2d} and rhombic C_{2v}. As for example, $[Cu^{II}Cl_4]^{2-}$ exhibits

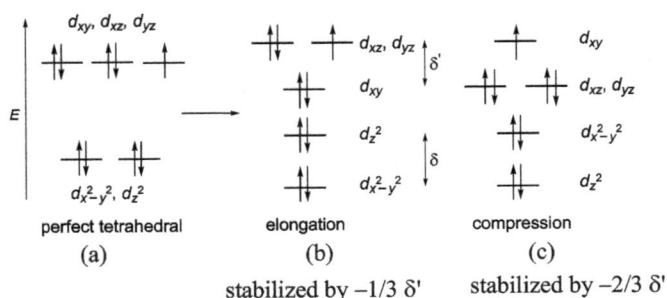

Fig. 6.13 The orbital splitting of $[Cu^{II}Cl_4]^{2-}$ in three situations: (a) perfect tetrahedral (hypothetical), (b) tetragonal elongation (Z-out), and (c) tetragonal compression (Z-in).

Table 6.2 Structural guide to decide which d^n ion in octahedral symmetry will give rise to perfect (no change in energy) and distorted geometry (Jahn Teller distortion).

d^n ion	z-out (octahedral)		z-in (octahedral)	
d^1	$-1/3\delta$		$-2/3\delta$	
d^2	$-2/3\delta$		$-1/3\delta$	
d^3	0		0	
	hs	ls	hs	ls
d^4	$-1/2\delta'$	$-1/3\delta$	$-1/2\delta'$	$-2/3\delta$
d^5	0	$-2/3\delta$	0	$-1/3\delta$
d^6	$-1/3\delta$	0	$-2/3\delta$	0
d^7	$-2/3\delta$	$-1/2\delta'$	$-1/3\delta$	$-1/2\delta'$
d^8	0		0	
d^9	$-1/2\delta'$		$-1/2\delta'$	

flattened tetrahedral (compression along z axis) structure (Fig. 6.13), which is basically a tetragonal D_{2d} distortion. Elongation along z axis results in rhombic C_{2v} distortion.

Table 6.2 acts as a quick structural guide to decide which d^n ion (remembering that d^4–d^7 ions in octahedral field may give rise to either hs or ls state, depending on the strength of ligands) will be perfect (no distortion), which d^n ion will prefer tetragonal elongation (z-out) and which d^n ion will prefer tetragonal compression (z-in). For octahedral geometry, partially occupied e_g orbitals (higher energy orbital set) lead to more pronounced distortions than partially occupied t_{2g} orbitals (lower energy orbital set).

> For octahedral geometry, the distortion is more pronounced when the degeneracy occurs in the e_g orbital set, as these orbitals point directly toward the ligands.

The Jahn Teller distortion can also be looked at from consideration of the point groups. If the symmetry of a complex is lowered, the complex will belong to a point group of lower symmetry.

Example 6.12 Bearing in mind the Jahn-Teller theorem, predict the structure of $[Cr(H_2O)_6]^{2+}$.

Answer As H_2O is a weak field ligand, the complex $[Cr^{II}(H_2O)_6]^{2+}$ is hs. In the case of hs-$[Cr^{II}(H_2O)_6]^{2+}$ (Table 6.2) the stabilization due to Jahn Teller effect is $-1/2\delta'$ for both z-out and z-in situation. However, in the case of z-out only two bonds along z-axes are disturbed but for z-in all four Cr–OH_2 bonds are disturbed. Thus, z-in requires spending of more energy and hence z-out is favored. Thus, the structure of $[Cr^{II}(H_2O)_6]^{2+}$ is tetragonally elongated.

Example 6.13 Comment on the structure (perfect, distorted etc.) of the following complexes: $[Cr(CN)_6]^{4-}$, $[Mn(H_2O)_6]^{3+}$, $[CoCl_4]^{2-}$.

Answer $[Cr^{II}(CN)_6]^{4-}$: As CN^- is a strong field ligand the d^4 octahedral complex will be ls. According to Fig. 6.5, the electron distribution is $t_{2g}^4 e_g^0$. According to Table 6.2, for this complex there will be minor (as uneven distribution is in the lower energy set) Z-out (tetragonal elongation) distortion.

$[Mn^{III}(H_2O)_6]^{3+}$: As H_2O is a weak field ligand the d^4 octahedral complex will be hs. According to Fig. 6.5, the electron distribution is $t_{2g}^3 e_g^1$. According to Table 6.2, for this complex there will be severe (as uneven distribution is in the higher energy orbital set, which directly faces the ligands) Z-out (tetragonal elongation) distortion.

$[Co^{II}Cl_4]^{2-}$: As Cl^- is a weak field ligand the d^7 complex will be tetrahedral. According to d-orbital splitting shown in Fig. 6.6, the electron distribution is $e^4 t_2^3$. For this complex there will not be any distortion, as electron distribution is spherically symmetric. It will have perfect tetrahedral geometry.

Example 6.14 Predict the type of distortion, if any, likely to be affecting the structure of the following complexes. Justify the answer.

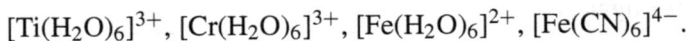

$$[Ti(H_2O)_6]^{3+}, [Cr(H_2O)_6]^{3+}, [Fe(H_2O)_6]^{2+}, [Fe(CN)_6]^{4-}.$$

Answer $[Ti^{III}(H_2O)_6]^{3+}$: According to Fig. 6.5, the electron distribution is $t_{2g}^1 e_g^0$. According to Table 6.2, for this d^1 octahedral complex there will be minor (as uneven distribution is in the lower energy set) Z-in (tetragonal compression) distortion.

$[Cr^{III}(H_2O)_6]^{3+}$: According to Fig. 6.5, the electron distribution for this d^3 octahedral complex is $t_{2g}^3 e_g^0$. According to Table 6.2, for this complex there will not be any distortion. It will have perfect octahedral geometry.

$[Fe(H_2O)_6]^{2+}$: As H_2O is a weak field ligand the complex will be hs. According to Fig. 6.5, the electron distribution for this d^6 hs octahedral complex is $t_{2g}^4 e_g^2$. According to Table 6.2, for this complex there will be minor (as uneven distribution is in the lower energy set) Z-in (tetragonal compression) distortion.

$[Fe^{II}(CN)_6]^{4-}$: As CN^- is a strong field ligand the complex will be ls. According to Fig. 6.5, the electron distribution for this d^6 ls octahedral complex is $t_{2g}^6 e_g^0$. According to Table 6.2, for this complex there will not be any distortion. It will have perfect octahedral geometry.

Example 6.15 In the ionic solid K_2CuF_4, four Cu–F bond distances (195 pm) are shorter than the other two Cu–F distances (208 pm). Explain.

Answer In the solid structure each Cu(II) ion attains six-coordination.

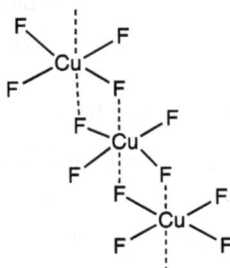

Due to Jahn Teller distortion of Cu(II) ion the observed fact is understandable.

(g) Crystal field splitting diagram for square planar symmetry

It is convenient to start with the regular octahedron (O_h) for the explanation in the formation of four-coordinate square planar (D_{4h}) complexes. A tetragonal distortion towards elongation (Fig. 6.12) reduces the energy of the dz^2 orbital and increases the energy of the dx^2-y^2 orbital. The distortion of a d^8 complex may be large enough to force the two e_g electrons to pair up in the dz^2 orbital. The extreme distortion of this kind may result in the complete loss of the ligands in the z-axis and formation of the d^8 square planar complex. The common occurrence of $4d^8$ and $5d^8$ complexes such as $[Pd^{II}Cl_4]^{2-}$ and $[Pt^{II}Cl_4]^{2-}$, respectively, correlates with the high values of the crystal field splitting parameter in these two series. This in turn gives rise to a high crystal field stabilization of the diamagnetic square planar complexes. Notably, a $3d^8$ complex such as $[Ni^{II}Cl_4]^{2-}$ is tetrahedral because Cl^- is a weak field ligand and the crystal field splitting parameter is quite small. In contrast, the complex $[Ni^{II}(CN)_4]^{2-}$ is square planar because CN^- is a strong field ligand and the crystal field splitting parameter is large enough to result in the formation of a square planar complex. Fig. 6.14 displays the d-orbital splitting pattern for square planar complexes. The sum of the three distinct orbital splitting are denoted as $\Delta_{SP}(= \Delta_1 + \Delta_2 + \Delta_3)$, which is greater than Δ_o. Simple theory predicts that $\Delta_{SP} \simeq 1.3\Delta_o$ for complexes of the same metal and ligands with the same M–L bond lengths.

(h) Crystal field splitting diagrams for trigonal bipyramidal and square pyramidal symmetry

In the trigonal bipyramidal (D_{3h}) geometry, the dz^2 orbital directly faces two ligands. Moreover, none of the other orbitals is face to face with any of the equatorial ligands. But the dxy and dx^2-y^2 orbitals are equally disposed with respect to the equatorial ligands, and thus form a doublet. The remaining dxz and dyz orbitals form the most stabilized doublet set. The d-orbital splitting is shown in Fig. 6.15(a).

In the square pyramidal (C_{4v}) geometry, the dx^2-y^2 orbital is the least stable as it is face to face with all four equatorial ligands. The d-orbital

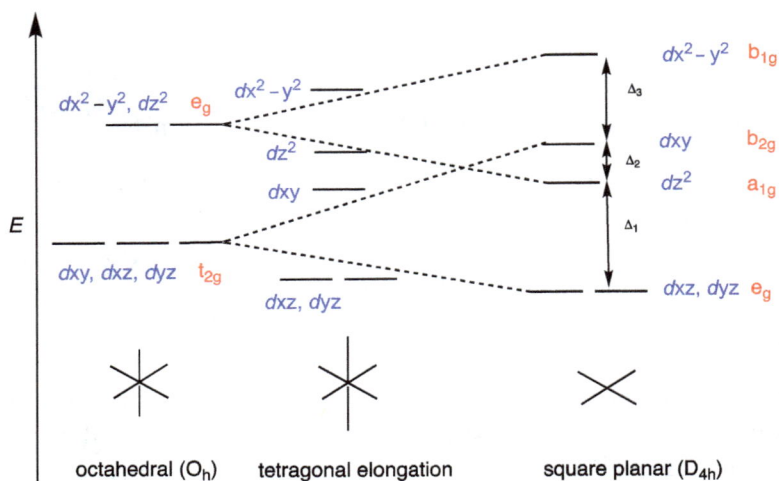

Fig. 6.14 Splitting of *d*-orbitals in square planar geometry, with symmetry representations.

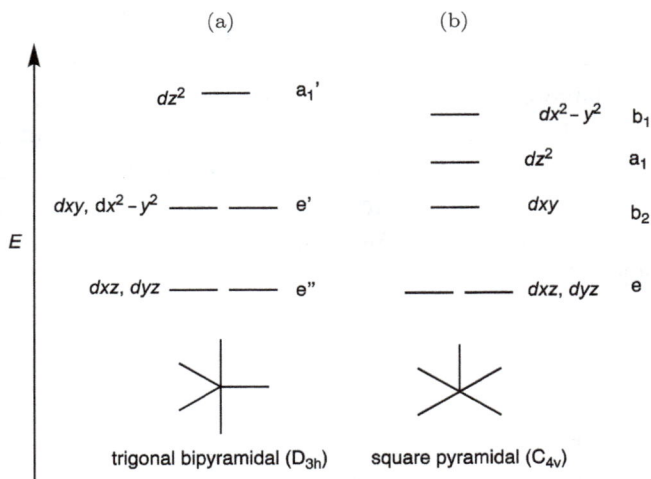

Fig. 6.15 Splitting of *d*-orbitals in trigonal bipyramidal and square pyramidal geometry, with symmetry representations.

splitting may be approximately that shown in Fig. 6.15(b). The relative positions of the dz^2 and dxy orbitals may get interchanged.

Example 6.16 In $[V^{IV}O(H_2O)_5]SO_4$ the V–O multiple covalent bond is very strong ($\sim$170 pm), whereas the equatorial V–OH$_2$ bond length is $\sim$230 pm. Comment on its d-orbital splitting diagram.

Answer In this complex the dz^2 orbital is the most destabilized, followed in order by the dx^2-y^2 orbital and the dxz, dyz orbital set. The dxy orbital is the most stabilized in the ordering. Hence, the energy ordering, in a tetragonal compression along the z-axis situation, is the following.

$$dz^2 > dx^2-y^2 > dxz, \ dyz > dxy$$

Example 6.17 Comment on the d-orbital splitting order in triangular planar and linear geometry.

Answer In a triangular planar geometry, the dx^2-y^2, dxy set of orbitals will be most destabilized, followed in order by the dxz, dyz set of orbitals. The dz^2 orbital is the most stabilized in the ordering. Hence, the energy ordering in a triangular planar geometry is,

$$dx^2-y^2, \ dxy > dxz, \ dyz > dz^2$$

In a linear geometry, the dz^2 orbital is the most destabilized, followed in order by the dxz, dyz set. The dx^2-y^2, dxy set of orbitals is the most stabilized in the ordering. Hence, the energy ordering in a linear geometry is,

$$dz^2 > dxz, \ dyz > dx^2-y^2, \ dxy$$

6.3 Splitting of levels and terms

A complete set of orbitals is a basis for an irreducible representation of the group (see Chapter 2). The s orbital is the totally symmetric one-dimensional representation [$(2l+1)$-fold set of functions; for s orbital it is $2 \times 0+1 = 1$], the p orbitals give a three-dimensional representation, d orbitals are five-dimensional, etc.

Although the full symmetry of the octahedron is O$_h$, we can gain all required information about the d orbitals by using only the pure rotational subgroup O because O$_h$ may be obtained from O by adding the inversion i.

However, we know that d orbitals are even to inversion, so that it is the only pure rotational operations of O which bring us new information. The character for rotation by the angle α is given by the equation,

$$\chi(\alpha) = \sin(l + 1/2)\alpha / \sin \alpha/2 (\alpha \neq 0)$$

For s orbital ($l = 0$) a twofold, threefold, and fourfold rotations, $\alpha = \pi$, $\alpha = 2\pi/3$, and $\alpha = \pi/2$ and hence the characters are,

$$\chi(C_2) = \sin\{(0 + 1/2)\pi\} / \sin(\pi/2) = \sin(\pi/2) / \sin(\pi/2) = 1$$

$$\chi(C_3) = \sin\{(0 + 1/2)2\pi/3\} / \sin(\pi/3) = \sin(\pi/3) / \sin(\pi/3) = 1$$

$$\chi(C_4) = \sin\{(0 + 1/2)\pi/2\} / \sin(\pi/4) = \sin(\pi/4) / \sin(\pi/4) = 1/1 = 1$$

For p orbital ($l = 1$) a twofold, threefold, and fourfold rotations, $\alpha = \pi$, $\alpha = 2\pi/3$, and $\alpha = \pi/2$ and hence the characters are,

$$\chi(C_2) = \sin\{(1 + 1/2)\pi\} / \sin(\pi/2) = \sin(3\pi/2) / \sin(\pi/2) = -1/1 = -1$$

$$\chi(C_3) = \sin\{(1 + 1/2)2\pi/3\} / \sin(\pi/3) = \sin \pi / \sin(\pi/3)$$

$$= 0 / \sin(\pi/3) = 0$$

$$\chi(C_4) = \sin\{(1 + 1/2)\pi/2\} / \sin(\pi/4) = \sin(3\pi/4) / \sin(\pi/4) = 1$$

For d orbital ($l = 1$) a twofold, threefold, and fourfold rotations, $\alpha = \pi$, $\alpha = 2\pi/3$, and $\alpha = \pi/2$ and hence the characters are,

$$\chi(C_2) = \sin\{(2 + 1/2)\pi\} / \sin(\pi/2) = \sin(5\pi/2) / \sin(\pi/2) = 1/1 = 1$$

$$\chi(C_3) = \sin\{(2 + 1/2)2\pi/3\} / \sin(\pi/3) = \sin(5\pi/3) / \sin(\pi/3)$$

$$= -\sin(\pi/3) / \sin(\pi/3) = -1$$

$$\chi(C_4) = \sin\{(2 + 1/2)2\pi/4\} \sin(\pi/4) = \sin(5\pi/4) / \sin(\pi/4) = -1$$

Referring to the character table for the group O and following the methods developed in Chapter 2, it is easily seen that the representations (characters) we have derived for C_2, C_3, and C_4 symmetry operations, respectively, for d orbital $(1, -1, -1)$ is reducible to $E + T_2$ [from Character Table of the point group O; E representation: $0, -1, 0$ and T_2 representation: $1, 0, -1$ lead to $1 (0 + 1), -1 (-1 + 0), -1 (0 + (-1))$]. Since the d wave functions are inherently g in their inversion property, in the group O_h

we have, $\Gamma_d = E_g + T_{2g}$. Thus, we see that the set of five d wave functions, degenerate in the free atom or ion (or, more precisely, under conditions of spherical symmetry) does not remain degenerate, when the atom or ion is placed in an environment with O_h symmetry. The wave functions are split into a triply degenerate set T_{2g} and a doubly degenerate set E_g. It is easy to apply the same treatment to electrons in other types of orbitals than d orbitals. The results obtained are collected in Table 6.3. It is seen that an s orbital is totally symmetric and has a_{1g} representation in the O_h environment. The set of three p orbitals remains unsplit, transforming as t_{1u}. All orbitals with higher values of the quantum number l, however, are split into two or more sets. This must be so, since the group O_h cannot allow any state to be more than threefold degenerate.

In a similar manner, we could determine the splitting of various sets of orbitals in environments of other symmetries. The results for a few point groups such as O_h, T_d, D_{4h} are in Table 6.4.

The results we have obtained so far for single electrons in variouh types of orbitals apply also to the behavior of terms arising from groups of electrons. For example, just as a single d electron in a free atom has a wave function which belongs to a fivefold degenerate set corresponding to the five values of the quantum number m may take, so a D state arising from any group of electrons has a completely analogous fivefold degeneracy because of the five values which the quantum number M may take. Exactly the same relationship exists between f orbital and F states, p orbitals and P states, and so on. Thus all of the results given in Table 6.3 for the splitting

Table 6.3 Splitting of one-electron levels in an octahedral environment

Type of level	l	$\chi(E)$	$\chi(C_2)$	$\chi(C_3)$	$\chi(C_4)$	Irreversible representations spanned
s	0	1	1	1	1	A_{1g}
p	1	3	-1	0	1	T_{1u}
d	2	5	1	-1	-1	$E_g + T_{2g}$
f	3	7	-1	1	-1	$A_{2u} + T_{1u} + T_{2u}$
g	4	9	1	0	1	$A_{1g} + E_g + T_{1g} + T_{2g}$
h	5	11	-1	-1	1	$E_u + 2T_{1u} + T_{2u}$
i	6	13	1	1	-1	$A_{1g} + A_{2g} + E_g + T_{1g} + 2T_{2g}$

Table 6.4   Splitting of one-electron levels in various symmetries

Type of level	O_h	T_d	D_{4h}
s	a_{1g}	a_1	a_{1g}
p	t_{1u}	t_2	$a_{2u} + e_u$
d	$e_g + t_{2g}$	$e + t_2$	$a_{1g} + b_{1g} + b_{2g} + e_g$
f	$a_{2u} + t_{1u} + t_{2u}$	$a_2 + t_1 + t_2$	$a_{2u} + b_{1u} + b_{2u} + 2e_u$
g	$a_{1g} + e_g + t_{1g} + t_{2g}$	$a_1 + e + t_1 + t_2$	$2a_{1g} + a_{2g} + b_{1g} + b_{2g} + 2e_g$
h	$e_u + 2t_{1u} + t_{2u}$	$e + t_1 + 2t_2$	$a_{1u} + 2a_{2u} + b_{1u} + b_{2u} + 3e_u$
i	$a_{1g} + a_{2g} + e_g + t_{1g} + 2t_{2g}$	$a_1 + a_2 + e + t_1 + 2t_2$	$2a_{1g} + a_{2g} + 2b_{1g} + 2b_{2g} + 3e_g$

of various sets of one-electron orbitals apply to the splitting of analogous Russell-Saunders terms.

In Table 6.4 the use of subscripts g and u is governed by the following rules. If the point group of the environment has no center of symmetry, then no subscripts are used. When the environment does have a center of symmetry, the subscripts are determined by the type of orbital. All AO's for which the quantum number l is even $(s,d,g,...)$ being centrosymmetric and hence of g character, and all AO's for which l is odd $(p,f,h,...)$ being antisymmetric to inversion and thus of u character.

In using Table 6.4 for term splittings the following rules apply. If the environment does not have a center of symmetry, then g and u subscripts are not used. For point groups having a center to which the inversion operation may be referred, the g or u character will be determined by the nature of the one-electron wave functions of the individual electrons making up the configuration from which the term is derived.

(i) Crystal field splitting of Russell-Saunders terms in octahedral symmetry

In Chapter 5 of this chapter we have considered free-ion terms, that means terms without the presence of a ligand field. Now we consider the influence of a crystal field (an O_h or a T_d) on a term. Terms are wavefunctions, just like orbitals, and therefore they behave like orbitals in a ligand field.

In an O_h field the d orbitals ($l = 2$) splits to give t_{2g} and e_g sets ($2l + 1 = 2 \times 2 + 1 = 5$; triply degenerate t and doubly degenerate e) (Fig. 6.5) and in a tetrahedral field the d orbitals splits to give e and t_2 sets (Fig. 6.6), and the ^2D term splits to give $^2T_{2g}$ (lower energy) and ^2E_g terms and ^2E (lower energy) and $^2T_2$ terms, respectively. The s orbital ($l = 0$) gives singly degenerate A state ($2l + 1 = 2 \times 0 + 1 = 1$). Similarly, the S term becomes A_{1g} terms (in O_h field) and A_2 (in T_d field). The p orbitals ($l = 1$) are triple-degenerate ($2l + 1 = 2 \times 1 + 1 = 3$) having T_{1g} symmetry in the point group O_h and T_2 symmetry in the point group T_d. The P terms do not split in energy but changes notation to T term; T_{1g} and T_2 terms in O_h and T_d field, respectively.

Free-ion terms split, according to symmetry. Splitting in octahedral symmetry is shown below in Table 6.5.

The splitting of a D term will be just the same as the splitting of the set of one-electron d orbitals.

The term symbols behave analogously to orbitals.

For d^n configurations, all will give g states in point groups possessing a center of symmetry.

Notably, the chemical environment does not interact directly with the electron spins, and thus all the states into which a particular term is split have the same spin multiplicity as the parent term.

Table 6.5   Splitting of Terms in Octahedral Symmetry

Term (multiplicity)	Irreducible representations
S ($L = 0$; multiplicity: $2 \times 0 + 1 = 1$)	A_{1g}
P ($L = 1$; multiplicity: $2 \times 1 + 1 = 3$)	T_{1g}
D ($L = 2$; multiplicity: $2 \times 2 + 1 = 5$)	$E_g + T_{2g}$
F ($L = 3$; multiplicity: $2 \times 3 + 1 = 7$)	$A_{2g} + T_{1g} + T_{2g}$
G ($L = 4$; multiplicity: $2 \times 4 + 1 = 9$)	$A_{1g} + E_g + T_{1g} + T_{2g}$
H ($L = 5$; multiplicity: $2 \times 5 + 1 = 11$)	$E_g + 2T_{1g} + T_{2g}$
I ($L = 6$; multiplicity: $2 \times 6 + 1 = 13$)	$A_{1g} + A_{2g} + E_g + T_{1g} + 2T_{2g}$

Splitting and ground state terms in octahedral and tetrahedral symmetry for d^1–d^9 are shown below.

d^n	R-S Term	Crystal field splitting: O_h (T_d)	Ground state: O_h (T_d)
d^1	^2D	$^2T_{2g} + {}^2E_g$ ($^2T_2 + {}^2E$)	$^2T_{2g}$ (^2E)
d^2	^3F	$^3T_{1g} + {}^3T_{2g} + {}^3A_{2g}$ ($^3T_1 + {}^3T_2 + {}^3A_2$)	$^3T_{1g}$ ($^3A_2$)
d^3	^4F	$^4T_{1g} + {}^4T_{2g} + {}^4A_{2g}$ ($^4T_1 + {}^4T_2 + {}^4A_2$)	$^4A_{2g}$ ($^4T_1$)
d^{4a}	^5D	$^5T_{2g} + {}^5E_g$ ($^5T_2 + {}^5E$)	^5E_g ($^5T_2$)
d^{5a}	^6S	$^6A_{1g}$ ($^6A_1$)	$^6A_{1g}$ ($^6A_1$)
d^{6a}	^5D	$^5T_{2g} + {}^5E_g$ ($^5T_2 + {}^5E$)	$^5T_{2g}$ (^5E)
d^{7a}	^4F	$^4T_{1g} + {}^4T_{2g} + {}^4A_{2g}$ ($^4T_1 + {}^4T_2 + {}^4A_2$)	$^4T_{1g}$ ($^4A_2$)
d^8	^3F	$^3T_{1g} + {}^3T_{2g} + {}^3A_{2g}$ ($^3T_1 + {}^3T_2 + {}^3A_2$)	$^3A_{2g}$ ($^3T_1$)
d^9	^2D	$^2T_{2g} + {}^2E_g$ ($^2T_2 + {}^2E$)	^2E_g ($^2T_2$)

^{a} high-spin.

Ground-state terms in octahedral symmetry for low-spin d^4–d^7 are shown below.

d^n	Ground state in O_h
d^4	$^3T_{1g}$
d^5	$^2T_{2g}$
d^6	$^1A_{1g}$
d^7	^2E_g

The ground-state term symbols for the grossly octahedral complexes $[Co^{II}(H_2O)_6]^{2+}$ and $[Ni^{II}(NH_3)_6]^{2+}$ are $^4T_{1g}$ and $^3A_{2g}$, respectively. The term symbols for $[Mn^{III}(CN)_6]^{3-}$, $[Fe^{II}(CN)_6]^{4-}$, and $[Fe^{III}(CN)_6]^{3-}$ are $^3T_{1g}$, $^1A_{1g}$, and $^2T_{2g}$, respectively.

The state labels (term symbols) also indicate the degeneracy of the electron configuration. As for example, for d^1 (t_{2g}^1), d^2 (t_{2g}^2), ls d^4 (t_{2g}^4), ls d^5 (t_{2g}^5), hs d^6 ($t_{2g}^4 e_g^2$), and hs d^7 ($t_{2g}^5 e_g^2$) it is T, which designates a triply degenerate asymmetrically occupied state. Similarly, hs d^4 ($t_{2g}^3 e_g^1$), ls d^7 ($t_{2g}^6 e_g^1$), and d^9 ($t_{2g}^6 e_g^3$) it is E, a doubly degenerate asymmetrically occupied state. For d^3 (t_{2g}^3), hs d^5 ($t_{2g}^3 e_g^2$), ls d^6 (t_{2g}^6), and d^8 ($t_{2g}^6 e_g^2$) it is A (or B), designates a non-degenerate state which is symmetrically occupied.

If the spectroscopic ground state of an octahedral complex is 'A', then there will not be any distortion (perfect octahedral geometry). For 'E' and 'T' ground states, there will be distortion (Fig. 6.12) and the extent of distortion will be severe for 'E' ground state and mild for 'T' ground state. It is understandable that for d^3, hs d^5, ls d^6, and d^8 electronic distribution (even distribution of electrons in t_{2g} and e_g set of orbitals) the term symbol will be 'A'. For d^1, d^2, ls d^4, ls d^5 the term symbol will be 'T'. Notably, only for hs d^4, ls d^7, and d^9 the term symbol will be 'E'. As the degeneracy occurs in the e_g orbital for hs d^4, ls d^7, and d^9 (uneven occupation of the e_g orbitals), the distortion is more pronounced in these systems.

Example 6.18 Assign the ground-state term symbols for the following six-coordinate (consider grossly octahedral geometry) complexes $[Ti(H_2O)_6]^{3+}$, $[V(H_2O)_6]^{3+}$, $[Cr(H_2O)_6]^{3+}$, $[Mn(H_2O)_6]^{3+}$, $[Fe(H_2O)_6]^{2+}$, $[Fe(H_2O)_6]^{3+}$, $[Co(H_2O)_6]^{3+}$, $[Co(CN)_6]^{4-}$, $[Ni(H_2O)_6]^{2+}$, and $[Cu(H_2O)_6]^{2+}$.

Answer $[Ti^{III}(H_2O)_6]^{3+}$: According to Fig. 6.5, the electron distribution for this d^1 complex is $t_{2g}^1 e_g^0$. The single electron in the t_{2g} set of orbitals can have three possibilities to reside $(dxy)^1$ or $(dxz)^1$ or $(dyz)^1$. This leads to triply degenerate T term. As it has only one unpaired electron this leads to spin multiplicity of 2 $(2 \times 1/2 + 1 = 2)$. Therefore, the ground-state term symbol for $[Ti(H_2O)_6]^{3+}$ is $^2T_{2g}$.

$[V^{III}(H_2O)_6]^{3+}$: According to Fig. 6.5, the electron distribution for this d^2 complex is $t_{2g}^2 e_g^0$. Two electrons in the t_{2g} set of orbitals can have three possibilities to reside $(dxy)^1(dxz)^1(dxz)^0$ or $(dxy)^1(dxz)^0(dyz)^1$ or $(dxz)^0(dxz)^1(dyz)^1$. This leads to triply degenerate T term. Two unpaired electrons lead to spin multiplicity of 3 $(2 \times 1 + 1 = 3)$. Therefore, the ground-state term symbol for $[V(H_2O)_6]^{3+}$ is $^3T_{1g}$.

$[Cr^{III}(H_2O)_6]^{3+}$: According to Fig. 6.5, the electron distribution for this d^3 complex is $t_{2g}^3 e_g^0$. Three electrons in the t_{2g} set of orbitals can have only one possibility to reside $(dxy)^1(dxz)^1(dyz)^1$. This leads to singly degenerate A term. Three unpaired electrons lead to spin multiplicity of 4 $(2 \times 3/2 + 1 = 4)$. Therefore, the ground-state term symbol for $[Cr(H_2O)_6]^{3+}$ is $^4T_{2g}$.

$[Mn^{III}(H_2O)_6]^{3+}$: H_2O is a weak field ligand and hence this d^4 complex is hs. According to Fig. 6.5, the electron distribution is $t_{2g}^3 e_g^1$. Four unpaired electrons in the t_{2g} and e_g set of orbitals can have only two possibilities to reside $\{(dxy)^1(dxz)^1(dyz)^1\}$ $\{(dx^2-y^2)^1(dz^2)^0\}$ or $\{(dxy)^1(dxz)^1(dyz)^1\}$ $\{(dz^2)^1(dx^2-y^2)^0\}$. This leads to doubly degenerate E term. Four unpaired electrons lead to spin multiplicity of 5 $(2 \times 2 + 1 = 5)$. Therefore, the ground-state term symbol for $[Mn(H_2O)_6]^{3+}$ is ^5E_g.

$[Fe^{II}(H_2O)_6]^{2+}$: H_2O is a weak field ligand and hence this d^6 complex is hs. According to Fig. 6.5, the electron distribution is $t_{2g}^4 e_g^2$. Four unpaired electrons in the t_{2g} and e_g set of orbitals can have only two possibilities to reside $\{(dxy)^2(dxz)^1(dyz)^1\}$ $\{(dx^2-y^2)^1(dz^2)^1\}$ or $\{(dxy)^1(dxz)^2(dyz)^1\}$ $\{(dx^2-y^2)^1(dz^2)^1\}$ or $\{(dxy)^1(dxz)^1(dyz)^2\}$ $\{(dx^2-y^2)^1(dz^2)^1\}$. This leads to triply degenerate T term. Four unpaired electrons lead to spin multiplicity of 5 $(2 \times 2 + 1 = 5)$. Therefore, the ground-state term symbol for $[Fe(H_2O)_6]^{2+}$ is $^5T_{2g}$.

$[Fe^{III}(H_2O)_6]^{3+}$: H_2O is a weak field ligand and hence this d^5 complex is hs. According to Fig. 6.5, the electron distribution is $t_{2g}^3 e_g^2$. Five unpaired electrons in the t_{2g} and e_g set of orbitals can have only one possibility to reside $\{(dxy)^1(dxz)^1(dyz)^1\}$ $\{(dx^2-y^2)^1(dz^2)^1\}$. This leads to singly degenerate A term. Five unpaired electrons lead to spin multiplicity of 6 $(2 \times 5/2 + 1 = 6)$. Therefore, the ground-state term symbol for $[Fe(H_2O)_6]^{3+}$ is $^6A_{1g}$.

$[Co^{III}(H_2O)_6]^{3+}$: Although H_2O is a weak field ligand but with Co(III) ion the d^6 complex is ls. According to Fig. 6.5, the electron distribution is $t_{2g}^6 e_g^0$. All six electrons are paired up in the t_{2g} set of orbitals. Hence there is only one possibility to distribute the electrons $\{(dxy)^2(dxz)^2(dyz)^2\}$ $\{(dx^2-y^2)^0(dz^2)^0\}$. This leads to singly degenerate A term. As there is no unpaired electron the spin multiplicity is 1 $(2 \times 0 + 1 = 1)$. Therefore, the ground-state term symbol for $[Co(H_2O)_6]^{3+}$ is $^1A_{1g}$.

$[Co^{II}(CN)_6]^{4-}$: CN^- is a strong field ligand and hence this d^7 complex is ls. According to Fig. 6.5, the electron distribution is $t_{2g}^6 e_g^1$. Single unpaired electron in the e_g set of orbitals can have only two possibilities to reside $\{(dxy)^2(dxz)^2(dyz)^2\}$ $\{(dx^2-y^2)^1(dz^2)^0\}$ or $\{(dxy)^2(dxz)^2(dyz)^2\}$ $\{(dz^2)^1(dx^2-y^2)^1\}$. This leads to doubly degenerate E term. Single unpaired electron leads to spin multiplicity of $2(2 \times 1/2 + 1 = 2)$. Therefore, the ground-state term symbol for $[Co^{II}(CN)_6]^{4-}$ is ^2E_g.

[$Ni^{II}(H_2O)_6]^{2+}$: According to Fig. 6.5, the electron distribution for this d^8 complex is $t_{2g}^6\, e_g^2$. Two unpaired electrons in the e_g set of orbitals can have only one possibility to reside $\{(dxy)^2(dxz)^2(dyz)^2\}\ \{(dx^2-y^2)^1(dz^2)^1\}$. This leads to singly degenerate A term. Two unpaired electrons lead to spin multiplicity of 3 ($2 \times 1 + 1 = 3$). Therefore, the ground-state term symbol for [$Ni(H_2O)_6]^{2+}$ is $^3A_{2g}$.

[$Cu^{II}(H_2O)_6]^{2+}$: According to Fig. 6.5, the electron distribution for this d^9 complex is $t_{2g}^6\, e_g^3$. One unpaired electron in the e_g set of orbitals can have only two possibilities to reside $\{(dxy)^2(dxz)^2(dyz)^2\}\ \{(dx^2-y^2)^2(dz^2)^1\}$ or $\{(dxy)^2(dxz)^2(dyz)^2\}\ \{(dz^2)^2(dx^2-y^2)^1\}$. This leads to doubly degenerate E term. One unpaired electron leads to spin multiplicity of 2 ($2 \times 1/2 + 1 = 2$). Therefore, the ground-state term symbol for [$Cu(H_2O)_6]^{2+}$ is ^2E_g.

Example 6.19 Predict the type of distortion, if any, likely to be affecting the structure of the following complexes. Justify answer.

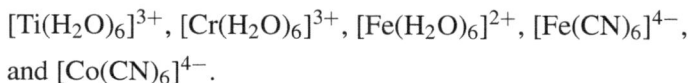

$$[Ti(H_2O)_6]^{3+}, [Cr(H_2O)_6]^{3+}, [Fe(H_2O)_6]^{2+}, [Fe(CN)_6]^{4-},$$
$$\text{and } [Co(CN)_6]^{4-}.$$

Answer The ground state term symbols for [$Ti^{III}(H_2O)_6]^{3+}$ (d^1 ion), [$Cr^{III}(H_2O)_6]^{3+}$ (d^3 ion), [$Fe^{II}(H_2O)_6]^{2+}$ (hs d^6 ion), [$Fe^{II}(CN)_6]^{4-}$ (ls d^6 ion), and [$Co^{II}(CN)_6]^{4-}$ (ls d^7 ion) are $^2T_{2g}$, $^4A_{2g}$, $^5T_{2g}$, $^1A_{1g}$, and ^2E_g, respectively. Hence, for [$Ti(H_2O)_6]^{3+}$ (mild distortion), [$Cr(H_2O)_6]^{3+}$ (no distortion), [$Fe(H_2O)_6]^{2+}$ (mild distortion), [$Fe(CN)_6]^{4-}$ (no distortion), and [$Co(CN)_6]^{4-}$ (severe distortion).

Example 6.20 The ground state for d^2 ion in O_h symmetry is $^3T_{1g}$. It is understandable that the ground-state will be a T term. But the question arises, why $^3T_{1g}$ and <u>NOT</u> $^3T_{2g}$?

Answer From O_h character table (see Further reading) the characters for T_{2g} irreducible representation are

E	$8C_3$	$6C_2$	$6C_4$	$3C_2 (= C_4^2)$	i	$6S_4$	$8S_6$	$3\sigma_h$	$6\sigma_d$
3	0	1	-1	-1	3	-1	0	-1	1

The character of the symmetric product,

$$[\chi(R)]^2 = 9 \quad 0 \quad 1 \quad 1 \quad 1 \quad 9 \quad 1 \quad 0 \quad 1 \quad 1$$

Symmetric product: $[\chi^2](R) = 1/2([\chi(R)]^2 + \chi(R^2))$

Antisymmetric product: $\{\chi^2\}(R) = 1/2([\chi(R)]^2 - \chi(R^2))$

We have further,

$E^2 = E$; $C_3^2 = C_3$; $C_2^2 = E$; $C_4^2 = C_2$; $C^2 = E$; $i^2 = E$; $S_4^2 = C_2$; $S_6^2 = C_3$; $\sigma_h^2 = E$; $\sigma_d^2 = E$

Hence,

$$\chi(R^2) = 3 \quad 0 \quad 3 \quad -1 \quad 3 \quad 3 \quad -1 \quad 0 \quad 3 \quad 3$$

We then get,

Symmetric product: $(9+3)/2 \ (0+0)/2 \ (1+3)/2 \ (1-1)/2 \ (1+3)/2$
$(9+3)/2 \ (1-1)/2 \ (0+0)/2 \ (1+3)/2 \ (1+3)/2$
$: 6 \ 0 \ 2 \ 0 \ 2 \ 6 \ 0 \ 0 \ 2 \ 2$

Antisymmetric product: $(9-3)/2 \ (0-0)/2 \ (1-3)/2 \ (1+1)/2 \ (1-3)/2$
$(9-3)/2 \ (1+1)/2 \ (0-0)/2 \ (1-3)/2 \ (1-3)/2$
$: 3 \ 0 \ -1 \ 1 \ -1 \ 3 \ 1 \ 0 \ -1 \ -1$

The symmetric representations are then found to be A_{1g}, E_g, and T_{2g} (by adding the characters together) and the antisymmetric representation is found to be T_{1g} (from the Character Table of O_h). If we also consider the spin functions, we have that the electronic configuration t_{2g}^2 can produce the states $^3T_{1g}$ and $^1A_{1g}$, ^1E_g, and $^1T_{2g}$. This latter result follows from the fact that the total wave function will be antisymmetric.

This is the reason why for d^2 in O_h symmetry the ground-state is $^3T_{1g}$.

(ii) Simplifying rules to construct CF diagrams

These simplifying rules arise out of *hole formalism* (hole equivalency) concept. As for example, d^2 ion in octahedral field is equivalent to reverse of d^8 ion in octahedral field i.e. in terms of energy the Orgel term-symbol splitting diagrams need to be reversed. It means that if T_{2g} term is stable in an octahedral field, T_2 term in a tetrahedral field is unstable. This is because d^2 electron (electron is negatively charged and hence is attracted by the nucleus) could be considered as d^8 positron (position carries positive charge and hence is repelled by the nucleus). For d^{2e-} and d^{8e+} the R-S terms are the same because electron-electron and positron-positron repulsions are the same. Therefore, where electron-electron repulsion leads to a stabilized set, position-electron interaction will lead to a destabilized set i.e. positrons are least stable where electrons are most stable. This calls for a reversal of the Orgel term-splitting diagrams.

Notably, if the Orgel term-splitting diagram of d^2 in octahedral symmetry is known one can construct the following diagrams due to the following reasons (i) tetrahedral splitting is reverse of octahedral splitting, (ii) high-spin d^{5+x} electron distribution is equivalent to d^x ($x = 1$–5) electron as the orbital angular momentum for high-spin d^5 situation is zero, and (iii) hole formalism:

$$d^2(\text{oct}) \equiv d^8(\text{tet}) \equiv \text{reverse of } d^8(\text{oct}) \equiv \text{reverse of } d^2(\text{tet}) \equiv d^7(\text{oct})$$

$$\equiv \text{reverse of } d^7(\text{tet}) \equiv \text{reverse of } d^3(\text{oct}) \equiv d^3(\text{tet})$$

6.4 Magnetic properties

Each electron has a magnetic moment (it is a quantity that describes the magnetic strength of a sample; the unit is emu in CGS unit) with one component associated with the spin angular momentum (s) of the electron and a second component associated with the orbital angular momentum (l) (except when the quantum number $l = 0$). For many complexes of first-row d-block metal ions we can ignore the second component and the magnetic moment μ [it is expressed in Bohr magneton $(\mu_B) = 9.27 \times 10^{-21}$ erg G^{-1} (G is Gauss) in CGS and 9.27×10^{-24} J T^{-1} in SI units; Danish physicist Niels Bohr (1885–1962), Nobel Prize in Physics in 1922; German mathematician and physicist Carl Friedrich Gauss (1777–1855)] can be regarded as being determined by the number of unpaired electrons, n. The total spin quantum number $S = ns = n \times 1/2$.

At first glance, there are two kinds of magnetic substances — *diamagnetic* and *paramagnetic*. Diamagnetic species are those where all electrons are spin-paired. If all the electrons in a molecule are paired there is no overall spin angular momentum ($M_s = 0$; $S = 0$) and therefore no inherent magnetic moment (μ) due to spin. Such a substance is weakly repelled by a magnetic field because the induced magnetic dipoles oppose the field. Diamagnetism is a property of all molecules. On the contrary, paramagnetism is associated with the unpaired electrons, owing to a resultant spin angular momentum ($S > 0$). The different paramagnetic behavior of the compounds of the same metal ion can be associated with a different number of unpaired electrons. Paramagnetic substances are strongly attracted into the magnetic field because the magnetic dipoles inherent in the substance align themselves in the direction of the applied field with a force proportional to its magnetic susceptibility χ. When the magnetic field is weak enough, χ is independent of H, such that one can write $M = \chi H$. It is the ratio of magnetization (M, volume magnetization = magnetic moment per unit volume; the unit is emu cm^{-3} or Oe in CGS unit) to magnetic field [the unit is Oersted (Oe) in CGS unit; unit of magnetic induction is Gauss (G) in CGS unit and T (Tesla) in SI unit; conversion factor: 1 T = 10^4 G] i.e. it is a measure of how magnetizable a substance can become in the presence of a magnetic field. In other words, it measures the ease of alignment of electron spins in an external magnetic field, which is related to the number of unpaired electron of the sample.

The extent of the paramagnetism due to spin angular momentum is much larger than that of diamagnetism and depends on the number of unpaired electrons. Thus, magnetic data, together with spectroscopic measurements, are used to establish oxidation states and bonding properties of transition metal complexes. For further detailed reading on magnetic properties of transition metal ions, which is outside the scope of this book chapter, the readers are encouraged to consult suggested references in the end of this chapter.

Diamagnetic substances are weakly repelled by a magnetic field. Paramagnetic substances are strongly attracted into the magnetic field.

Magnetic susceptibility (χ) indicates the degree of magnetization (M) of a material in response to an applied magnetic field (H) and it is the ratio of magnetization to magnetic field. (Volume) susceptibility is a dimensionless quantity (emu cm^{-3} or Oe).

Magnetic moment or more precisely magnetic dipole moment is the measure of the object's tendency to align with a magnetic field.

For a magnetic substance, at first one determines experimentally gram or mass magnetic susceptibility [$\chi_g = \chi_v$ (volume susceptibility)/ρ, where ρ is the density of the magnetic substance (the unit of χ_g is cm^3 g^{-1} $\equiv$ emu g^{-1} $\equiv$ emu g^{-1} Oe^{-1} in CGS unit; 1 emu = 1 erg G^{-1} and

1 emu $= 1$ cm^3). Next one calculates the molar magnetic susceptibility (χ_M) by multiplying the χ_g value with molar mass (g mol^{-1}) of the sample ($\chi_M = \chi_g \times$ M, where M is the molecular mass; the unit is emu mol$^{-1} \equiv$ emu mol^{-1} Oe^{-1}). Then one calculates the corrected molar susceptibility (χ_M^{corr} in cm^3 mol$^{-1} \equiv$ emu mol$^{-1} \equiv$ emu mol^{-1} Oe^{-1}), which takes into consideration the molar susceptibility (χ_M) and diamagnetic properties of the paired electrons present (χ_{dia} is a negative quantity) in the transition metal complex: $\chi_M^{corr} = \chi_M - \chi_{dia}$.[2]

In the ideal case, $\chi_{para} = \{N_A \mu_B^2 g^2 / 3k_B T\} S(S+1)$

Since $\mu = g[S(S+1)]^{1/2}$, $\mu^2 = g^2 S(S+1)$, $\chi_{para} = N_A \mu_B^2 \mu^2 / 3k_B T$

Therefore, $\mu = (3k_B / N_A \mu_B^2)^{1/2} (\chi_{para} T)^{1/2}$ in units of Bohr Magneton (1 emu $\equiv 1$ erg G^{-1}).

Owing to spin-orbit coupling, the ideal case is rarely attained. In real systems, the effective magnetic moment is given by

$$\mu_{eff} = (3k_B / N_A \mu_B^2)^{1/2} (\chi_M^{corr} T)^{1/2} = 2.828 (\chi_M^{corr} T)^{1/2}$$

($N_A \mu_B^2 / 3k_B = 0.125$ cm^3 K mol^{-1} and $\chi_M^{corr} T =$ cm^3 mol^{-1} K in CGS unit; see below).

Any deviations from ideal behavior are embodied in the g value, which becomes an experimental parameter in molecules rather than a fundamental constant as it is for free electrons. Thus, the susceptibility is related to magnetic moment by the expression,[3]

$$\chi_M^{corr} = \{N_A \mu_B^2 / 3k_B T\} \mu_{eff}^2$$

where k_B is Boltzmann's constant (1.381×10^{-16} erg K^{-1} in CGS unit), T is temperature in K, N_A is Avogadro's number (6.022×10^{23} mol^{-1}), μ_B is the Bohr magneton (9.274×10^{-21} erg G^{-1} in CGS unit), and μ_{eff} is the experimentally determined effective magnetic moment expressed in Bohr magneton (μ_B).

Finally, the spin-only effective magnetic moment (μ_{eff}, experimentally obtained) can be estimated, which assumes that orbital angular momentum contributions to the magnetic moment are absent.

The expression for calculation of spin-only magnetic moment,

$$\mu_{so} = g\{S(S+1)\}^{1/2} = \{4S(S+1)\}^{1/2} \mu_B$$

---

[2] G. A. Bain and J. F. Berry, *J. Chem. Educ.* **2008**, *85*, 532.
[3] R. L. Carlin, *J. Chem. Educ.* **1966**, *43*, 521.

since for a free electron, $g = 2$. $\mu_{so} = \{n(n+2)\}^{1/2}\mu_B$ (see Chapter 5), where n is the number of unpaired electron. Given that hs electronic configurations are typically found for metal(II) hexa-aqua complexes $[M^{II}(H_2O)_6]^{2+}$, the μ_{so} values are ~ 3.87 and ~ 2.83 μ_B for Co(II) $(S = 3/2)$ and Ni(II) $(S = 1)$, respectively. The observed values (μ_{eff}) are, however, in the range 4.7–5.2 μ_B and 2.8–3.4 μ_B, respectively. For tetrahedral Co(II) and Ni(II) complexes the μ_{eff} values are normally found in the range 4.3–4.7 μ_B and 3.5–4.2 μ_B, respectively. Thus, orbital angular momentum also contributes to paramagnetism (see below). For many (but not all) first-row metal ions, spin-orbit coupling constant λ (see Chapter 5), is very small and the spin and orbital angular momenta of the electrons operate independently.

The Curie [Pierre Curie (1859–1906) was a French physicist; Nobel Prize in Physics in 1903] law expresses the relation between the magnetic susceptibility of a paramagnetic substance to its temperature. It follows that χ should increase as temperature decreases. This is because, at the lower temperature, there is less thermal motion opposing the alignment of the magnetic dipoles with the applied magnetic field. Normal paramagnetic substances obey the Curie Law,

$$\chi = \{Ng^2\mu_B^2/3kT\}S(S+1) = (N\mu_B^2/3k)g^2S(S+1)/T = C/T$$

where C $(= (N\mu_B^2/3k)g^2S(S+1) = 0.125g^2S(S+1)$ cm^3 K mol^{-1}) is the Curie constant. It is a material-dependent property.

$N\mu_B^2/3k = 0.125$ erg G^{-2} K mol^{-1} in CGS unit $= 0.125$ cm^3 K mol^{-1} in CGS unit, since 1 erg $= g$ cm^2 s^{-2} and 1 G $= g^{1/2}$ cm$^{-1/2}$ s^{-1}.

Then, G$^2 = g$ cm^{-1} s^{-2} $=$ erg cm^{-2} $\times$ cm^{-1} $=$ erg cm^{-3}. Hence, 1 erg $=$ G^2 cm^3 (i.e. G$^2 =$ erg cm^{-3})

The Curie law is obeyed if a plot of $1/\chi$ versus T gives a straight line passing through the origin (0 K) of slope $1/C$. Many transition metal compounds satisfy the Curie law, although deviations are known to occur. Many paramagnetic substances do give a straight line which often intercepts either just a little above 0 K or just a little below 0 K. These substances are said to obey the Curie-Weiss Law,

$$\chi = C/(T - \Phi)$$

where C is the Curie constant and Φ is known as the Weiss constant. The intercept with the T axis yields both the sign and the value of Φ

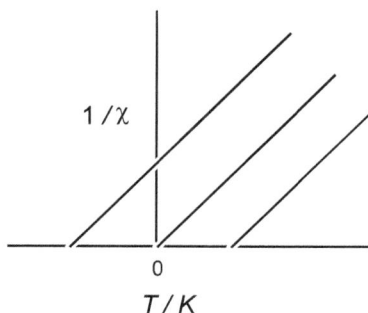

Fig. 6.16 Plots for $1/\chi_M^{corr}$ versus T for paramagnetic complexes.

(see Fig. 6.16). The substances which give rise to positive value of Φ are called ferromagnetic and which give rise to negative value of Φ are called antiferromagnetic.

(a) Orbital contribution to magnetic moment

From a quantum mechanics viewpoint, the magnetic moment is dependent on both spin and orbital angular momentum contributions.

For an electron to have orbital angular momentum, the orbital it occupies must be able to transform into an entirely equivalent and degenerate orbital by a simple rotation. It is the rotation of the electrons that induces the orbital contribution.

In an octahedral complex, the three t_{2g} orbitals (dxy, dxz, dyz) can be interconverted by rotations through $90°$. Thus, an electron in a t_{2g} orbital has orbital angular momentum. The e_g orbitals (dx^2-y^2, dz^2) cannot be interconverted by rotation about any axis as the orbital shapes are different. Therefore an electron in the e_g set does not contribute to the orbital angular momentum and is said to be quenched. Electrons in the t_{2g} set do not always contribute to the orbital angular momentum. For example, in the d^3 (t_{2g}^3) case, an electron in one of the three d orbitals cannot by rotation be placed in the other orbital as the orbital already has an electron of the same spin quantum number as the incoming electron. This process is also called quenching. If all the t_{2g} orbitals are doubly occupied, electron rotation is also impossible.

Tetrahedral complexes can be treated in a similar way with the exception that we fill the e orbitals first, and the electrons in these e orbitals do not contribute to the orbital angular momentum.

In hs O_h complexes, orbital contributions to the magnetic moment are important only for the configurations $(t_{2g})^1$, $(t_{2g})^2$, $(t_{2g})^4 (e_g)^2$, and $(t_{2g})^5 (e_g)^2$.

For T_d complexes, it is similarly the configurations that give rise to an orbital contribution are $(e)^2 (t_2)^1$, $(e)^2 (t_2)^2$, $(e)^4 (t_2)^4$, and $(e^4) (t_2)^5$.

These results lead us to the conclusion that an d^7 hs O_h complex should have a magnetic moment greater than the spin-only value of 3.87 μ_B but a d^7 T_d complex should not. However, the observed values of μ_{eff} for $[Co^{II}(H_2O)_6]^{2+}$ and $[Co^{II}Cl_4]^{2-}$ are 5.0 and 4.4 μ_B, respectively, i.e. both complexes have magnetic moments greater than μ (spin-only). The third factor involved is spin–orbit coupling (see Chapter 5).

The spin-only eqn $\mu_{so} = g\{S(S+1)\}^{1/2} = \{4S(S+1)\}^{1/2} = \{n(n+2)\}^{1/2}$ μ_B can be modified to include the orbital angular momentum. The expression (van Vleck formula) is

$$\mu_{L+S} = \{4S(S+1) + L(L+1)\}^{1/2}$$

Strictly, the eqn for μ_{L+S} applies to free ions but, in a complex ion, the crystal field partly or fully quenches the orbital angular momentum.

The following table gives a list of all d^1 to d^9 configurations including hs and ls complexes and a statement of whether or not a direct orbital contribution is expected.

| d^n ion | Octahedral | | Tetrahedral |
	high-spin	low-spin	high-spin
d^1	Yes	—	No
d^2	Yes	—	No
d^3	No	—	Yes
d^4	No	Yes	Yes
d^5	No	Yes	No
d^6	Yes	No	No
d^7	Yes	No	No
d^8	No	—	Yes
d^9	No	—	Yes

These are explained by mixing of the ground state with the excited state.

(b) Spectroscopic term symbols and orbital contribution

An orbital angular momentum contribution is expected when the ground term is triply degenerate i.e. a T state. For such complexes one uses van Vleck formula. These systems show temperature dependence as well.

The configurations corresponding to the A_1 (from free ion term S), E (from free ion term D), or A_2 (from free ion term F) (see Chapter 5) do not directly contribute to the orbital angular momentum.

For the A_2 and E terms there is always a higher T term of the same multiplicity as the ground term. This T term can affect the magnetic moment through 'mixed in' effect or in other words, as a result of a mixing of states. The relevant phenomenological expression is,

$$\mu_{eff} = \mu_{so}(1 - \alpha\lambda/\Delta)$$

where α is a constant (4 for an A_2 term and 2 for an E term), λ is the spin-orbit coupling constant (see Chapter 5) which is generally only available for the free ion but this does give important information since the sign of the value varies depending on the orbital occupancy (number of d electrons), Δ is the crystal field splitting energy. The parameter Δ again is often not available for complexes.

For less than half-filled d-electrons (d^1 to d^4) the value of λ (in cm^{-1}) is positive [for d^1 Ti(III) (155), d^2 V(III) (105), d^3 Cr(III) (90), d^4 Mn(III) (88)], hence μ_{eff} is less than μ_{so}.

For more than half-filled d-electrons (d^6 to d^9) the value of λ (in cm^{-1}) is negative [d^6 Fe(II) (-102), d^7 Co(II) (-172), d^8 Ni(II) (-315), d^9 Cu(II) (-830)], hence μ_{eff} is greater than μ_{so}.

Example 6.21 For [Fe(CN)$_6$]$^{3-}$ experimentally determined μ_{eff} value is much higher than its μ_{so} value. Justify.

Answer CN$^-$ is a strong field ligand and hence d^5 [Fe^{III}(CN)$_6$]$^{3-}$ is a ls complex with one unpaired electron. The values of L and S are 2 and 1/2, respectively. For this complex the ground-state spectroscopic symbol is $^2T_{2g}$. For T state the relevant magnetic expression to follow is

$$\mu_{so} = \{n(n+2)\}^{1/2} = (1 \times 3)^{1/2} = 1.73 \ \mu_B$$

$$\mu_{eff}(\mu_{L+S}) = \{4S(S+1)+L(L+1)\}^{1/2} \ \mu_B$$

$$= \{4 \times 1/2(1/2+1)+2(2+1)\}^{1/2}$$

$$= \{4 \times 1/2 \times 3/2) + 2 \times 3\}^{1/2}$$

$$= (3+6)^{1/2} = 3 \ \mu_B$$

Example 6.22 Tetrahedral Co(II) complexes have room-temperature effective magnetic moments that range from 4.3–4.7 μ_B. For example, $\mu_{eff} = 4.59 \ \mu_B$ for $[CoCl_4]^{2-}$ (Given: λ for d^7 free ion $= -172 \ cm^{-1}$ and Δ for the complex $= 3100 \ cm^{-1}$).

Answer The ground-state term for the d^7 free ion is ^4F. The term ^4F splits into $^4A_2 + {}^4T_1 + {}^4T_2$. The ground-state term for d^7 Co(II) in tetrahedral field is $^4A_2$. The complex has three unpaired electrons and hence $\mu_{so} = (3 \times 5)^{1/2} = 3.87 \ \mu_B$. For an A term the constant $\alpha = 4$. Using λ for the free ion $-172 \ cm^{-1}$, which we can use as an approximation, $\Delta = 3100 \ cm^{-1}$, and $\mu_{so} = 3.87 \ \mu_B$.

Hence, applying $\mu_{eff} = \mu_{so} \ (1 - \alpha\lambda/\Delta)$, $\mu_{eff} = 3.87 \times \{1 - (4 \times -172)/3100\} = 4.72 \ \mu_B$.

Thus this eqn rationalizes the experimental result.

Example 6.23 Calculate the magnetic moment of Ce(III) and Nd(III) complexes. Atomic numbers of Ce and Nd are 58 and 60, respectively. Given, $g = 1 + \{J(J+1) + S(S+1) - L(L+1)\}/2J(J+1)$.

Answer The electronic configurations are

Ce(III): [Pd] $4f^1 5s^2 5p^6$
Nd(III): [Pd] $4f^3 5s^2 5p^6$
The ground state: $^{2S+1}L_J$
Ce(III): $L = 3$, $S = 1/2$, $J = 5/2$; ground state $^2F_{5/2}$ and $g = 6/7$
Nd(III): $L = 6$, $S = 3/2$, $J = 9/2$; ground state $^4I_{9/2}$ and $g = 8/11$
Applying, $\mu_J = g\{J(J+1)\}^{1/2} \ \mu_B$ (since for lanthanides J is a good quantum number)

For Ce(III): $6/7 \times \{(5/2 \times 7/2)\}^{1/2} = 0.86 \times 2.96 = 2.55 \; \mu_B$
For Nd(III) complex: $8/11 \times \{9/2 \times 11/2\}^{1/2} = 0.73 \times 4.97 = 3.63 \; \mu_B$

(c) Forms of paramagnetism

The paramagnetic centers may interact with neighboring center(s). This interaction leads to ferromagnetism (in the case where the neighbouring magnetic dipoles are aligned in the same direction) and antiferromagnetism (where the neighbouring magnetic dipoles are aligned in alternate directions). These two forms of paramagnetism show characteristic variations of the magnetic susceptibility with temperature. The magnetic substance that displays "normal" (Curie law) paramagnetic behavior is displayed in Fig. 6.17(a). In the case of ferromagnetism, above the Curie temperature (T_C) the magnetic substance displays "normal" (Curie law) paramagnetic behavior. Below the Curie temperature the material displays strong magnetic properties. The magnetic behavior is displayed in Fig. 6.17(b). For antiferromagnetism, above the Néel (Louis Néel; French physicist who in 1936 successfully explained antiferromagnetism) temperature the substance displays "normal" paramagnetic behavior. Below the Neel temperature the material displays weak magnetic properties which at lower temperatures can become essentially diamagnetic. The magnetic behavior is displayed in Fig. 6.17(c).

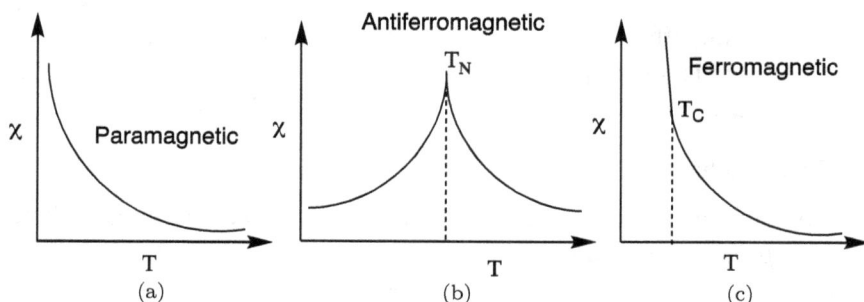

Fig. 6.17 Plots of χ vs. T for (a) paramagnetic, (b) antiferromagnetic, and (c) ferromagnetic substances.

(d) Magnetic field dependence

The magnetic susceptibility of diamagnetic and paramagnetic compounds are independent of external magnetic field strength. A plot of magnetization (M) vs. applied field (H) is linear and the slope of the plot give the susceptibility at that temperature.

Superparamagnetic materials show a nonlinear behavior. Since χ is the slope of this plot, the susceptibility changes at each field strength. Superparamagnetism is identified by the size of the susceptibility, which at low fields is typically larger than paramagnetism. Further, when the field is reversed the magnetization is reversible.

Ferromagnetic materials show bulk magnetic behavior and when M is plotted vs. H there is a sharp nonlinear rise in M but this asymptotes to some maximum value called the saturation magnetization. When the field is reversed the saturation value is maintained to below $H = 0$ and then eventually drops to the same saturation magnetization but with a negative value. This is known as hysteresis.

Ferromagnetism arises because below some transition temperature, the Curie Temperature (T_C), the microscopic magnetic moments all align in the same direction to give a bulk moment: ↑ ↑ ↑ ↑. Antiferromagnetism is similar to ferromagnetism except that the microscopic moments align in an antiparallel fashion: ↑ ↓ ↑ ↓. The consequence of this is that the field dependence is nonlinear and hysteretic but the magnetic moment is much smaller, close to 0.

Related to this is ferrimagnetism, where there are two microscopic magnetic sites with different spin quantum numbers. The microscopic moments can align parallel or antiparallel: ↑ ↓ ↑ ↓. The net effect is a bulk ferromagnetic response, although the saturation moment can be much smaller.

(e) Super-exchange phenomena

The effective magnetic moment (μ_{eff}) for copper(II) acetate monohydrate is ~ 1.4 μ_B at room temperature (μ_{so} value for one unpaired spin is 1.73 μ_B). X-ray crystallographic analysis revealed that the compound is actually a dimer (Fig. 6.18). Considering tetragonal elongation at the Cu(II) centers, the d orbital splitting is $(dxz)^2$, $(dyz)^2 < (dxy)^2 < (dz^2)^2 < (dx^2 - y^2)^1$. There is some direct overlap between the two $dx^2 - y^2$ orbitals (see metal-metal bonding section).

[Cu^{II}$_2$(O$_2$CMe)$_4$(H$_2$O)$_2$]

copper(II) acetate hydrate dimer

[(H$_2$O)Cu^{II}(CH$_3$COO)$_4$Cu^{II}(OH$_2$)]

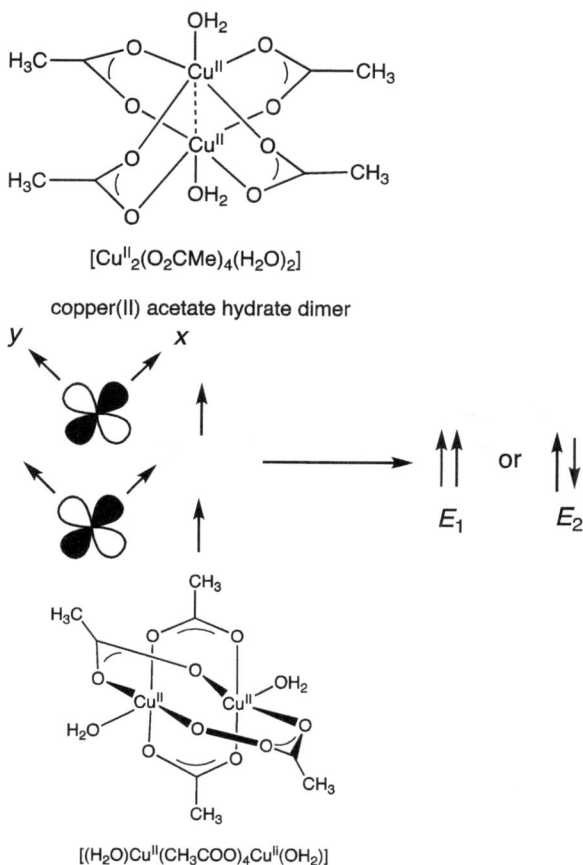

Fig. 6.18 Structures of Copper(II) acetate hydrate dimer and orbital interactions for magnetic-exchange.

The two magnetic exchange interactions have energies of E_1 (ferromagnetic) and E_2 (antiferromagnetic) (Fig. 6.18). At a particular temperature (Néel temperature), E_2 may be less than E_1, and the magnetic interaction will be antiferromagnetic.

The oxovanadium(IV) complex of a tridentate Schiff base ligand is actually a dimer (Fig. 6.19). Moreover, the V=O distance is quite short ($\sim$1.67 Å). Therefore, tetragonal compression is expected at the metal center. The d orbital splitting is $dxy < dxz, dyz < dx^2-y^2 < dz^2$. Through the phenolate bridges the two oxovanadium(IV) units are brought so close

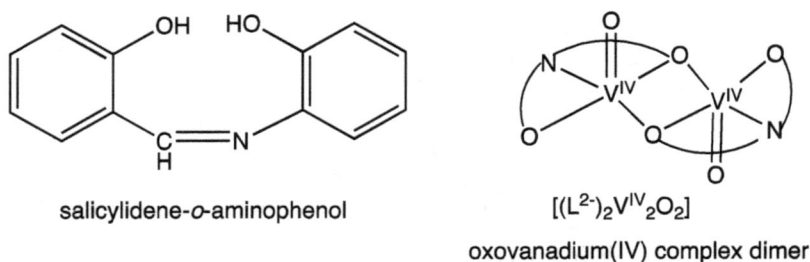

salicylidene-*o*-aminophenol

$[(L^{2-})_2V^{IV}_2O_2]$

oxovanadium(IV) complex dimer

Fig. 6.19   Structure of oxovanadium(IV) dimer.

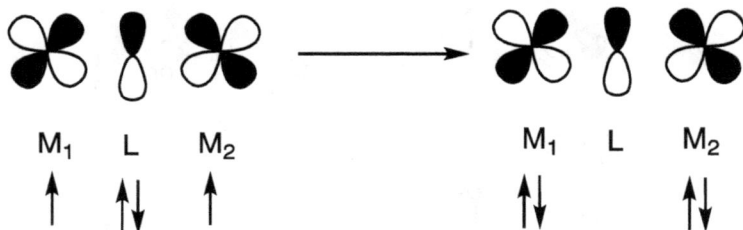

Fig. 6.20   Magnetic interaction between two half-filled *dxz/dyz* orbitals and a filled *pπ* orbital.

together that the two *dxy* orbitals carrying the unpaired spins can effectively overlap resulting in lowering of spin (antiferromagnetic coupling).

There should be an overlap of a filled ligand orbital with two partially filled orbitals of two metal ions (Fig. 6.20).

When there is overlap of a *pπ*-type ligand orbital with two half-filled orbitals of two different metal ions (*dxx, dyz* orbitals) then an interesting situation arises. Since the metal ion orbitals are only half-filled a situation may arise when one of the electrons of the filled π orbitals of the ligand may tend to move into the half-filled π orbitals of the metal ion. A paired spin will have to result in order to obey Pauli's exclusion principle. The other electron of the ligand π orbital may simultaneously move into the π orbital of the other metal ion, again a paired spin would result. If this reorganization of spin becomes of lower energy at a particular temperature the system will be stabilized. Further lowering of temperature may enhance this stability. This spin-spin neutralization mechanism is possible due to interaction of a filled ligand orbital in between two half-filled metal ion orbitals. In other words, such antiferromagnetic pathway gives us a strong evidence in favor of overlap of metal orbitals and ligand orbitals.

(f) Temperature-independent paramagnetism (TIP)

The magnetic susceptibility of a molecule in which all the electrons are paired depends on the relative magnitude of a diamagnetic contribution χ_D (χ_{dia}) and a paramagnetic contribution χ_T. In contrast to the paramagnetism due to unpaired electron spins (χ_{para}), χ_T does not depend on the temperature and if it is larger than χ_D the molecule is said to show temperature-independent paramagnetism (TIP). In some substances TIP can make the molecules paramagnetic, although all the electrons in the ground state are paired, if there are low-lying excited states to which electrons can readily move. Many molecules fall into this category, particularly transition-metal complexes possessing paired d electrons. The examples are $Mn^{VII}O_4^-$, $Cr_2^{VI}O_7^{2-}$ etc.

(g) Spin-crossover complexes

Octahedral complexes of d^4–d^7 ions may be either hs (maximum number of unpaired electrons) or ls (maximum number of paired up electrons), depending on the strength of crystal field (Δ_o) of the ligand and the magnitude of the mean pairing energy (P) of the metal ion. For strong-field ligands the spin-pairing occurs and hence ls complexes are formed. For weak-field ligands the resulting complexes are hs. The spin-transition or spin-crossover (SC) phenomenon requires Δ_o to be of the same order of magnitude as P. The SC compounds are one of the representative examples of molecular bistability.[4] The hs and ls states of an SC compound are interconvertible by several different physical perturbations such as temperature,

---

[4](a) Y.-S. Meng and T. Liu, *Acc. Chem. Res.* **2019**, *52*, 1369; (b) G. Molnár, S. Rat, L. Salmon, W. Nicolazzi, and A. Bousseksou, *Adv. Mater.* **2018**, *30*, 17003862. (c) K. Senthil Kumar and M. Ruben, *Coord. Chem. Rev.* **2017**, *346*, 176. (d) P. Gütlich, A. B. Gaspar, and Y. Garcia, *Beilstein J. Org. Chem.* **2013**, *9*, 342. (e) P. Gütlich and H. A. Goodwin, *Top. Curr. Chem.* **2004**, *233*, 1. (f) P. Gütlich, Y. Garcia, and H. A. Goodwin, *Chem. Soc. Rev.* **2000**, *29*, 419.

pressure, light-irradiation as well as chemical modifications such as variation in the nature of associated counteranion and the degree of solvation in the crystallized salts.

Many mononuclear six-coordinate iron(II) complexes with heterocyclic nitrogen donor ligands and α-diimine ligands have been found to exhibit SC behavior.[4] In order for thermally-induced spin-transition to occur the difference in the Gibb's free energies for the two spin states involved must be on the order of thermal energy, $k_B T$. An increase in temperature favors the hs state, while lowering the temperature favors the ls state. Thus, a possibility exists of identifying both the spin states. Thermally-driven spin-state transitions between the hs and ls state occur because in such systems two electronic states of differing multiplicity (consider $Fe^{II}N_6$ complexes (see below): hs $(S = 2)$ and ls $(S = 0)$) become nearly equienergetic and a thermal population of the excited state is possible.

For a ls $\rightleftharpoons$ hs equilibrium in an $Fe^{II}N_6$ complex, thermodynamic parameters for the dynamic equilibrium

$$\text{ls-Fe(II)}(^1A_{1g}; S = 0) \rightleftharpoons \text{hs-Fe(II)}(^5T_{2g}; S = 2)$$

may be readily evaluated using the following equations,

$$m_{(\text{hs})} = \{\chi_M - \chi(\text{ls})\}/\{\chi(\text{hs}) - \chi(\text{ls})\}$$
$$m_{(\text{ls})} = \{\chi(\text{hs}) - \chi_M)\}/\{\chi(\text{hs}) - \chi(\text{ls})\}$$
$$K_{\text{eq}} = m_{(\text{hs})}/m_{(\text{ls})} = \{\chi_M - \chi(\text{ls})\}/\{\chi(\text{hs}) - \chi_M)\} \tag{1}$$

where χ_M is the corrected molar susceptibility at a given temperature, $\chi(\text{hs})$ and $\chi(\text{ls})$ are the corrected molar susceptibilities of the pure hs and ls component, respectively, and $m_{(\text{hs})}$ and $m_{(\text{ls})}$ are the corresponding mole fractions. The pure hs behavior can be evaluated from the data for hs and the pure ls behavior can be determined from each system, separately, from the low-temperature data. Thermodynamic parameters can be evaluated from the equilibrium constants using $\ln K$ vs. $1/T$ plots (see below). In the solid state, these plots have a pronounced curvature at the low-temperature end. Only the linear portion of the plots is used in evaluating the parameters. The curvature in these plots is probably due either to small inaccuracies in the values of $\chi(\text{ls})$ and $\chi(\text{hs})$ states or results from a small amount of impurity. In either case, its effect is considered to be negligible.

As χ is additive but μ is not but μ^2 is, we can rewrite the final form as

$$K_{eq} = m(\text{hs})/m(\text{ls}) = (\mu_{\text{exptl}}^2 - \mu_{\text{ls}}^2)/(\mu_{\text{hs}}^2 - \mu_{\text{exptl}}^2) \qquad (2)$$

$$K_{eq} = m(\text{hs})/m(\text{ls}) = m(\text{hs})/\{1 - m(\text{hs})\} \ (\text{since } m(\text{hs}) + m(\text{ls}) = 1) \ (3)$$

Again, $\Delta G = -RT\ln K_{eq}$; $\Delta H - T\Delta S = -RT\ln K_{eq}$; $\ln(K_{eq}) = -\{\Delta H - T\Delta S/RT\}$

Therefore, $K_{eq} = \exp\{(-\Delta H + T\Delta S)/RT\} = \exp\{-\Delta H/RT + \Delta S/R\}$ (4)

Combining eq 3 and eq 4,

$$m(\text{hs})/\{1 - m(\text{hs})\} = \exp\{-\Delta H/RT + \Delta S/R\}$$

$$\text{Therefore, } m(\text{hs}) = 1/\exp\{(\Delta H/RT - \Delta S/R) + 1\} \qquad (5)$$

The results of a system discovered from the author's laboratory are presented (Fig. 6.21) and briefly discussed below.[5]

The $\chi_M T$ value is equal to $3.36\,\text{cm}^3\,\text{mol}^{-1}\,\text{K}$ (a value which corresponds to μ_{eff} value of $5.18\ \mu_B$) at 300 K and it decreases when cooling to reach a value of 0.59 ($\mu_{\text{eff}} = 2.18\ \mu_B$) at 70 K, and remains almost constant down to 30 K [$\chi_M T = 0.57\,\text{cm}^3\,\text{mol}^{-1}\,\text{K}$ ($\mu_{\text{eff}} = 2.14\ \mu_B$)]. Thereafter, it decreases slowly down to 5.2 K [$\chi_M T = 0.38\,\text{cm}^3\,\text{mol}^{-1}\,\text{K}$ ($\mu_{\text{eff}} = 1.74\ \mu_B$)]. The magnetic moment value determined at 300 K clearly indicates that the attainment of hs ($^5T_2$) state is complete at the

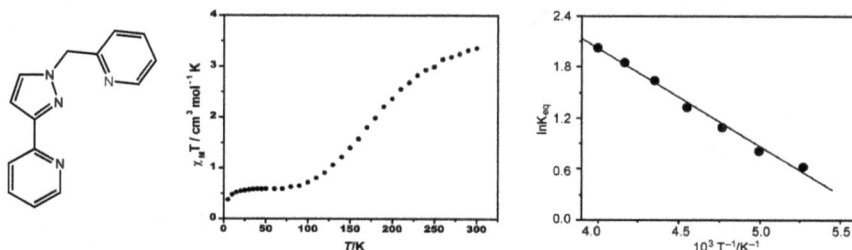

Fig. 6.21 (a) Temperature dependence of the $\chi_M T$ (cm^3 mol^{-1} K) product for powdered samples of [Fe(L)$_2$](ClO$_4$)$_2$ and (b) Plot of ln K_{eq} vs. $1/T$ between 190–250 K.

[5]S. Singh, V. Mishra, J. Mukherjee, N. Seethalekshmi and R. Mukherjee, *Dalton Trans.* **2003**, 3392.

highest temperature. The data at the lowest temperature is significantly higher than the temperature-independent paramagnetism (TIP; see Section 6.4(f)) value (0.6 μ_B), expected for diamagnetic ls-Fe(II) complexes. The $\chi_M T$ value at 5.2 K indicates the presence of $\sim$10% hs species in the ls region. In these calculations μ_{hs} was taken as 5.2 μ_B and μ_{ls} was taken as 0.6 μ_B to take account of the TIP. The $\ln K_{eq}$ vs. $1/T$ relationship is linear to a good approximation over the temperature range 190–250 K (Fig. 6.21). The thermodynamic parameters derived from this straight line were $\Delta H = 9.7$ kJ mol^{-1} and $\Delta S = 56$ J K^{-1} mol^{-1}. The source of the ΔH is the Fe–N bond length increase, which accompanies the transition from ls to hs, *i.e.*, reorganization of the inner coordination sphere. Measured values of the entropy changes at ls $\rightleftharpoons$ hs transitions for complexes of Fe^{II}N$_6$ vary between about 48 and 86 J K^{-1} mol^{-1}. This is considerably more than the value expected for a change solely by the spin multiplicity change ($R \ln 5 = 13.4$ J K^{-1} mol^{-1}). The excess value no doubt arises from the changes in the molecular vibrational contributions.

6.5 Nephelauxetic effect

An analysis of the electronic spectral transitions in a complex (Section 6.6; see below) allows one to make an estimate of the interelectronic repulsion, among the electrons in the d-orbitals. It is denoted by the parameter B, called the Racah parameter. The energy separation between the states with maximum possible multiplicity (e.g., between the ^{3}F and ^{3}P states of a d^2 ion and ^{4}F and ^{4}P states of a d^3 ion; see below) is represented by 15B. The value of B in a complex is always smaller than its value for the free ion. The value of B (free ion) is a fixed quantity. However, in a complex we have an experimental measure of B (analysis of electronic spectra). Thus, B (complex) < B (free ion). This is explained as due to 'cloud expansion'. In other words, when the metal d orbitals and the ligand orbitals overlap (in contrary to the basic postulate of CFT), the ligand electrons enter into the metal d orbitals, screening the d electrons from the nucleus. This in turn reduces the effective nuclear charge of the nucleus felt by the d electrons, leading to an expansion of the d orbitals i.e. 'cloud expansion'. As a result, the interelectronic repulsion decreases. The reduction in B from its free ion value is expressed in terms of the nephelauxetic parameter

β ($=$ B (complex)/B (free ion)). The value of β is a measure of covalency. The values of β depend on the identity of the metal ion and the ligand, and vary along the nephelauxetic series:

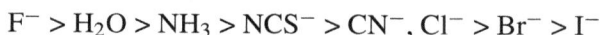

$$F^- > H_2O > NH_3 > NCS^- > CN^-, Cl^- > Br^- > I^-$$

A small β indicates enhanced d electron delocalization on to the ligands and hence a significant covalent character in the complex. Thus, the value of β is a measure of covalency. The smaller the value, the greater is the covalency between the metal ion and the ligands. The softer the ligand, the smaller is the value of β. Notably, since the σ-overlap of e_g set is larger than the π-overlap of t_{2g} (see below), the cloud expansion is larger in the former case.

6.6 Deficiencies of crystal field theory

 (i) The placement of CO next to CN^- in the spectrochemical series, since CO is neutral the electrostatic repulsion is not expected to be large.
 (ii) The Nephelauxetic effect
(iii) The electronic spectra (d–d transitions) of transition metal complexes (see below)

6.7 Electronic spectral properties

The electronic absorption spectra arise due to transition of an electron from a lower energy level to a higher energy level of the frontier orbitals. A characteristic feature of the transition elements and their complexes is their color, which are extremely varied. Notably, they absorb light in the visible region. An attempt to explain this, at least in qualitative fashion, is worthwhile. Such an exercise provides a good opportunity to correlate the relationship between the color of a complex and the position of its absorption band, and the relationships of these in turn to the transitions between electronic energy levels. The colors of transition metal complexes are usually attributed to electronic transitions involving predominantly metal d orbitals. Such transitions are termed as d–d transition. In this type of transition, there is little shift of electron density from any atom or another. There is another kind of transition the charge transfer (CT) transition. In this case, an electron is transferred from a MO centered mainly on the ligand(s) to one centered mainly on the metal atom, or vice versa. In this type of transition, the atomic charges in the initial and final states are appreciably different.

It is now possible to make quantitative predictions of the spectral properties of these complexes, as well as to find a theoretical rationale for much of the experimental data. Studies of electronic spectra of metal complexes provide information about structure and bonding.

The spectrum of a colored solution can be measured quite easily using an absorption spectrophotometer. Absorption bands are described in terms of λ_{max} (usually given in nm, but it may also be expressed in terms of wavenumbers $\bar{\upsilon}$ in cm^{-1} ($\bar{\upsilon} = 1/\lambda = v/c$, c is velocity of light) corresponding to the absorption maximum A_{max}. The molar extinction coefficient (or molar absorptivity) ε_{max} (in $M^{-1} cm^{-1}$) indicates how intense an absorption is and reflects the probability of the corresponding electronic transition. The ε_{max} is related to A_{max} ($= \varepsilon_{max}cl$; c is the concentration of the solution expressed in moles/liter and l is the path length (in cm) of the spectrometer cell/cuvette). The ε_{max} values in the range $\approx 10-200\ M^{-1} cm^{-1}$ are observed for d–d transitions. The CT transitions are intense. In some spectra, CT absorptions mask bands due to d–d transitions, although CT absorptions (as well as ligand-centered n–π* and π–π* bands) often occur at higher energies than d–d absorptions. The ε_{max} values in the range $\geq 10^3\ M^{-1} cm^{-1}$ are observed for CT transitions.

Electronic spectral bands are usually broad because the electronic transitions are coupled to vibrational transitions. Each broad band consists of an envelope of transitions from the various occupied vibrational levels of the electronic ground state to various vibrational levels of the electronic excited state. The absorption of a photon of light occurs in $\approx 10^{-18}$ s whereas molecular vibrations and rotations occur more slowly. Thus, the electronic spectrum will record a range of energies corresponding to different vibrational states, associated with the electronic levels.

Absorption of electronic transition is accompanied by a change in the electric dipole moment of the molecule for such electric dipole transitions. The absorption bands for transition metal complexes belong to this class. Certain requirements of the ground and excited states — *selection rules* — must be followed for a transition to occur. If a selection rule is obeyed, the transition is said to be allowed and if the selection rule is violated, the transition is said to be forbidden. The selection rules deal with specific idealized descriptions of the ground and excited states. We often observe forbidden transitions. Nevertheless, the observed intensities for forbidden transitions are low (lower values of ε_{max}) and hence have lower probability.

Selection Rules

Two selection rules for the occurrence of an electronic transition are as follows.

(i) Transitions which involve a change in the orbital angular momentum quantum number $\Delta l = \pm 1$ are allowed. This is known as Laporte selection rule. According to this selection rule, d–d transitions are not allowed, since $\Delta l = 0$. They are still observed because of relaxation in the Laporte rule. On the other hand, the CT transitions are allowed, as this selection rule is valid (here metal-ligand orbital overlap is considered; see below).

(ii) Transitions involving states of the same spin multiplicity i.e. ΔS (precisely, $2S + 1$) $= 0$ are allowed.

The significant measure of the intensity of the electronic transitions, and hence the probability of the electronic transition, is not ε_{max} but rather the area under the band, the integrated intensity I $(= \int \varepsilon \, dv)$. A useful theoretical measure of intensity is the *oscillator strength* (f). The oscillator strength can be calculated from ε_{max} as an approximation.

$$f \propto \int \varepsilon \, dv, \quad \text{then } f \propto \varepsilon_{max} \Delta v_{1/2}$$

where $\Delta v_{1/2}$ is the half-width, the width at $\varepsilon_{max}/2$ (Fig. 6.22). Moreover, the oscillator strength (f) is proportional to the dipole moment (D), which is the square of the *transition moment integral* (M). The term $D^{1/2} = \int \Psi_i \mu_e \Psi_f \, d\tau$, where Ψ_i and Ψ_f are two wave functions between which

$$f \propto D = M^2$$

the transition is to occur [i is the initial lower energy state (Ψ_i) and f is the final higher energy state (Ψ_f)], μ_e $(= -er)$ is the *electric dipole*

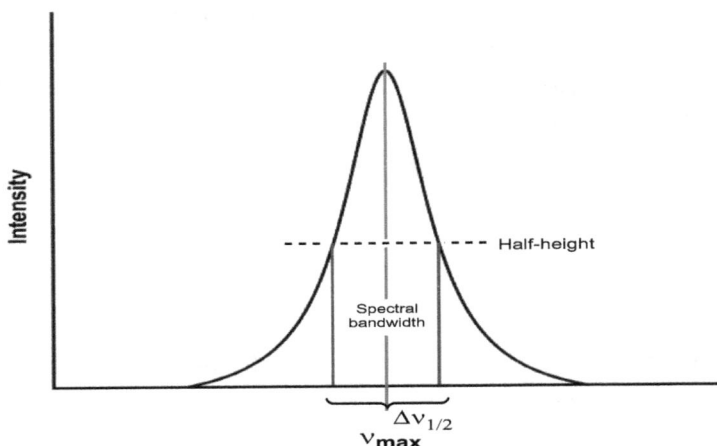

Fig. 6.22   The half band-width of an electronic spectral band.

moment operator, and $d\tau$ is a volume element. This applies for *electric-dipole transitions*. The integral can be regarded as corresponding roughly to charge displacement during the transition. This results in a change in electric-dipole moment. For a single atom, the dipole moment is the sum of the products of the charge on the electron times the distance of the electron from the nucleus. The summation for all electrons and all nuclei in a molecule is the electric-dipole moment vector μ_e. Moreover, the oscillator strength (intensity) is proportional to the dipole moment D, which is the square of the *transition moment integral* M. Since the integral of an odd function vanishes (the positive and negative portions cancel when integrated over all space), the intensity will be nonzero only for an even function.

Hence, for an electronic transition to take place, the integral must be nonzero. Since *r* is a vector quantity, it changes sign under inversion. Thus, μ is inherently odd and therefore has u parity. The integral will also change sign under inversion (u parity), if Ψ_i and Ψ_f have the same parity (either g or u), because g × u × g = u and u × u × u = u and hence becomes zero. However, if Ψ_i and Ψ_f have opposite parity, the integral does not change sign under inversion (g parity) of the coordinates because g × u × u = g and hence becomes nonzero.

The consequence of this analysis is noteworthy. In a centrosymmetric complex such as octahedral, crystal-field/ligand-field/d–d transitions are therefore forbidden (g × u × g = u; d orbitals have g parity). The relatively weakness of d–d transitions (low ε_{max} values) in octahedral complexes accounts for their forbidden nature. However, for a tetrahedral complex, which has no center of symmetry, the Laporte selection rule is silent and therefore exhibits enhanced intensity, compared with those in octahedral complexes. This is the basis for the symmetry selection rule.

The question remains why d–d transitions occur at all, even though they are weak. The d–d transitions occur due to relaxation of Laporte selection rule. First, a complex is distorted from perfect centrosymmetric nature, imposed by crystal packing forces or asymmetry in the metal coordination environment due to chelating ligand structure. Laporte selection rule may be relaxed for metal complexes due to *vibronic mechanism*. This can be explained in the following manner. An odd vibration which destroys the center of inversion distorts the regular octahedron so that g and u parity is no more applicable. Therefore, there could be mixing of metal d and ligand p orbitals. The vibrational and electronic transitions are coupled (see above). The transitions involve excited vibrational states of the electronic ground and/or excited states. This contributes to the broadening of bands.

The transition-moment integral M has two parts dealing with orbital and spin wave functions. Moreover, M also includes a Franc-Condon factor dependent on the nuclear coordinates. Ignoring Franc-Condon factor (during an electronic transition there is no change in the position of the nucleus), an integral dealing with the orbital and spin wave functions is,

$$M = \int \Psi_i\, \mu_e\, \Psi_f d\tau_e \text{ (orbital wave function)}$$

$$\times \int \Psi_i\, \mu_e\, \Psi_f d\tau_s \text{ (spin wave function)}$$

The operator μ_e does not operate on the spin wave functions (dipole moment is independent of spin of the system). M is zero if one of the integrals is zero. If M is zero, then so are f and ε_{max}. The transition is forbidden. The transitions that are forbidden, in terms of the approximations we use, sometimes occur, but with low probability and hence low intensity.

Spin Selection Rule. The integral of the spin wave functions determines the spin selection rule. Wave functions of different spin multiplicity are orthogonal, so the integral is zero unless the i and f states have the same spin. Transitions that involve a change in the number of unpaired electrons are spin-forbidden. A spin-allowed electronic transition involves promotion of an electron without change in its spin. For a singlet to a triplet transition, we go from two electrons with opposite spins in the initial state to two with the same spin occupying different orbitals in the final state. Hence, the transition is forbidden.

a) Orgel diagrams for octahedral and tetrahedral symmetry

Spectral terms (see Chapter 5) represent electronic energy states. Each spectral term represents an array of microstates. Different terms of a configuration have different energies on account of the repulsion between electrons. Term symbols give a detailed description of an electron configuration. Electron-electron repulsions are difficult to estimate, but the discussion is simplified by considering two extreme cases.

In the weak field limit, the crystal field/ligand field in O_h, as measured by Δ_o, is so weak that only electron-electron repulsions are important. In such a situation the relative energies of the terms can be expressed in terms of Racah parameter B (see Section 6.5). In the strong field limit, the ligand field is so strong that electron-electron repulsions can be ignored. In this situation the energies of the terms can be expressed solely in terms of Δ_o. Then, with the two extremes established, one can consider intermediate cases by drawing a correlation diagram between the two situations. We shall illustrate this by considering two simple cases d^1 and d^2 electron configuration.

The 10 microstates for the d^1 configuration are represented by the ^2D $[(2S + 1) \times (2L + 1) = (2 \times 1/2 + 1) \times (2 \times 2 + 1) = 10; L = 2$ for D] spectroscopic term. In an O_h crystal field the d orbitals split giving rise to t_{2g} and e_g sets, the familiar CF splitting (see Fig. 6.5). Similarly, the ^2D term splits to give $^2T_{2g}$ and ^2E_g terms (see Section 6.3).

The way orbitals split under the influence of a crystal field, the term symbols split the same way.

In the strong field limit, only two discrete electronic configurations $(t_{2g})^1$ (lower energy) and $(e_g)^1$ (higher energy) results. Six microstates for

$(t_{2g})^1$ (the electron can be on any of the three orbitals with spin either up or down), corresponding to $^2T_{2g}$ term and four microstates for $(e_g)^1$ (the electron can be on any of the two orbitals with spin either up or down), corresponding to ^2E_g term. For the one-electron case, there is a direct correspondence between the electron configuration (d^1, $(t_{2g})^1$ or $(e_g)^1$) and the energy state because d^1 gives only ^2D and $(t_{2g})^1$ and $(e_g)^1$ give only $^2T_{2g}$ and ^2E_g terms, respectively. The splitting of the energy states is obtained by plotting the energies of the spectral terms (here it is only ^2D), as obtained from gas-phase atomic spectroscopy, on one vertical axis ($Dq = 0$) and the energies of the strong field configurations along the other vertical axis (Fig. 6.23(a)). The relative energies of these configurations can be expressed as the CFSE, $-4Dq$ for $(t_{2g})^1$ and $+6Dq$ for $(e_g)^1$. As the O_h crystal field is gradually increased on the ^2D term, the energies of $^2T_{2g}$ and ^2E_g diverge. From the other extreme, as we relax the strong field the energy separation of the states arising from $(t_{2g})^1$ and $(e_g)^1$ decreases. Since the same states arise from the two limiting cases, a correlation between the two is expected. If we draw lines, connecting the states, then we obtain a correlation diagram for d^1, or an energy-level splitting diagram. Figure 6.23 is called the Orgel diagram for d^1 electron configuration. It should be remembered that Orgel diagrams consider only states with the same spin multiplicity as that of the ground state and plots the energy levels of the states as CFSE. For d^1 case, the energy separation between $^2T_{2g}$ and ^2E_g states is $10Dq/\Delta_o$. Here only one electronic transition ($^2T_{2g} \rightarrow {}^2E_g$) is possible.

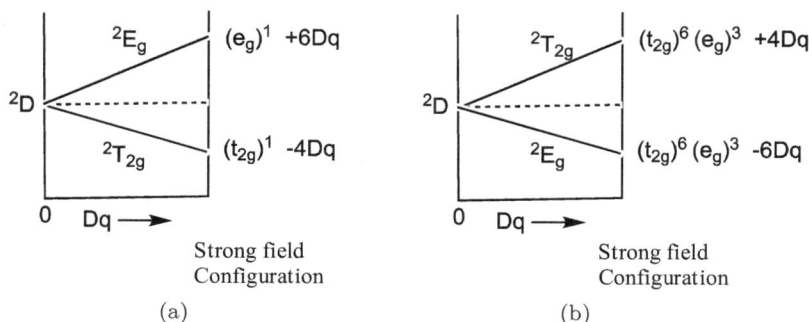

Fig. 6.23 Correlation diagrams for (a) d^1 configuration and (b) d^9 configuration in an octahedral field.

We know that the d^9 configuration gives only a ^2D term. This is a one-hole case $[(d^{10})^{e-} + (d^1)^{e+}]$. This is a consequence of *hole formalism*. The strong field configurations and the corresponding energy states are $(t_{2g})^6$ $(e_g)^3$, ^2E_g (lower energy; CFSE = $-6Dq$) and $(t_{2g})^5$ $(e_g)^4$, $^2T_{2g}$ (higher energy; CFSE = $+4Dq$). These correspond to those for d^1 case, except the energies are inverted (Fig. 6.23(b)).

A similar reasoning applies to hs d^6 [hs d^5 $(L = 0) + d^1$ i.e. effectively hs d^6 will behave as that of d^1 case] octahedral case. The free ion term ^5D splits in an octahedral field giving $^5T_{2g}$ and ^5E_g terms. The hs strong field configuration $(t_{2g})^4$ $(e_g)^2$ (CFSE = $-4Dq$) gives only $^5T_{2g}$ and hs $(t_{2g})^3$ $(e_g)^3$ (CFSE = $+6Dq$) gives only ^5E_g. The only difference is that now the states are quintets, while for d^1 the states are doublet. The same correlation diagram (Fig. 6.23(a)) can be used, with ^5D instead of ^2D. The splitting will be $^5T_{2g}$ instead of $^2T_{2g}$ and ^5E_g instead of ^2E_g. For this hs d^6 case also only one transition is possible ($^5T_{2g} \rightarrow {}^5E_g$) because the five electrons of one spin set are fixed and only the one of opposite spin is promoted.

Similarly, hs d^4 $[(d^6)^{e-} + (d^2)^{e+}]$ octahedral is the hole counterpart of hs octahedral d^6, and the only hs term is ^5D. In an octahedral field the free ion term ^5D splits giving ^5E_g $(t_{2g})^3$ $(e_g)^1$ and $^5T_{2g}$ $(t_{2g})^2$ $(e_g)^2$. Since ^5E_g is lower in energy, the splitting is opposite that of the hs d^6 case and the same as for d^9 (see above).

Let us now consider the situation in T_d crystal field. We have seen that the d orbital splitting diagram for T_d is opposite that for O_h (see Fig. 6.6). For d^1, $(e)^1$ (^2E) is lower in energy than $(t_2)^1$ ($^2T_2$). The correlation diagram corresponds to the left side of Fig. 6.24. Similarly, we find that d^9 and d^4 are represented by the right side for T_d and d^1 and d^6 by the left side. It should be remembered that in T_d geometry, g and u have no meaning.

Let us now consider the cases of O_h crystal field for d^2, d^3, hs d^5, hs d^7, and d^8 electronic configurations. We start with the case for d^2 electronic configuration. There are two terms of highest spin multiplicity ^3F and ^3P, and three strong field configurations increasing in energy in the order $(t_{2g})^2 < (t_{2g})^1$ $(e_g)^1 < (e_g)^2$. The ^3P term becomes $^3T_{1g}$ in an octahedral field, with no splitting. The ^3F term splits giving rise to $^3T_{1g}$, $^3T_{2g}$, and $^3A_{2g}$ terms (see Section 6.3). We obtain the order of energies for these states from the correlation diagram. To construct the correlation diagram,

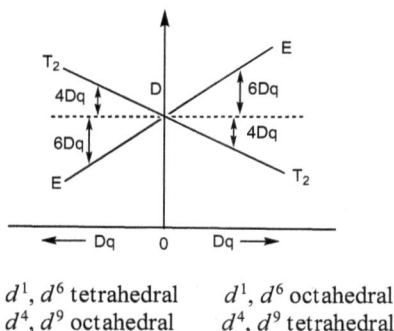

d¹, d⁶ tetrahedral   d¹, d⁶ octahedral
d⁴, d⁹ octahedral   d⁴, d⁹ tetrahedral

Fig. 6.24 Orgel diagram for d^1, d^9 and high-spin d^4, d^6 configurations.

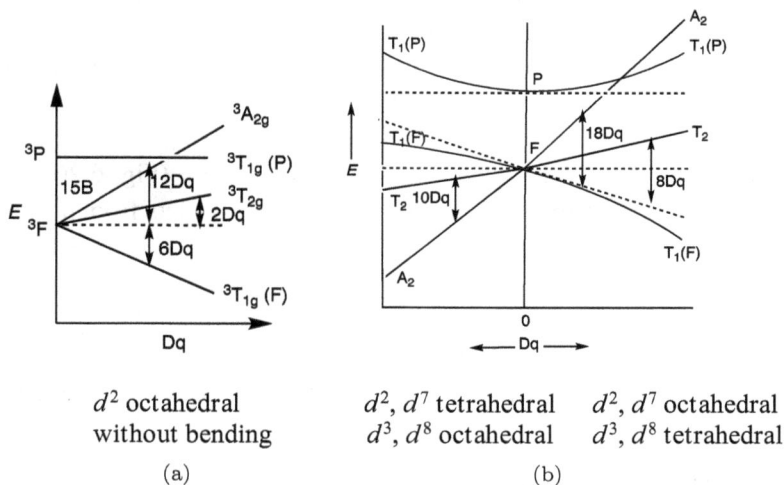

d^2 octahedral d^2, d^7 tetrahedral d^2, d^7 octahedral
without bending d^3, d^8 octahedral d^3, d^8 tetrahedral

(a)   (b)

Fig. 6.25 Orgel diagrams for (a) d^2 octahedral without bending and (b) d^2, d^3, high-spin d^7, and d^8 in O_h and T_d fields.

we must sort out the singlets and triplets arising from the strong field configurations. Let us consider the Orgel diagram for d^2 case as shown in
Fig. 6.25.

As we have seen for the d^1 and d^6 cases, the diagram applies also to
the hs d^7 (hs d^5 $(L = 0) + d^2$ i.e. effectively hs d^7 will behave as that of
d^2 case); the reverse splitting applies for d^3 [(hs d^5)$^{e-}$ + $(d^2)^{e+}$] and for

d^8 $[(d^{10})^{e-} + (d^2)^{e+}]$. The diagram Fig. 6.25 also applies to tetrahedral complexes.

For the general diagram (Fig. 6.25(b)), the spin multiplicities and g and u subscripts are dropped. In the diagram for d^3 and d^8 (O_h) and d^2 and d^7 (T_d), we find that the line for $T_1(F)$ (T_1 of F parentage), increases in energy with increasing Dq while that for $T_1(P)$ does not change with Dq. The extrapolation of the lines would cross (dotted lines) (on the left side of the diagram Fig. 6.25(b)). As Dq increases, some lines are curved on account of the mixing of terms of the same symmetry type because of configuration interaction. Terms of the same symmetry obey the *noncrossing rule*. The rule states that if the increasing ligand field causes two weak field terms of the same symmetry to approach, they do not cross but bend away from each other (Fig. 6.25(b)). On the right side of the diagram (Fig. 6.25(b)), the lines for the two T_1 states also bend away from one another at high field strength. In this diagram we consider only the states with the same spin multiplicity as the ground state, ignoring, for the present states of higher multiplicities.

Let us consider the correlation diagram for d^2 case (Fig. 6.26). From this diagram, we find the following correlations for triplets from d^2.

The free ion term ^3F splits to give $^3T_{1g}$, $^3T_{2g}$, and $^3A_{2g}$ in an octahedral crystal field. Where a state occurs only once ($^3T_{2g}$ and $^3A_{2g}$) we can obtain the relative energy of the state from the energy of the configuration from which it is derived. For example, $+2Dq$ for $^3T_{2g}$ from $(t_{2g})^1$ $(e_g)^1$ (CFSE = $-4Dq + 6Dq$) and $+12Dq$ for $^3A_{2g}$ from $(e_g)^2$ ($2 \times 6Dq$). The energy of

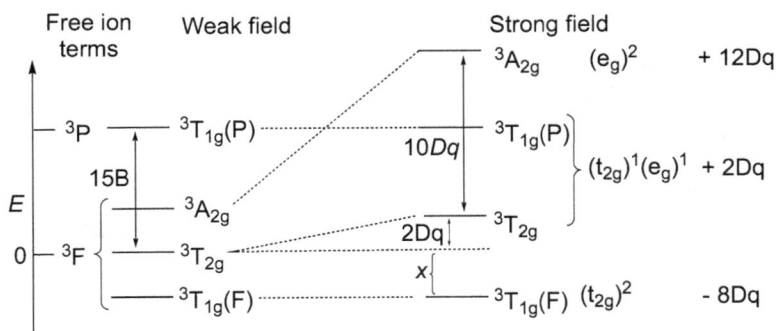

Fig. 6.26 The correlation diagram for d^2.

the third state $(^3T_{1g})$ from ^3F is derived from the unchanged "baricenter" rule, making use of the orbital degeneracy (the dimensionality of the representations, 1 for A and 3 for T terms):

$$1 \times 12Dq + 3 \times 2Dq + 3 \times xDq = 0$$
$$^3A_{2g} \qquad ^3T_{2g} \qquad ^3T_{1g}(F)$$
$$x = -6Dq \text{ for } ^3T_{1g}(F).$$

In the Orgel diagram these splittings are relative to the "unsplit" energy (dashed horizontal line). The energy separations shown are relative to the ground state, without allowing for bending (Fig. 6.26).

The energy separation between F and P terms for d^2, d^3, d^7, and d^8 is 15B, where B is interelectronic repulsion parameter (see Section 6.5). B is evaluated for the free ion. The value of B in complexes B′ is always less than the value for the free ion because of the decreased interelectronic repulsion resulting from electron delocalization (covalency). B′ is usually about 0.7B to 0.9B, depending on the extent of metal-ligand bond covalency. A reasonable value of B′ can be estimated from analysis of electronic spectra (see below). In the diagram (Fig. 6.26) B is interelectronic repulsion parameter for a free ion and B′ is for the d^2 ion in an octahedral crystal field.

b) Tanabe-Sugano diagrams

In Tanabe-Sugano (Japanese scientists Y. Tanabe and S. Sugano, 1954) diagrams the term energies of the levels of a d^n ion are plotted as the vertical coordinate (Y-axis) in units of the interelectronic repulsion parameter (Racah parameter) B and crystal field strength is the horizontal coordinate (X-axis) in units of Δ_o/B. Thus the term energies are expressed as E/B and the crystal field splitting Δ_o is expressed as Δ_o/B. These diagrams require two parameters B and C $(C \approx 4B)$ to describe the interelectronic repulsions for a d electron system. The diagrams can only be drawn when the ratio C/B is specified.

Example 6.24 The complex $[V(H_2O)_6]^{3+}$, doped into alumina (Al_2O_3), displays two CF transitions $^3T_{1g}(F) \rightarrow {}^3T_{2g}$ at 17 400 cm^{-1} (v_1) and $^3T_{1g}(F) \rightarrow {}^3T_{1g}(P)$ at 25 400 cm^{-1} (v_2). The transition v_1 corresponds to Δ_o. Calculate Δ_o and B using a Tanabe-Sugano diagram.

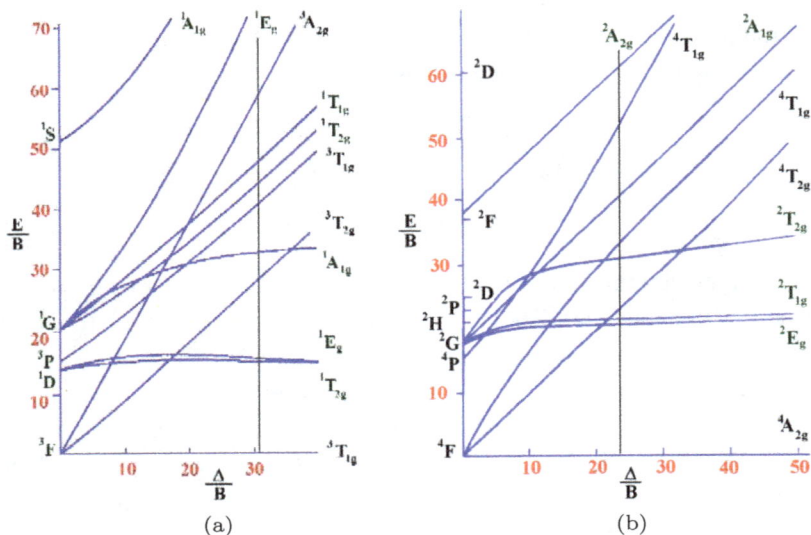

Fig. 6.27 Tanabe-Sugano diagram for (a) d^2 and (b) d^3 ion in an O_h field.

Answer From the given data, the ratio of v_2/v_1 is 1.46. According to the Tanabe-Sugano diagram for this d^2 complex $[V^{III}(H_2O)_6]^{3+}$ (Fig. 6.27(a)) the only point where this energy ratio is satisfied is with the value of $\Delta_0/B = 30.9$ (shown by a vertical line in Fig. 6.27(a)). Interpolation of the graph to find the Y-axis values for the spin-allowed transitions gives $v_1/B = 28.78$ and $v_2/B = 41.67$.

For v_1, $E/B = 28.78$, $E = 17\ 400\ \text{cm}^{-1} = 28.78B$, then $B = 17\ 400/28.78$; $B \simeq 605\ \text{cm}^{-1}$. The value of $\Delta_0/B = 30.9$. Hence, $\Delta_0 = 30.9B = 30.9 \times 605 \simeq 18\ 690\ \text{cm}^{-1}$. Similarly, for v_2, $E/B = 41.67$, $E = 25\ 400\ \text{cm}^{-1} = 41.67B$, $B \simeq 610\ \text{cm}^{-1}$. The value of B' is less than that of the free-ion (V(III): $860\ \text{cm}^{-1}$) because the expansion of d-electron charge on complexation reduces by 30% the interelectronic repulsions. This indicates a strong Nephelauxetic effect.

Example 6.25 The complex $[Cr(H_2O)_6]^{3+}$ displays two CF transitions $^4A_{2g} \rightarrow {}^4T_{2g}$ at $17\ 400\ \text{cm}^{-1}$ (v_1) and $^4A_{2g} \rightarrow {}^4T_{1g}$ at $24\ 500\ \text{cm}^{-1}$ (v_2). Calculate Δ_0 and B using a Tanabe-Sugano diagram.

Answer For this d^3 complex $[Cr^{III}(H_2O)_6]^{3+}$ the transition v_1 corresponds to Δ_0. First, we calculate the ratio of v_2/v_1, which is 1.412. The only point where this energy ratio is satisfied is with the value of $\Delta_0/B = 24.0$ (shown by arrows in Fig. 6.27(b)). Interpolation of the graph to find the Y-axis values for the spin-allowed transitions gives $E/B = v_1/B = 24.0$ and $E/B = v_2/B = 34.0$. Since $v_1 = 17\,400\,\text{cm}^{-1}$, for the first spin-allowed transition, $E/B = 17\,400/B = 24.0$, then $B = 17\,400/24.0$; $B = 725\,\text{cm}^{-1}$. For v_2, $E/B = 34.0$. Since $v_2 = 24\,500\,\text{cm}^{-1}$, for the second spin-allowed transition, $E/B = 24\,500/B = 34.0$, then $B = 24\,500/34.0$; $B \simeq 721\,\text{cm}^{-1}$.

The value of $\Delta_0/B = 24.0$. Hence, $\Delta_0 = 24B = 24 \times 721 \simeq 17\,300\,\text{cm}^{-1}$.

The value of B' is less than that of the free-ion (Cr(III): $1030\,\text{cm}^{-1}$) because the expansion of d-electron charge on complexation reduces by 30% the interelectronic repulsions. This indicates a strong nephelauxetic effect.

Following the equations provided by Dou[6] and Lever[7] for A_2 ground state ions the values of Δ_0 and B can also be obtained.

Example 6.26 The complex $[Ni(H_2O)_6]^{2+}$ displays three CF transitions $^3A_{2g}(F) \rightarrow {}^3T_{2g}(F)$ at $8700\,\text{cm}^{-1}$ (v_1), $^3A_{2g}(F) \rightarrow {}^3T_{1g}(F)$ at $14\,500\,\text{cm}^{-1}$ (v_2), and $^3A_{2g}(F) \rightarrow {}^3T_{1g}(P)$ at $25\,300\,\text{cm}^{-1}$ (v_3). Calculate Δ_0 and B using a Tanabe-Sugano diagram.

Answer For this d^8 complex $[Ni^{II}(H_2O)_6]^{2+}$ the transition v_1 corresponds to $\Delta_0 = 8700\,\text{cm}^{-1}$. Now we calculate the ratio of v_2/v_1, which is 1.74. The only point where this energy ratio is satisfied is with the value of $\Delta_0/B = 9.8$ (Fig. 6.28). Interpolation of the graph to find the Y-axis values for the spin-allowed transitions gives $E/B = v_2/B = 16.0$ and $E/B = v_3/B = 27.85$. Since $v_2/B = 16.0$, $B = v_2/16 = 14\,500/16 = 906\,\text{cm}^{-1}$. From the Tanabe-Sugano diagram $\Delta_0 \simeq 8900\,\text{cm}^{-1}$. The value of B' is less than that of the free-ion (Ni(II): $1040\,\text{cm}^{-1}$) because the expansion of d-electron charge on complexation reduces by 13% the interelectronic repulsions. This justifies the nephelauxetic effect.

---

[6]D. R. Brown and R. R. Pavlis, *J. Chem. Educ.* **1985**, *62*, 807.
[7](a) Y.-s. Dou, *J. Chem. Educ.* **1990**, *67*, 134. (b) A. B. P. Lever, *J. Chem. Educ.* **1968**, *45*, 711.

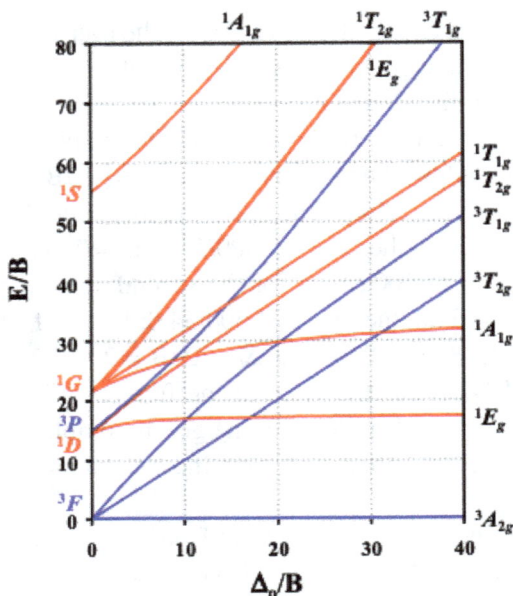

Fig. 6.28   The Tanabe-Sugano energy diagram for d^8 ion in an O_h field.

Example 6.27   The spin-allowed optical transitions for $[Co^{II}(H_2O)_6]$-$(CF_3SO_3)_2$ ($S = 3/2$) are observed at $8350\,cm^{-1}$ (1198 nm) due to $^4T_{1g}(F)$ $\rightarrow$ $^4T_{2g}(F)$ (ν_1), at $16\,000\,cm^{-1}$ (625 nm) due to $^4T_{1g}(F) \rightarrow {^4}A_{2g}(F)$ (ν_2), and at $19\,400\,cm^{-1}$ (shoulder) (515 nm) due to $^4T_{1g}(F) \rightarrow {^4}T_{1g}(P)$ (ν_3). Justify its orange appearance.

Answer   The lowest energy transition (ν_1) lies in the near-infrared region and is outside the experimental range (usually 200–1100 nm). The other two overlap in the visible region of the spectrum with the ν_2 absorption dominating the ν_3 (shoulder) one. These factors ultimately lead the orange appearance of $[Co^{II}(H_2O)_6](CF_3SO_3)_2$.

For d^4–d^7 ions the possibility exists for isolating both hs and ls complexes with appropriate ligands. The Tanabe-Sugano diagrams of these ions in O_h field have a vertical line, cutting X-axis at $\Delta_o \approx P$. The left of which presents the behavior of a hs situation and the right of which signifies ls situation. The case of d^6 and d^7 are displayed in Fig. 6.29(a) and Fig. 6.29(b), respectively.

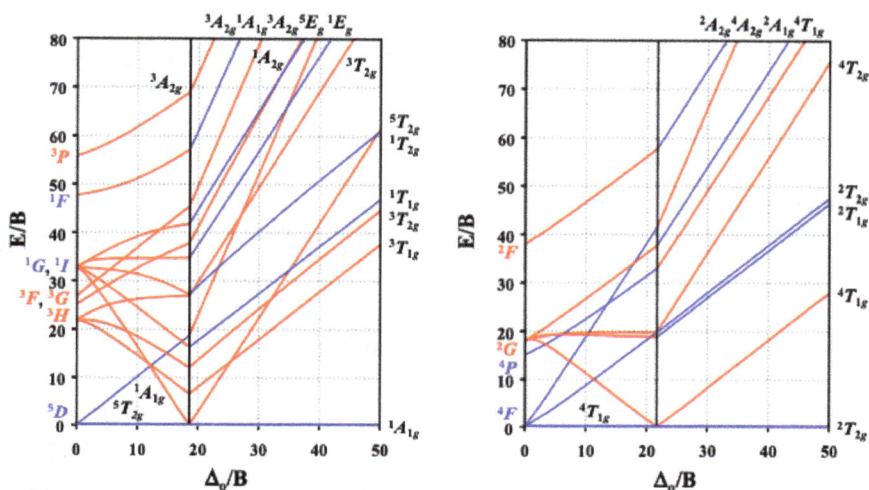

Fig. 6.29   The Tanabe-Sugano energy diagrams for d^6 and d^7 in O_h field.

Example 6.28   The spectra of ls d^6 complex $[Co^{III}(en)_3]^{3+}$ (en = 1,2-diaminoethane) displays two absorptions (CF transitions) $^1A_{1g} \rightarrow {}^1T_{1g}$ at 21 400 cm^{-1} (v_1) and $^1A_{1g} \rightarrow {}^1T_{2g}$ at 29 400 cm^{-1} (v_2). Calculate Δ_o and B using a Tanabe-Sugano diagram.

Answer   Let us consider the Tanabe-Sugano diagram for d^6 ion in O_h field (Fig. 6.29(a)). The ratio v_2/v_1 of these two energies from Y-axis is 1.37 for $\Delta_o/B = 40$. The only point where this energy ratio is satisfied is with the value of $\Delta_o/B = 38$ from X-axis. Now $38B = 21\ 400$ cm^{-1} and it gives B = 563 cm^{-1}. Again, $\Delta_o/B = 40$, hence $\Delta_o = 40B = 40 \times 563 = 22\ 520$ cm^{-1}. The value of B' is well below that of the free-ion (Co(III): 1400 cm^{-1}) because the expansion of d-electron charge on complexation reduces by 40% the interelectronic repulsions (nephelauxetic effect).

Following the equations provided by Dou[6] and Lever[7] for A_1 ground state ions the values of Δ_o and B can also be obtained.

For further discussion on electronic absorption spectra, see the advanced texts listed in the end of chapter reading list and in particular the book by Figgis.

Observed color/Color seen (Complementary color)	Color of light absorbed	Wavelength of light absorbed in nm (approx.)
Green	Purple	680–780
Bluish green	Red	620–680
Blue	Orange	580–620
Violet	Yellow	520–580
Purple	Green	500–520
Red	Bluish green	470–500
Orange	Blue	440–470
Yellow	Violet blue	420–440
Greenish yellow	Violet	380–420

The color that we see is the color that is not absorbed but is transmitted. The transmitted light is the complement of the absorbed light.

Color absorbed	Color observed
Violet	YellowGreen
Blue Violet	Yellow
Blue	Orange
BlueGreen	Red
Green	Purple
YellowGreen	Violet
Yellow	Blue Violet
Orange	Blue
Red	Bluegreen

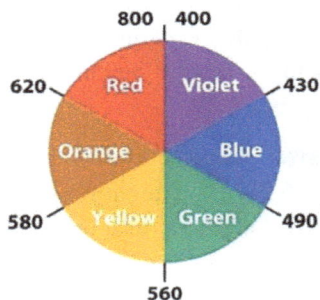

Numbers are in nm

Example 6.29 For hs $[Co(NH_3)_6]^{2+} \Delta_o = 10\ 100\ cm^{-1}$ and $B = 920\ cm^{-1}$. Given this information, how many electronic spectral transitions are expected and at what energies?

Answer For $[Co^{II}(NH_3)_6]^{2+}$ we should consider d^7 Tanabe-Sugano diagram (Fig. 6.29). We find that $\Delta_o/B \simeq 11$. According to Fig. 4B.x we expect three transitions in the UV-vis region: $^4T_{1g}(F) \rightarrow {}^4T_{2g}(E_1)$,

$^4T_{1g}(F) \rightarrow {}^4A_{2g}(E_2)$, $^4T_{1g}(F) \rightarrow {}^4T_{1g}(P)(E_3)$. In the order of increasing energy $E_1/B \simeq 10$, $E_2/B \simeq 18$, and $E_3/B \simeq 23$. Then, $E_1 = 10 \times 920\,\mathrm{cm}^{-1} = 9200\,\mathrm{cm}^{-1}$, $E_2 = 18 \times 920\,\mathrm{cm}^{-1} = 16\ 500\,\mathrm{cm}^{-1}$, and $E_3 = 23 \times 920\,\mathrm{cm}^{-1} = 21\ 200\,\mathrm{cm}^{-1}$.

Example 6.30 In the electronic spectrum of $[Ti(H_2O)_6]^{3+}$ the main absorption peak is at $20\ 300\,\mathrm{cm}^{-1}$. Calculate Δ_o for this complex in kJ mol^{-1}.

Answer We know that $1\,\mathrm{kJ\,mol}^{-1} = 83.7\,\mathrm{cm}^{-1}$.
  Then $20\ 300\,\mathrm{cm}^{-1}$ value of Δ_o is $20\ 300/83.7 = 242.5\,\mathrm{kJ\,mol}^{-1}$.

Example 6.31 In the electronic spectrum of $[Ti(H_2O)_6]^{3+}$ an absorption peak at $20\ 300\,\mathrm{cm}^{-1}$ with a shoulder is observed. Justify the observation.

Answer The complex $[Ti^{III}(H_2O)_6]^{3+}$ ion is a d^1 system. For d^1 perfect octahedral complex an electronic transition $^2T_{2g} \rightarrow {}^2E_g$ ($t^1_{2g} \rightarrow e^1_g$) is expected. Notably, for $[Ti^{III}(H_2O)_6]^{3+}$ both the ground state and the excited state are responsive to Jahn-Teller distortion. Ground state (t^1_{2g}) distortion leads to tetragonal compression (z-in) (Fig. 6.12). Excited state (e^1_g) distortion is of the same type. Thus, the spectrum of the ion shows the result of this splitting: $(dxy)^1 \rightarrow (dx^2 - y^2)^1$ (v_1) and $(dxy)^1 \rightarrow (dz^2)^1$ (v_2). This explains the origin of observing two transitions — a peak and a shoulder.

Example 6.32 The Δ_o values of $[Cr(H_2O)_6]^{3+}$ and $[Co(NH_3)_6]^{3+}$ are 17 400 cm^{-1} and 22 900 cm^{-1}, respectively. Predict the colors of the ions.

Answer We know, $E = hv$; $v = c/\lambda$; $\bar{v} = 1/\lambda$).
  For $[Cr^{III}(H_2O)_6]^{3+}$ $\lambda = 1/\bar{v} = 1/\Delta_o = 1/17\ 400\,\mathrm{cm}^{-1} \simeq 575 \times 10^{-7}$ cm $= 575$ nm; an absorption band is expected at 575 nm, which corresponds to the absorption of yellow light (see color wheel). The expected color is violet. However, the actual color is green. In this case an additional absorption occurs at $24\ 700\,\mathrm{cm}^{-1}$ ($\simeq 405$ nm). Hence, consideration of Δ_o alone the color for $[Cr^{III}(H_2O)_6]^{3+}$ cannot be predicted.

For $[Co^{III}(NH_3)_6]^{3+}$ $\lambda = 1/\bar{v} = 1/\Delta_o = 1/22\ 900\,\mathrm{cm}^{-1} \simeq 437 \times 10^{-7}$ cm $= 437$ nm; an absorption band is expected at ~ 440 nm, which corresponds to the absorption of blue light. The observed color of the complex is orange, as expected.

Example 6.33 Assuming that light of a specific wavelength is absorbed by a coordination complex, (i) predict the color of the complex as observed, upon absorption of 42.9 kcal per mole of light energy and (ii) predict the color of the visible light absorbed by the complex, if it is found that the frequency of light corresponding to the observed color of the complex is 5.3×10^{14} sec^{-1}.

Answer We know $E = N_o h\nu = N_o hc/\lambda = N_o hc\bar{\nu}$

(i) The energy absorbed per mol $= 42.9$ kcal $= 42.9 \times 10^3 \times 4.18 \times 10^7$ erg $= 42.9 \times 4.18 \times 10^{10}$ erg

    Then, $\lambda = N_o hc/E = 6.023 \times 10^{23} \times 6.626 \times 10^{-27}$ erg sec $\times 3.0 \times 10^{10}$ cm sec$^{-1}/42.9 \times 4.18 \times 10^{10}$ erg $= 6.68 \times 10^{-5}$ cm $= 668 \times 10^{-7}$ cm $= 668$ nm

    The color absorbed is red and the complex will be seen as green.

(ii) The frequency of light corresponding to the observed color $= 5.30 \times 10^{14}$ sec^{-1}

    Again, $\lambda = c/\nu = 3.0 \times 10^{10}$ cm sec$^{-1}/5.3 \times 10^{14}$ sec$^{-1} = 566 \times 10^{-7}$ cm $= 566$ nm.

    The color absorbed is violet, so the complex will appear yellow.

Example 6.34 In the electronic spectrum of CoF_6^{3-} two distinct peaks are observed. Account for this observation.

Answer For the $Co^{III}F_6^{3-}$ ion (both $^5T_{2g}$ and ^5E_g are Jahn-Teller active), the splitting of the two bands is sufficient to cause two distinct peaks (at $\sim$11 000 cm^{-1} and at $\sim$15 000 cm^{-1}) to be found in its absorption spectrum.

c) Charge transfer transitions

In addition to transitions between d-orbitals (d-d transitions), transitions from ligand-based orbitals to metal d-orbitals (ligand-to-metal) and metal d-orbitals to ligand-based orbitals (metal-to-ligand) are also possible. Such transitions are called charge-transfer in character, since quite a lot of charge transfer (electron transfer) occurs from the metal to the ligand orbital and vice versa. While for ligand-to-metal charge-transfer (LMCT) transition,

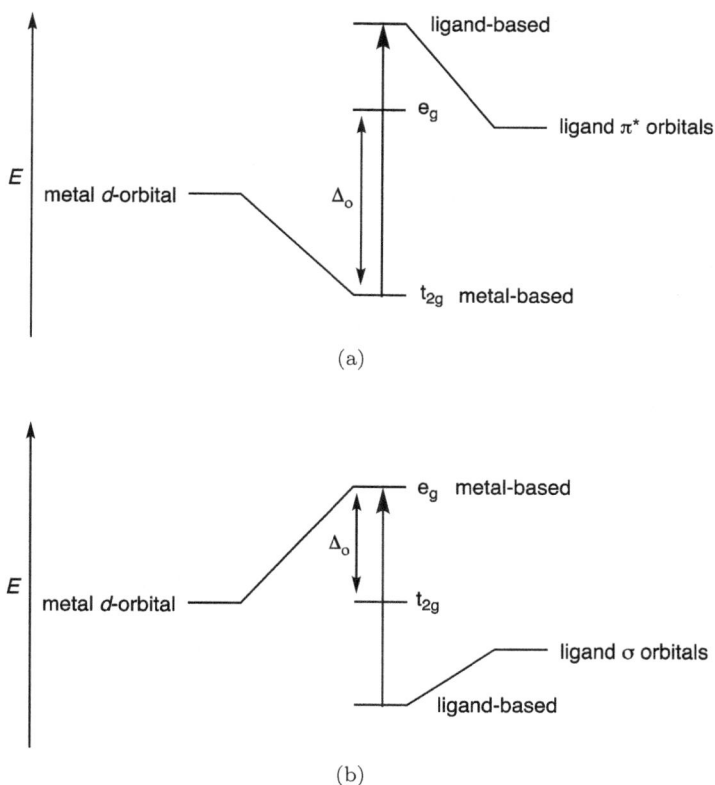

Fig. 6.30 Schematic representation of (a) MLCT and (b) LMCT transitions.

ligand is oxidizable and metal is reducible and for metal-to-ligand charge-transfer (MLCT) transition, metal is oxidizable and the ligand is reducible. Schematic representations of such transitions are shown in Fig. 6.30. These transitions are quite intense ($\varepsilon \approx 10^3$ to $10^6 \, M^{-1} \, cm^{-1}$) as they are both Laporte and spin-allowed. For charge-transfer (CT) transitions to occur there may or may not have d electron. As for example, even though $Mn^{VII}O_4^-$ does not have any d electron it is strongly colored (violet) and absorbs at 524 nm with $\varepsilon \approx 2240 \, M^{-1} \, cm^{-1}$. The Mn(VII) ion in MnO_4^- is surrounded by four oxide (O^{2-}) ions in a tetrahedral arrangement. Here electronic charge is transferred from O^{2-} to Mn(VII) ion and hence it is a case of LMCT transition (Fig. 6.30(b)).

Similarly, the intense orange color of acidic aqueous solutions of $Cr_2O_7^{2-}$ [$(O^{2-})_3Cr^{VI}$-O^{2-}-$Cr^{VI}(O^{2-})_3$]$^{2-}$ ion is due to transfer of charge from O^{2-} ions to Cr(VI) ion.

Intense red color of [Fe(bpy)$_3$]$^{2+}$ (bpy = 2,2'-bipyridine) is due to transfer of filled ls d^6 Fe(II) electron to low-lying (easily accessible) empty antibonding orbital (π^* LUMO) of bpy (MLCT transition) (Fig. 6.30(a)).

6.8 Molecular orbital treatment (theory)

The MO treatment for metal complexes is the same as for AB_n molecules, discussed in Chapter 3.

(a) ML_6 complexes (say, $M = Cr^0$, $L = CO$).

M–L σ-Bonding

We consider with only σ bonds. For the time being we disregard π bonding. To consider the MO energy level diagram for this ML_6 complex, we should consider which orbitals of the metal will be available for σ interaction with ligand group orbitals (LGO's) of six CO's (see Chapter 3).

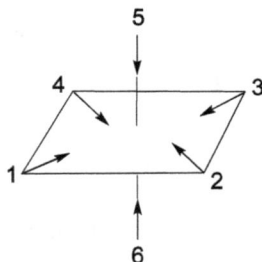

Let us consider now the effects of the symmetry operations of the O point group on the σ orbitals of an octahedral complex:

O	E	$6C_4^z$	$3C_2^z(=C_4^2)$	$8C_3$	$6C_2'$	
A_1	1	1	1	1	1	$x^2+y^2+z^2$
A_2	1	-1	1	1	-1	
E	2	0	2	-1	0	$(2z^2-x^2-y^2, x^2-y^2)$
T_1	3	1	-1	0	-1	$(R_x, R_y, R_z), (x, y, z)$
T_2	3	-1	-1	0	1	(xy, xz, yz)

O	E	$(C_4)^z$	$(C_2)^z$	C_3	C_2'	i
$\Gamma\sigma$	6	2	2	0	0	0

$$a_{A1} = 1/24[1 \times 6 \times 1 + 6 \times 2 \times 1 + 3 \times 2 \times 1 + 8 \times 0 \times 1 + 6 \times 0 \times 1]$$

$$= 1/24 \times 24 = 1$$

$$a_E = 1/24[1 \times 6 \times 2 + 6 \times 2 \times 0 + 3 \times 2 \times 2 + 8 \times 0 \times -1 + 6 \times 0 \times 0]$$

$$= 1/24 \times 24 = 1$$

$$a_{T1} = 1/24[1 \times 6 \times 3 + 6 \times 2 \times 1 + 3 \times 2 \times -1 + 8 \times 0 \times 0 + 6 \times 0 \times -1]$$

$$= 1/24 \times 24 = 1$$

$$a_{T2} = 1/24[1 \times 6 \times 3 + 6 \times 2 \times -1 + 3 \times 2 \times -1 + 8 \times 0 \times 0 + 6 \times 0 \times 1]$$

$$= 1/24 \times 0 = 0$$

The results of the transformation of the σ vectors for this complex are shown above. The representation for the σ vectors reduces to $A_1 + E + T_1$ in the O point group. Since there will be an A_{1g} LGO and the character for the center of inversion for Γ_σ is zero, the representations must be $A_{1g} + E_g + T_{1u}$ for O_h point group.

$$\Gamma_\sigma = A_1 + E + T_1(O)$$

$$\Gamma_\sigma = A_{1g} + E_g + T_{1g}(O_h)$$

M orbitals	Symmetry representation
s	A_{1g}
px, py, pz	T_{1u}
dz^2, dx^2-y^2	E_g
dxy, dxz, dyz	T_{2g}

$$L_{A1g}(s) = 1/\sqrt{6}(-p\sigma_1 - p\sigma_2 + p\sigma_3 + p\sigma_4 - p\sigma_5 + p\sigma_6) \text{ (Fig. 6.31)}$$

A_{1g} is the symmetry representation for s orbital. All six p orbitals contribute equally i.e. each has equal weightage to the MO. The ligand orbitals, with positive lobes (the sign of the wave function) are shaded and point to the positive direction of the axes, are written with a '+' sign.

Three p orbitals (px, py, and pz) of the metal are triply degenerate and their symmetry representation is T_{1u}. The LGO's for interacting with three *p* orbitals are as follows (Fig. 6.31).

$$L_{T1u}(px) = 1/\sqrt{2}(-p\sigma_1 - p\sigma_3)$$

$$L_{T1u}(py) = 1/\sqrt{2}(-p\sigma_2 - p\sigma_4)$$

$$L_{T1u}(pz) = 1/\sqrt{2}(+p\sigma_5 - p\sigma_6)$$

Now let us identify the LGO's to interact with metal $d_{x^2-y^2}$ and d_{z^2} orbitals.

$$L_{Eg}(dx^2-y^2) = 1/\sqrt{4}(-p\sigma_1 + p\sigma_2 + p\sigma_3 - p\sigma_4).$$

E_g is doubly degenerate and the symmetry representation for $d_{x^2-y^2}$ and d_{z^2} orbitals. From orbital interaction point of view they are Fig. 6.31(f) and (g), respectively.

We should remember that the total participation of a ligand orbital must be equal to 1. Keeping this in mind we can estimate the coefficients (x) of $p\sigma_1$, $p\sigma_2$, $p\sigma_3$ or $p\sigma_4$ as x and the coefficients (y) of $p\sigma_5$ or $p\sigma_5$ for symmetry matching interactions of LGO's with metal d_z^2 orbital.

LGO	(Coefficient of $p\sigma_1$)^2	(Coefficient of $p\sigma_5$)^2
L_{A1g} (s)	1/6	1/6
L_{T1u} (px)	1/2	0
L_{T1u} (py)	0	0
L_{T1u} (pz)	0	1/2
L_{Eg} (dx^2-y^2)	1/4	0
L_{Eg} (d_z^2)	x^2	y^2

Then, $x^2 + 1/6 + 1/2 + 1/4 = 1$; $x^2 = 1/12$; $x = 1/\sqrt{12}$

$y^2 + 1/6 + 1/2 = 1$; $y^2 = 2/6 = 4/12$; $y = 2/\sqrt{12}$

$$L_{Eg}(d_z^2) = 1/\sqrt{12}(p\sigma_1 + p\sigma_2 - p\sigma_3 - p\sigma_4 - 2p\sigma_5 + 2p\sigma_6)$$

The MO energy level diagram for ML$_6$ (only σ bonding) is displayed in Fig. 6.32.

The total number of valence electron for [Cr(CO)$_6$] is 6 (Cr^0) + 6 × 2 (each CO molecule contributes 2 electron) = 18. The electronic distribution: $(a_{1g})^2$ $(t_{1u})^6$ $(e_g)^4$ $(t_{2g})^6$. Number of bonding electron: 12; number of nonbonding electron: 6.

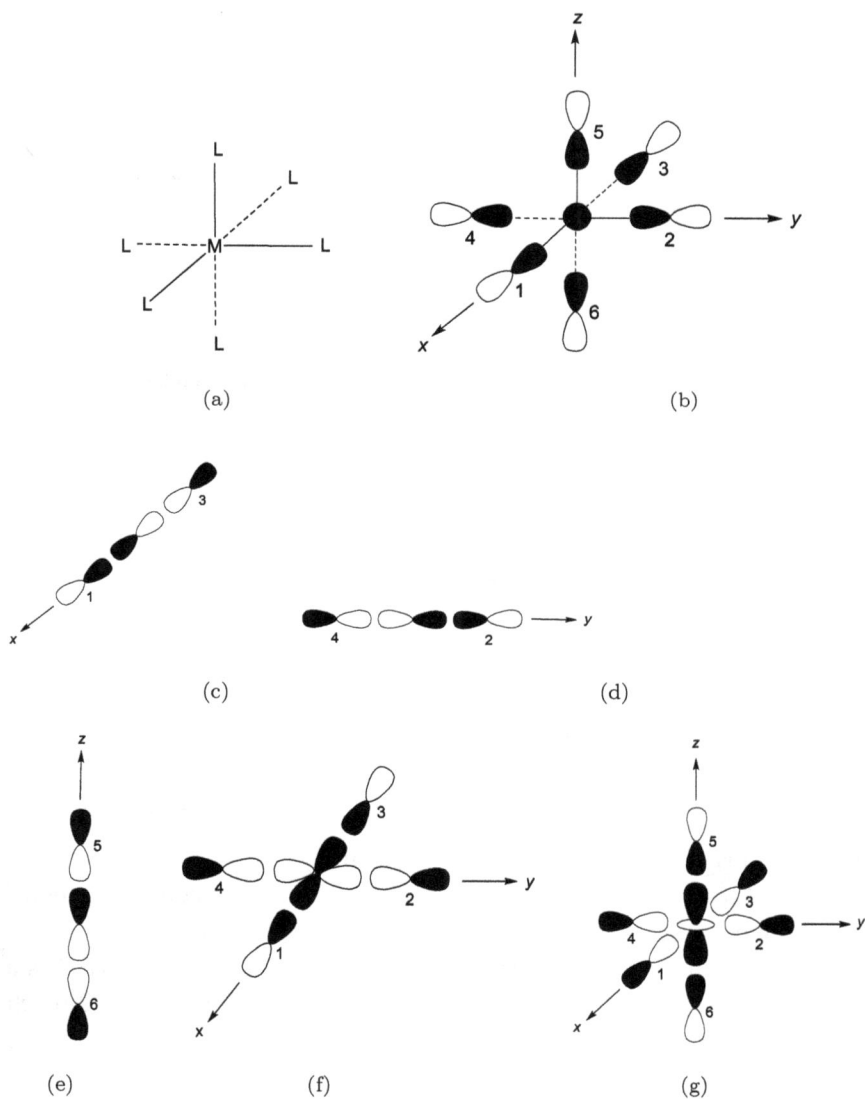

Fig. 6.31 The σ-bonding MO's (as LGO's) for octahedral ML$_6$.

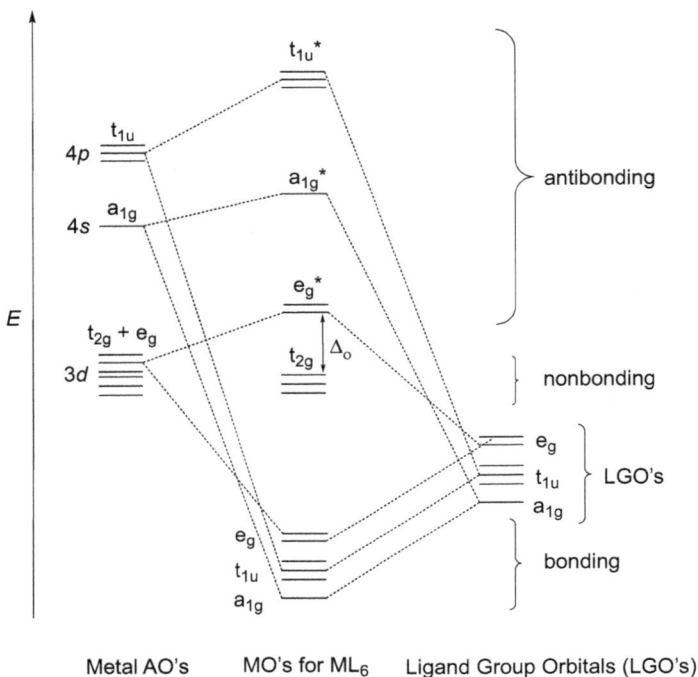

Fig. 6.32   MO energy level diagram for ML_6 (only σ bonding).

The total number of valence electron for $[Cr^{III}(H_2O)_6]^{3+}$ is 3 (Cr^{III}) + 6×2 (each H_2O molecule contributes 2 electron) = 15. The electronic distribution: $(a_{1g})^2 (t_{1u})^6 (e_g)^4 (t_{2g})^3$. Number of bonding electron: 12; number of nonbonding electron: 3; number of unpaired electron: 3.

The total number of valence electron for hs-$[Fe^{III}(H_2O)_6]^{3+}$ (H_2O is a weak field ligand and hence it is a hs complex) is 5 (Fe^{III}) + 6×2 (each H_2O molecule contributes 2 electron) = 17. The electronic distribution: $(a_{1g})^2 (t_{1u})^6 (e_g)^4 (t_{2g})^3 (e_g^*)^2$. Number of bonding electron: 12; number of nonbonding electron: 5; number of unpaired electron: 5.

The total number of valence electron for ls-$[Co(NH_3)_6]^{3+}$ (NH_3 is a strong field ligand and hence it is a ls complex) is 6 (Co^{III}) + 6×2 (each NH_3 molecule contributes 2 electron) = 18. The electronic distribution: $(a_{1g})^2 (t_{1u})^6 (e_g)^4$ (number of bonding electron: 12) and $(t_{2g})^6$ (number of nonbonding electron: 6). Number of unpaired electron: 0.

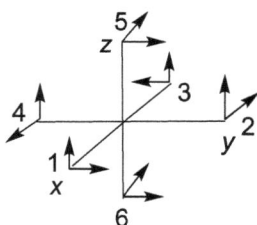

O	E	$6C_4^z$	$6C_2^z$	$8C_3$	$3C'_2$	i (O_h)
$\Gamma\pi$	12	0	0	0	-4	0

Fig. 6.33 Orientation of the vectors for π bonding.

M–L π-bonding and effects of π bonding

In this section we consider M–L π-bonding and its effect in an octahedral complex. The π orbitals of the six ligands in an octahedral complex can be represented by a set of vectors orientated as shown in Fig. 6.33. In this case, all 12 vectors belong to the same set since they can be interchanged by some operation of the group. The result of operating on the 12π vectors with one operation of each group for O is in a Table shown above (Fig. 6.33). The irreducible representations are $2T_1 + 2T_2$ (see below). The character is zero for i (O_h), so the representations are $T_{1g} + T_{2g} + T_{1u} + T_{2u}$. Since there are no metal t_{1g} or t_{2u} orbitals, these LGO's must be nonbonding. The t_{1u} (p) metal orbitals are used for σ bonding and since they give better σ overlap, they are expected to participate particularly in σ bonding. The t_{2g} orbitals are more important for π bonding. The t_{1u} and t_{2g} MO's are illustrated in Fig. 6.34, as combinations of the metal and LGO's. Sketches describing the LCAO's are obtained easily using the metal orbitals as templates. Ligand π orbitals are drawn in with the signs matching those of the metal orbital. The ligand orbitals are shown as p orbitals; they can be the π^* orbitals of a ligand such as CN^-.

We consider irreducible representations for five d orbitals.

$$a(A_1) = 1/24[1 \times 12 \times 1 + 6 \times 0 \times 1 + 6 \times 0 \times 1 + 8 \times 0 \times 1 + 3 \times -4 \times 1]$$

$$= 1/24 \times 0 = 0$$

$$a(E) = 1/24[1 \times 12 \times 2 + 6 \times 0 \times 0 + 6 \times 0 \times 0 + 8 \times 0 \times -1 + 3 \times -4 \times 2]$$
$$= 1/24 \times 0 = 0$$
$$a(T_1) = 1/24[1 \times 12 \times 3 + 6 \times 0 \times 1 + 6 \times 0 \times -1 + 8 \times 0 \times 0 + 3 \times -4 \times -1]$$
$$= 1/24 \times 48 = 2$$
$$a(T_2) = 1/24[1 \times 12 \times 3 + 6 \times 0 \times -1 + 6 \times 0 \times 1 + 8 \times 0 \times 0 + 3 \times -4 \times -1]$$
$$= 1/24 \times 48 = 2$$

From the character table it is obvious that,

$$\Gamma_\pi = 2T_1 + 2T_2(O)$$

$$\text{Then, } \Gamma_\pi = T_{1g} + T_{2g} + T_{1u} + T_{2u}(O_h).$$

Metal-ligand π bonding can involve either partner as the π donor (the ligands such as Cl^-, O^{2-}, OH^- etc.). If the metal is the donor, the metal t_{2g} electrons go into the bonding t_{2g} MO's. If the metal is the acceptor, the ligand electrons fill the bonding t_{2g} MO's and any metal t_{2g} electrons must go into the antibonding $t_{2g}*$ orbitals. Metals early in each transition series and those in high oxidation states have few d electrons and can serve as π acceptors. Ligands such as F^-, Cl^-, OH^- and O^{2-} are good π donors. Since 10Dq (Δ_o) is the separation between the t_{2g} orbitals occupied by metal electrons (this is now $t_{2g}*$) and the metal e_g orbitals (really e_g* with respect to the σ bonding), ligand-to-metal π donation decreases Δ_o (Fig. 6.34). If the metal has filled, or nearly filled, t_{2g} orbitals and the ligands have empty low-energy orbitals, then metal-to-ligand donation is favored. Metals late in each transition series and those in low oxidation states can serve as π donors with ligands having empty low-energy orbitals, such as the d orbitals of P or S or the $\pi*$ orbitals of the CN^-, CO, NO^+, 2,2'-bipyridine (bipy) or 1,10-phenanthroline (phen). Metal-to-ligand π bonding increases the M–L bond order and bond strength. It increases 10Dq (Δ_o), and in the case of $\pi*$ acceptors such as CN^-, it lowers the bond order for the C–N bond. Since CO is a weak Lewis base, the π acceptor role is important in placing CO high in the spectrochemical series. The lowering of CO stretching frequency has been used as an indication of the extent of M–L π bonding.

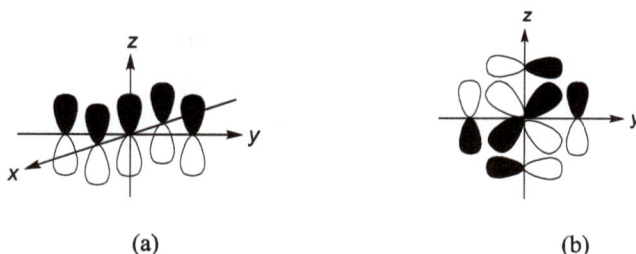

(a)

(b)

one of the t_{1u} π orbitals (*pz* orbital) other two identical orbitals are along the *x* (*px* orbital) and *y* (*py* orbital) directions

one of the t_{2g} orbitals (*dyz* orbital) other two identical orbitals are in the *xy* (*dxy* orbital) and *xz* (*dxz* orbital)

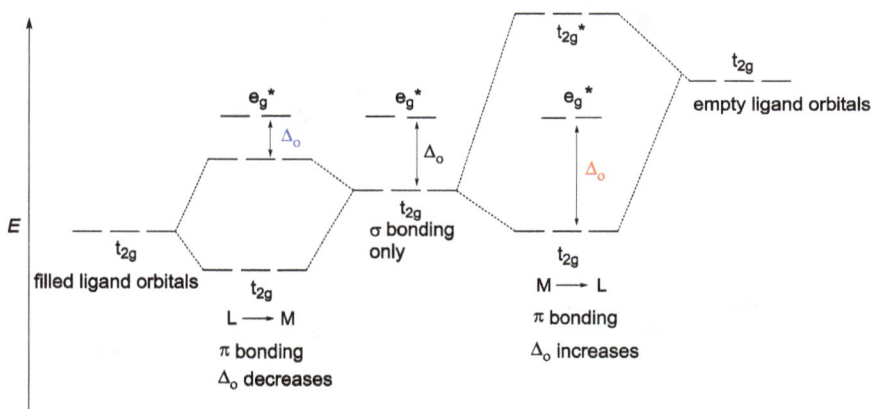

Fig. 6.34 Effects of π bonding using (a) filled low-energy π ligand-orbitals for L → M donation and (b) empty ligand-orbitals of π symmetry for M → L donation.

(b) ML$_4$ complex (square planar geometry)

Let us now consider another important structure in transition metal chemistry. It is the square planar arrangement of ligands about a metal. Sketches of the orbital interaction between relevant metal *d* AO's and matched LGO's are displayed in Fig. 6.35. To describe the MO energy level diagram for this geometry we start with reducible representation ($\Gamma\sigma$) associated with σ bonding. The σ orbitals of the four ligands can be represented by a set of vectors orientated as shown below.

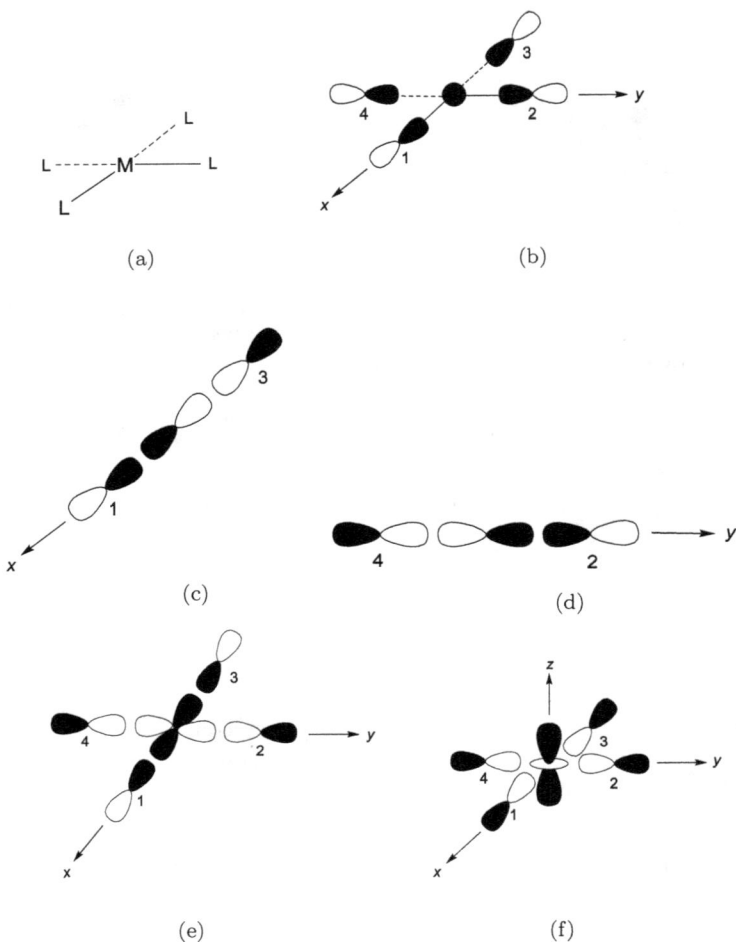

Fig. 6.35 Orbital interactions (only σ bonding).

The result of operating on the 4σ vectors for D_{4h} is shown below.

D_{4h}	E	C_4	C_2	C'_2	C''_2	i	S_4	σ_h	σ_v	σ_d
$\Gamma\sigma$	4	0	0	2	0	0	0	4	2	0

From the character table for D_{4h} point group (Table 3.5), it is obvious that the irreducible representations for reducible representation ($\Gamma\sigma$)

are A_{1g} $(s, dz^2) + B_{1g}$ $(dx^2-y^2) + E_u$ (px, py). Sketches describing the LCAO's are obtained easily using the metal orbitals as templates. Ligand π orbitals are drawn in with the signs matching those of the metal orbital.

$$L_{A1g}(s) = 1/\sqrt{4}(-p_{x1} - p_{y2} + p_{x3} + p_{y4})$$
$$= 1/2(-p\sigma_1 - p\sigma_2 + p\sigma_3 + p\sigma_4)$$

A_{1g} is the symmetry representation for s orbital. All four p orbitals contribute equally i.e. each has equal weightage to the MO. The ligand orbitals, with positive lobes (the sign of the wave function) are shaded and point to the positive direction of the axes, are written with a '+' sign.

Two p orbitals (p_x and p_y) of the metal are triply degenerate and their symmetry representation is E_u. LGOs for interacting with three p orbitals are as follows.

$$L_{Eu}(px) = 1/\sqrt{2}(-p_{x1} - p_{x3})$$
$$L_{Eu}(py) = 1/\sqrt{2}(-p_{y2} - p_{y4})$$
$$L_{B1g}(dx^2-y^2) = 1/\sqrt{4}(-p_{x1} + p_{y2} + p_{x3} - p_{y4})$$
$$= 1/2(-p\sigma_1 + p\sigma_2 + p\sigma_3 - p\sigma_4)$$
$$L_{A1g}(dz^2, s) = 1/\sqrt{4}(p_{x1} + p_{y2} - p_{x3} - p_{y4})$$
$$= 1/2(p\sigma_1 + p\sigma_2 - p\sigma_3 - p\sigma_4)$$

Now let us consider reducible representation (π) associated with π bonding. The result of operating on the four vectors for D_{4h} point group (see Table 2.3) is shown below.

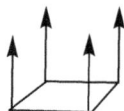

D_{4h}	E	C_4	C_2	C'_2	C''_2	i	S_4	σ_h	σ_v	σ_d
$\Gamma\pi$(II)	4	0	0	-2	0	0	0	-4	2	0

$$a(A_{1g}) = 1/16[1 \times 4 \times 1 + 2 \times 0 \times 1 + 1 \times 0 \times 1 + 2 \times -2 \times 1$$
$$+ 2 \times 0 \times 1 + 1 \times 0 \times 1 + 2 \times 0 \times 1 + 1 \times -4 \times 1$$
$$+ 2 \times 2 \times 1 + 2 \times 0 \times 1] = 1/16 \times 0 = 0$$

$$a(A_{2g}) = 1/16[1 \times 4 \times 1 + 2 \times 0 \times 1 + 1 \times 0 \times 1 + 2 \times -2 \times -1$$
$$+ 2 \times 0 \times -1 + 1 \times 0 \times 1 + 2 \times 0 \times 1 + 1 \times -4 \times 1$$
$$+ 2 \times 2 \times -1 + 2 \times 0 \times -1] = 1/16 \times 0 = 0$$

$$a(B_{1g}) = 1/16[1 \times 4 \times 1 + 2 \times 0 \times -1 + 1 \times 0 \times 1 + 2 \times -2 \times 1$$
$$+ 2 \times 0 \times -1 + 1 \times 0 \times 1 + 2 \times 0 \times -1 + 1 \times -4 \times 1$$
$$+ 2 \times 2 \times 1 + 2 \times 0 \times -1] = 1/16 \times 0 = 0$$

$$a(B_{2g}) = 1/16[1 \times 4 \times 1 + 2 \times 0 \times -1 + 1 \times 0 \times 1 + 2 \times -2 \times -1$$
$$+ 2 \times 0 \times 1 + 1 \times 0 \times 1 + 2 \times 0 \times -1 + 1 \times -4 \times 1$$
$$+ 2 \times 2 \times -1 + 2 \times 0 \times 1] = 1/16 \times 0 = 0$$

$$a(E_g) = 1/16[1 \times 4 \times 2 + 2 \times 0 \times 0 + 1 \times 0 \times -2 + 2 \times -2 \times 0$$
$$+ 2 \times 0 \times 0 + 1 \times 0 \times 2 + 2 \times 0 \times 0 + 1 \times -4 \times -2$$
$$+ 2 \times 2 \times 0 + 2 \times 0 \times 0] = 1/16 \times 16 = 1$$

$$a(A_{1u}) = 1/16[1 \times 4 \times 1 + 2 \times 0 \times 1 + 1 \times 0 \times 1 + 2 \times -2 \times 1$$
$$+ 2 \times 0 \times 1 + 1 \times 0 \times -1 + 2 \times 0 \times -1 + 1 \times -4 \times -1$$
$$+ 2 \times 2 \times -1 + 2 \times 0 \times -1] = 1/16 \times 0 = 0$$

$$a(A_{2u}) = 1/16[1 \times 4 \times 1 + 2 \times 0 \times 1 + 1 \times 0 \times 1 + 2 \times -2 \times -1$$
$$+ 2 \times 0 \times -1 + 1 \times 0 \times -1 + 2 \times 0 \times -1 + 1 \times -4 \times -1$$
$$+ 2 \times 2 \times 1 + 2 \times 0 \times 1] = 1/16 \times 16 = 1$$

$$a(B_{1u}) = 1/16[1 \times 4 \times 1 + 2 \times 0 \times -1 + 1 \times 0 \times 1 + 2 \times -2 \times 1$$
$$+ 2 \times 0 \times -1 + 1 \times 0 \times -1 + 2 \times 0 \times 1 + 1 \times -4 \times -1$$
$$+ 2 \times 2 \times -1 + 2 \times 0 \times 1] = 1/16 \times 0 = 0$$

$$a(B_{2u}) = 1/16[1 \times 4 \times 1 + 2 \times 0 \times -1 + 1 \times 0 \times 1 + 2 \times -2 \times -1$$
$$+ 2 \times 0 \times 1 + 1 \times 0 \times -1 + 2 \times 0 \times 1 + 1 \times -4 \times -1$$
$$+ 2 \times 2 \times 1 + 2 \times 0 \times -1] = 1/16 \times 16 = 1$$

$$a(E_u) = 1/16[1 \times 4 \times 2 + 2 \times 0 \times 0 + 1 \times 0 \times -2 + 2 \times -2 \times 0$$
$$+ 2 \times 0 \times 0 + 1 \times 0 \times -2 + 2 \times 0 \times 0 + 1 \times -4 \times 2$$
$$+ 2 \times 2 \times 0 + 2 \times 0 \times 0] = 1/16 \times 0 = 0$$

From D_{4h} character table (see Table 2.3) it is obvious that,

$$\Gamma_\pi(\parallel) = A_{2u} + B_{2u} + E_g$$

Checking the representations of the orbitals of M, we see that of the parallel set (in-plane π bonding) A_{2u} and E_g are of the proper symmetry for π bonding and B_{2u} is nonbonding. For the perpendicular set (out of plane π bonding), B_{2g} (M dxy orbital) and E_u (M px and py orbitals) have proper symmetry for π bonding and A_{2g} is nonbonding. The bonding and nonbonding orbitals are sketched from the LCAO's in Fig. 6.36. The non-bonding orbitals are sketched from their symmetry properties as given in the character table. The pz, dxy, and (dxz, dyz) orbitals on the central atom were found to be σ nonbonding so these can participate only in π bonding. The d orbitals give very good overlap. The (px, py) orbitals have proper symmetry for both σ and π bonding. The p orbitals provide much better overlap for σ bonding than for the sidewise π interaction, so they are expected to participate primarily in σ bonding.

Here we have dealt with the π-type orbitals on the ligands as vectors. These could be p orbitals, or in the case of ligands such as CN^-, π^* orbitals. The π bonding is important in many square planar $[M(CN)_4]^{n-}$ complexes involving donation from the filled M orbitals into the empty π^* orbitals of CN^-.

The MO energy level diagram for ML_4 (D_{4h}; $\sigma + \pi$ bonding) is displayed in Fig. 6.37.

6.9   Metal-metal bonding

As orbital interaction between atoms from the viewpoint of energy-matching and symmetry-matching lead to MO's forming σ and π bonds, then conceptually one can think of a δ bond, involving interactions between appropriate d orbitals.[8] In this section we will discuss metal-metal bonding concepts. The presence of M–M bonding and the essence of orbital interactions are shown in Fig. 6.38 and Fig. 6.39, respectively. The overlap of the metal d_{z^2} orbitals leads to σ-bond formation, while d_{xz} and d_{yz} orbitals on two metals overlap giving rise to a degenerate pair of

---

[8]F. A. Cotton, *J. Chem. Educ.* **1983**, *60*, 713.

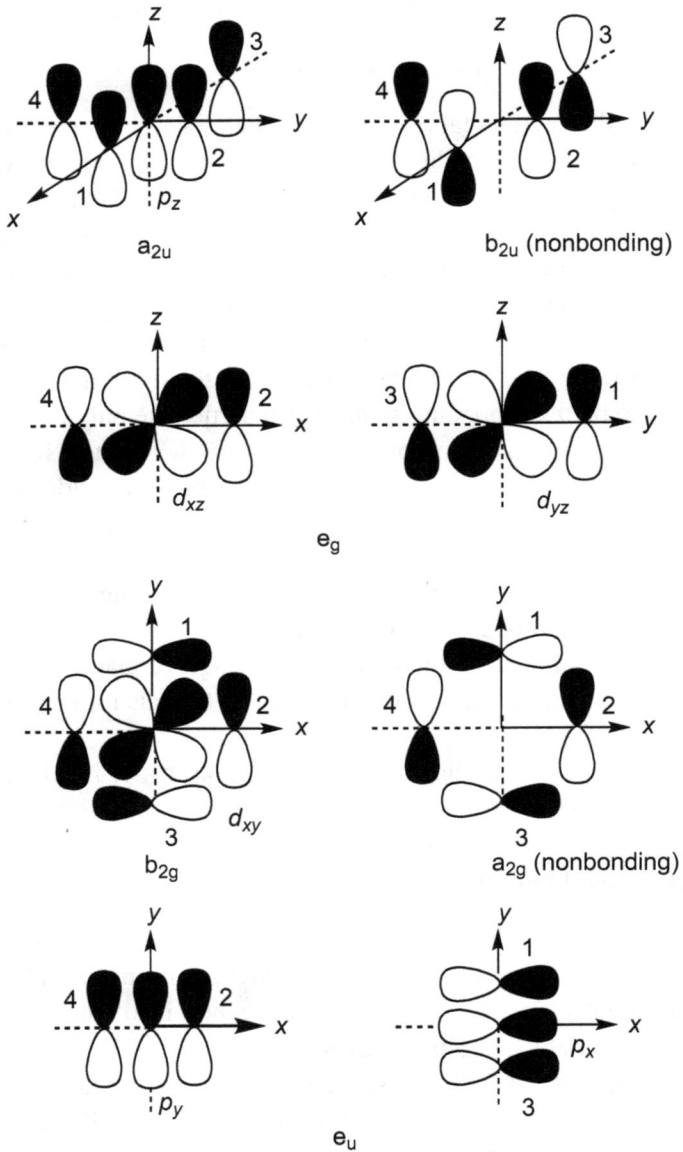

Fig. 6.36 Bonding antibonding MO's for ML_4 (D_{4h}).

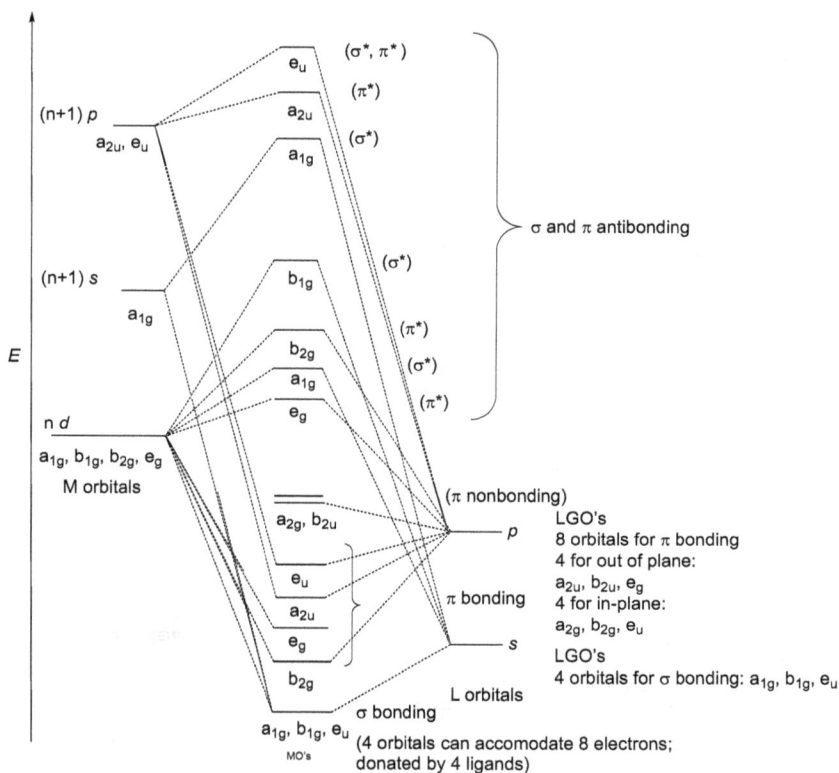

Fig. 6.37 MO energy level diagram for ML$_4$ (D$_{4h}$; $\sigma + \pi$ bonding).

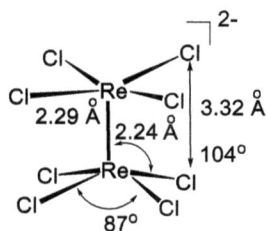

Fig. 6.38 ChemDraw* structure of [Re$_2$Cl$_8$]$^{2-}$ ion.

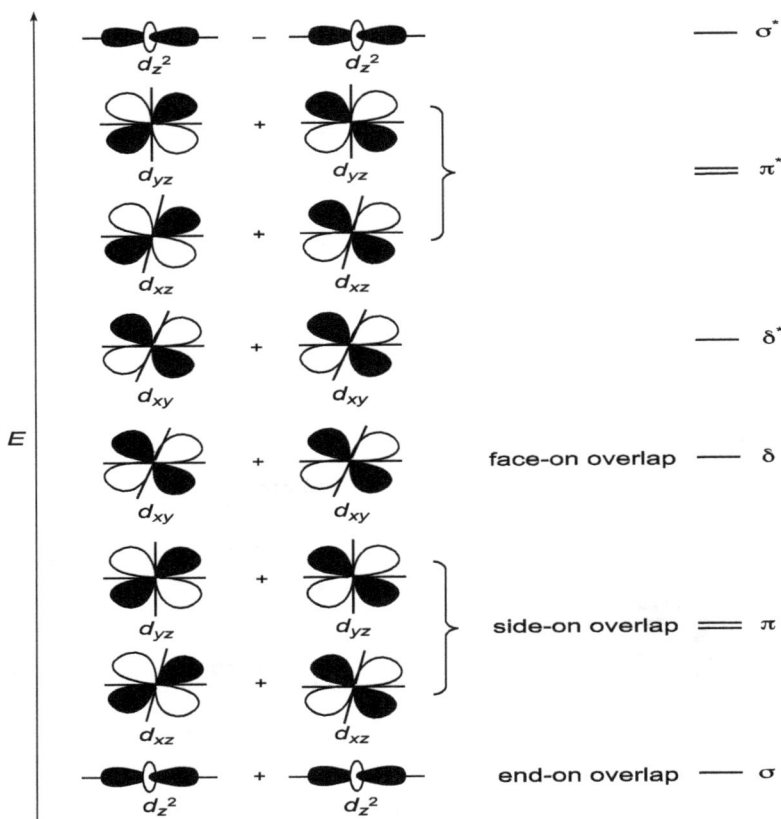

Fig. 6.39 The d-orbital overlap scheme for the formation of metal-metal bond between two MX_4 units.

π-orbitals. Finally, overlap of the two d_{xy} orbitals gives rise to a δ-bond. The degree of overlap follows the order $\sigma > \pi > \delta$. Fig. 6.39 also shows an approximate energy-level diagram for the σ, π, δ, δ^*, π^*, σ^* (in the order of increasing energy) MO's.

Let us consider the well-known compound $[Re_2^{III}Cl_8]^{2-}$ ion (Fig. 6.38). The four Cl^- ligands to each Re(III) ion are coordinated in a square fashion in each $ReCl_4^-$ ion, using the metal dx^2-y^2 orbitals. This causes dx^2-y^2 orbitals to go high in energy, as they become mainly Re–Cl antibonding in character. Two such $ReCl_4^-$ units form $[Re_2^{III}Cl_8]^{2-}$ ion. It is clear that we are left with four bonding orbitals - one σ, two π, and one δ and four corresponding antibonding orbitals - one σ^*, two π^*, and one δ^*. In $[Re_2^{III}Cl_8]^{2-}$ ion we have 8 metal electrons to accommodate in these bonding and antibonding orbitals.

Thus, when only eight electrons ($\sigma^2\pi^4\delta^2$ configurations) are present we have quadrupole bonds, for six electrons ($\sigma^2\pi^4$ configurations) we have triple bonds, for four electrons ($\sigma^2\pi^2$ configurations) we have double bonds (also for twelve electron system). Addition of electrons beyond the eight needed for a quadruple bond gives rise to triple bonds (also for six and ten electron systems), and bonds of still lower order, as we move toward the right side of the transition series. Thus, $\sigma^2\pi^4\delta^2\delta^{*2}$ configuration also provides a triple bond, and for the Rh_2^{4+} and Pt_2^{6+} species the $\sigma^2\pi^4\delta^2\delta^{*2}\pi^{*4}$ configurations give rise to net single bonds. Obviously, for sixteen electron systems no M–M bond is expected.

Throughout this range of compounds, bonds of half-integral order are also quite common, an example being the $[Rh_2(O_2CCF_3)_4]^+$ ion with thirteen electrons ($\sigma^2\pi^4\delta^2\delta^{*2}\pi^{*3}$), with a bond order of 1.5. The number of electrons present controls the bond order or in other words the number of M–M bonds.

Further reading

R. L. Dutta, *Inorganic Chemistry: Part-I Principles*, 6th edition, The New Book Stall, Kolkata (2009)

C. E. Housecroft and A. G. Sharpe, *Inorganic Chemistry*, 2nd edition, Person Education Ltd. (2005)

D. F. Shriver and P. W. Atkins, *Inorganic Chemistry*, 3rd edition, Oxford University Press (1999)

J. D. Lee, *Concise Inorganic Chemistry*, 5th edition, Blackwell Science Limited (1996)

J. E. Huheey, E. A. Keiter, and R. L. Keiter, *Inorganic Chemistry: Principles of Structure and Reactivity*, 4th edition, Addison-Wesley Publishing Company (1993)

W. L. Jolly, *Modern Inorganic Chemistry*, 2nd edition, McGraw-Hill Inc., McGraw-Hill International Editions Chemistry Series (1991)

F. A. Cotton and G. Wilkinson, *Basic Inorganic Chemistry*, Wiley Eastern Limited (1988)

K. F. Purcell and J. C. Kotz, Inorganic Chemistry, Saunders Golden Sunburst Series, W. B. Saunders Company, Holt-Saunders Japan (1985)

B. E. Douglas, D. H. McDaniel, and J. J. Alexander, Concepts and Models of Inorganic Chemistry, 2nd edition, John Wiley & Sons, Inc. (1983)

B. E. Douglas and C. A. Hollingsworth, *Symmetry in Bonding and Spectra — An Introduction*, Academic Press, Inc. (1985)

B. N. Figgis, *Introduction to Ligand Fields*, Wiley Eastern Limited (1976)

C. J. Ballhausen, *Introduction to Ligand Field Theory*, McGraw-Hill book Company, Inc. (1962)

O. Kahn, Molecular Magnetism, VCH (1993)

R. L. Carlin, *Magnetochemistry*, Springer-Verlag (1986)

Exercises

6.1 Both *trans*-$[Co(NH_3)_4Cl_2]^+$ and $[Cu(NH_3)_6]^{2+}$ have tetragonally distorted octahedral structure. Explain.

6.2 Although d^8 tetrahedral complexes are somewhat stabilized by flattening of tetrahedron, similar but opposite statements apply to d^9 tetrahedral complexes. Explain by consideration of energy level diagram.

6.3 Consider the molecule ML_6, which has O_h symmetry. Changing only one or more M–L bond lengths show how one reduces the symmetry to C_{4v}, D_{2h}, C_{3v}, and C_{2v}.

6.4 The spin-only magnetic moment of an octahedral Co(II) complex is 3.87 μ_B. Write its electronic configuration and calculate the CFSE involved.

6.5 Determine whether $MnFe_2O_4$ is normal or inverse spinel.

6.6 Explain the fact that $[CoCl_4]^{2-}$ is blue in color while $[Co(H_2O)_6]^{2+}$ is pink.

6.7 Discuss MO diagram for a tetrahedral ML_4 complex (σ bonding).

6.8 The $[Mn(H_2O)_6]^{2+}$ ion in its absorption spectrum exhibits many very weak sharp peaks. Explain.

Chapter 7

Reactions of *d*-Block Complexes

This chapter is devoted to discussions of the reactions of transition metal complexes. The rates (kinetics) and mechanisms of two most important reaction types of inorganic complexes: substitution and electron transfer (redox) reactions are the focus of this discussion. Mechanistic understanding of chemical reactions drives research and guides teaching of reactivity in chemistry.

7.1 Labile and inert metal ions

Complexes that are thermodynamically unstable but survive for long periods (at least a minute) in solution are called *inert*. Complexes that undergo more rapid equilibration are called *labile*.

General classification:

$$[M^{III}(H_2O)_6]^{3+} + H_2O^* \xrightarrow{k} [M^{III}(H_2O)_{6-n}(H_2O^*)_n]^{3+} + H_2O$$

The rates of such reactions can be followed by using isotopic tracers.

Rate law (rate of disappearance of $[M^{III}(H_2O)_6]^{3+}$):

$$-d/dt\,[M^{III}(H_2O)_6]^{3+} = k\,[M^{III}(H_2O)_6^{3+}]\,[H_2O^*]$$

$[H_2O^*]$ is constant, as it is present in large excess, $k \simeq 10^{-6}$ to $10^{10}\ \mathrm{s}^{-1}$.

Remembering, 10^{10} is diffusion-controlled limit (diffusion-controlled reactions are those in which the reaction rate is equal to the rate of transport of the reactants through the reaction medium, usually a solution).

Lifetime / residence time of coordinated H_2O: $k^{-1} \simeq \tau = 10^{-10}$ to $10^6\ \mathrm{s}$

251

Consideration of the characteristic rate constants (k) and hence life-times (τ) for the exchange of water molecules in octahedral complexes of the important aqua metal ions, the following generalizations are made that help us to differentiate between the *labile* and the *inert* complexes:

Class I. Exchange of water is very fast and is essentially diffusion-controlled $(k \gtrsim 10^8 \, s^{-1})$. The class encompasses ions of Groups IA, IIA (except Be^2 and Mg^{2+}), and IIB (except Zn^{2+}), plus Cr^{2+} and Cu^{2+}.

Class II. Rate constants are in the range from 10^4 to $10^8 \, s^{-1}$. This class includes most of the first-row bivalent transition metal ions (the exceptions are V^{2+}, Cr^{2+}, and Cu^{2+}) and Mg^{2+}, and the trivalent lanthanides.

Class III. Rate constants are roughly in the range 1 to $10^4 \, s^{-1}$. This class includes Be^{2+}, Al^{3+}, V^{2+}, and some of the first-row trivalent transition metal ions.

Class IV. Rate constants are roughly in the range from 10^{-3} to $10^{-6} \, s^{-1}$. The following ions fall within this class: Cr^{3+}, Co^{3+}, Rh^{3+}, Ir^{3+}, and Pt^{2+}.

Metal ions which fall in Classes I–III are labile and Class IV ions are inert.

Across the $3d$ series, complexes of hs-M(II) ions are moderately labile, with distorted Cu(II) complexes among the most labile.

Complexes of M(III) ions are less labile. Inertness is quite common among the complexes of the 2nd and 3rd transition series, reflecting their large CFSE and strength of the metal–ligand bonding.

d^3 and d^6 complexes of 1st transition series, such as $[Cr^{III}(NH_3)_6]^{3+}$, $[Co^{III}(H_2O)]^{3+}$, and particularly of strong field ligands such as ls-$[Fe^{II}(CN)_6]^{4-}$, ls-$[Fe^{II}(bpy)_3]^{2+}$ (bpy $= 2,2'$-bipyridine), and ls-$[Fe^{II}(phen)_3]^{2+}$ (phen $= 1,10$-phenanthroline) are generally inert. In the 1st transition series, the least labile M(II) and M(III) ions are those with large CFSE.

The bivalent $d^{0,10}$ complexes, such as Zn(II), Cd(II), and Hg(II), are highly labile.

The trivalent ions of the f-block are all very labile.

Generalizations of kinetic behavior:

(i) All ions of 1st transition series and lanthanides are labile ($k \geq 10^{-2}\,s^{-1}$) except Cr(III) and ls-Co(III).

(ii) Low-spin complexes tend to be less labile than high-spin complexes in 1st transition series.

(iii) All ions of 2nd and 3rd transition series are inert ($(k \leq 10^{-2}\,s^{-1})$) except $d^{0,10}$ ions.

(iv) Labile vs. inert behavior as established by water exchange rates apply to majority of other complexes if, for a given metal ion, the oxidation and spin state are the same.

Labile and inert refer to kinetic and not thermodynamic stability!

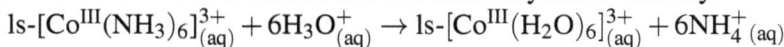

$$ls\text{-}[Co^{III}(NH_3)_6]^{3+}_{(aq)} + 6H_3O^+_{(aq)} \rightarrow ls\text{-}[Co^{III}(H_2O)_6]^{3+}_{(aq)} + 6NH_4^+{}_{(aq)}$$

$$K_{eq} \sim 10^{25}, \Delta G^{\circ}_{298\,K} \sim -35\,kcal/mol$$

Therefore, $ls\text{-}[Co^{III}(NH_3)_6]^{3+}$ (d^6 ion) will exist for days in acid solution because of its kinetic inertness.

7.2 Ligand substitution reactions

Substitution reactions are defined as due to replacement of one ligand by another, usually without change in the oxidation state of the metal ion.

Important distinctions:

Transition state ($\ddagger$) represents the species of fleeting existence corresponding to a high point on a potential energy curve (Fig. 7.1).

Intermediates are species of finite lifetime corresponding to a local minimum in high energy portion of potential energy curve.

(1) Octahedral complexes

Generalized substitution reaction of an octahedral complex:

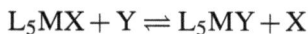

$$L_5MX + Y \rightleftharpoons L_5MY + X$$

Mechanisms are distinguished by the nature of the transition state or intermediate.

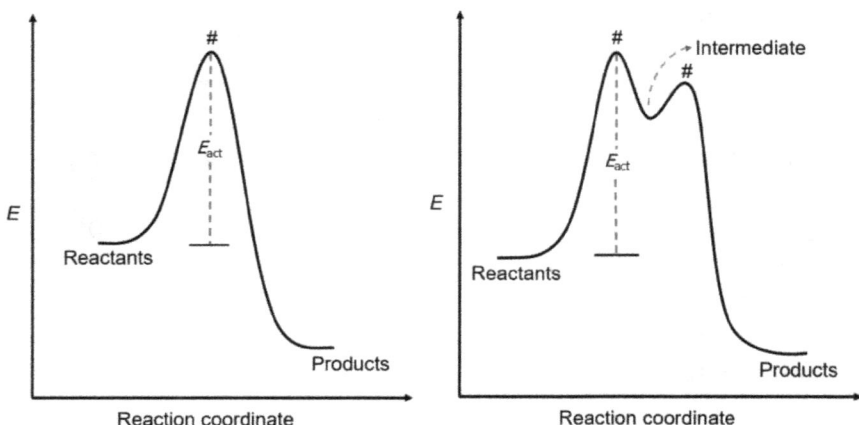

Fig. 7.1 Plots for activation energy.

- Associative mechanisms

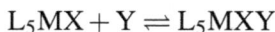

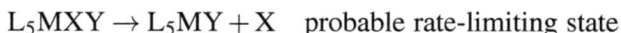

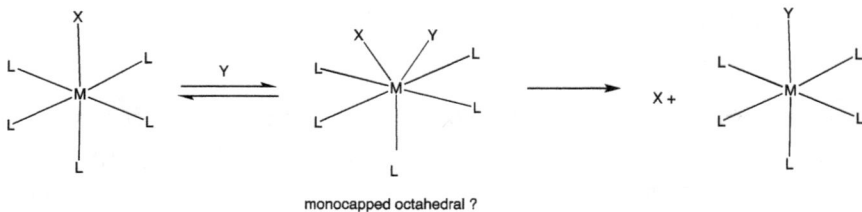

$$L_5MX + Y \rightleftharpoons L_5MXY$$

$$L_5MXY \rightarrow L_5MY + X \quad \text{probable rate-limiting state}$$

monocapped octahedral ?

A mechanism $\equiv$ S_N2 limiting (S_N2 lim.)

The transition state or intermediate may exist with well-formed M–X and M–Y bonds prior to dissociation of X, or M–Y may form as M–X is weakening. In S_N2 limiting, an intermediate with higher coordination number is formed when the entering ligand bonds with the metal. For S_N2, no such intermediate step is observed.

$$[Y-M \cdots X]^{\ddagger} \quad I_a \text{ (associative interchange) mechanism}$$

The departing ligand X is assumed to have a fairly weak bond with the metal M that the incoming Y ligand can easily disrupt. M–Y bond making and M–X bond breaking,

Associative mechanism → a → A (S_N2 lim.) or I_a (S_N2)

For octahedral complexes virtually no well-established 'a mechanism' is known.

• Dissociative mechanisms

$$L_5MX \rightleftharpoons L_5L + X$$

$$L_5M + Y \rightarrow L_5MY$$

The complex accumulates enough energy to break completely the M–X bond, giving rise to a five-coordinate intermediate (Fig. 7.2).

It is to be noted how easily trigonal bipyramid (TBP) and square pyramid (SP) are interconverted (Fig. 7.3). Theoretical and experimental results

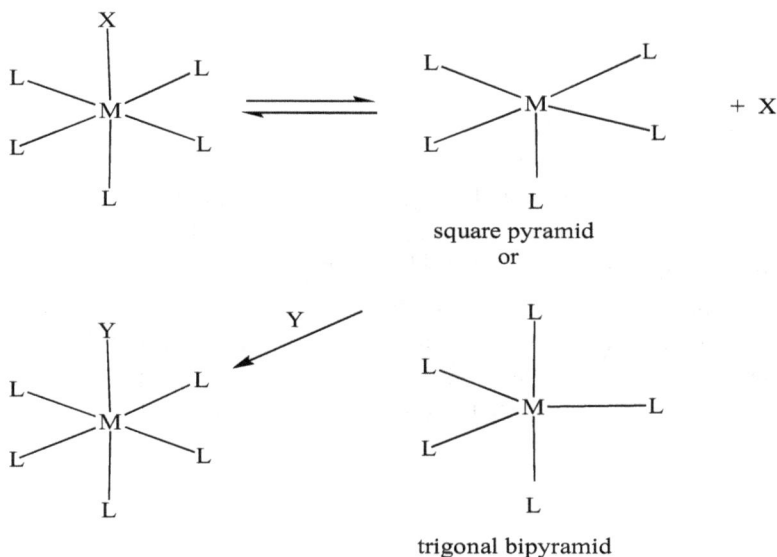

D mechanism ≡ S_N1 limiting (S_N1 lim.)

Fig. 7.2   Dissociative mechanism.

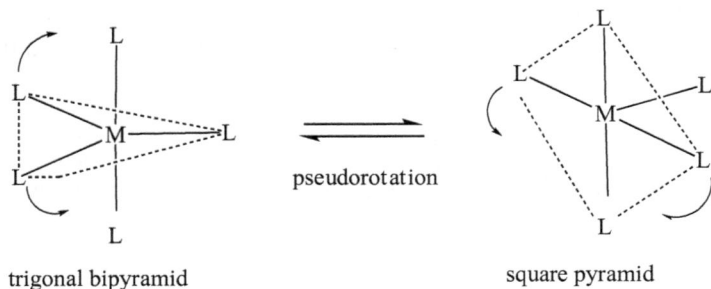

Fig. 7.3   Interconversion between SP and TBP geometry.

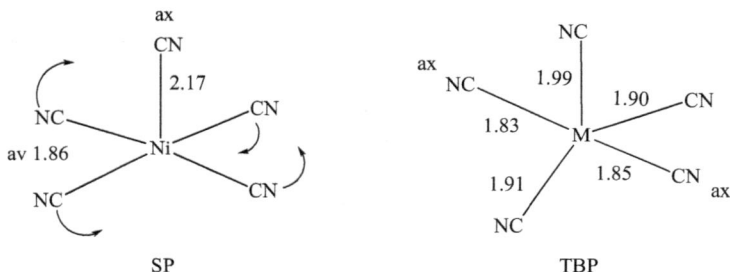

Fig. 7.4   Showing small structural displacements for SP → TBP interconversion.

show that, in general, TBP and SP geometry have very small ($\lesssim 5$ kcal/mol) energy difference (Berry pseudorotation: see Chapter 1).

The crystal lattice of $\{[Cr^{III}(en)_3]^{3+}[Ni^{II}(CN)_5]^{3-}\}$ (en = 1,2-diaminoethane) contains both five-coordinate (square pyramidal SP and trigonal bipyramidal TBP) structures (distances are in Å) (Fig. 7.4).

Let us consider the case where ML_5 has no appreciable lifetime. M begins to form a bond with Y before M–X is completely broken.

$$[X\text{–}M\cdots Y]^{\ddagger}$$

$M\cdots Y$ bond making and M–X bond breaking,

 $L_5MX + Y \rightleftharpoons (L_5MX, Y)$ rapid equilibrium

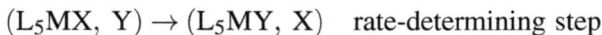 $(L_5MX, Y) \rightarrow (L_5MY, X)$ rate-determining step

 $(L_5MY, X) \rightarrow L_5MY + X$ fast

Dissociative mechanisms → d → D (S_N1 lim.) or I_d (S_N1)

I_d (dissociative interchange)

Dissociative mechanisms are prevalent in the substitution reactions of octahedral complexes.

Now, let us examine some experimental data, in relation to the following predictions.

(a) Anation of Co(III) complexes (leaving group constant)

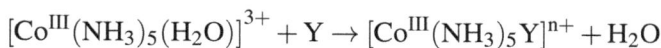

$$[Co^{III}(NH_3)_5(H_2O)]^{3+} + Y \rightarrow [Co^{III}(NH_3)_5Y]^{n+} + H_2O$$

$$\text{rate} = -d/dt\ [Co^{III}(NH_3)_5(H_2O)]^{3+} = k\ [Co^{III}(NH_3)_5(H_2O)]^{3+}\ [Y]$$

Y is used in large excess; hence [Y] is $\sim$ constant.

The kinetic data:

Y	$10^6 \times k\ (s^{-1})$; 45°C
H_2O	100
N_3^-	100
SO_4^{2-}	24
Cl^-	21
NCS^-	16

Rate constants are not very sensitive to the bond being made. On the other hand, relatively minor dependence on entering group Y. Therefore, it is a dissociative mechanism; most likely I_d.

(b) Substitution of Ru(III) complexes

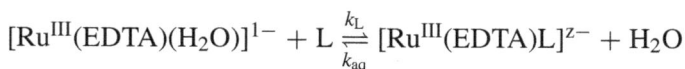

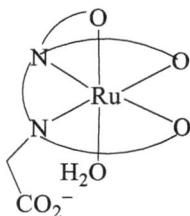

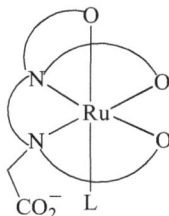

$$[Ru^{III}(EDTA)(H_2O)]^{1-} + L \underset{k_{aq}}{\overset{k_L}{\rightleftharpoons}} [Ru^{III}(EDTA)L]^{z-} + H_2O$$

$$\text{rate} = -d/dt\,[Ru^{III}(EDTA)(H_2O)]^{1-} = k_L[Ru^{III}(EDTA)(H_2O)]^{1-}\,[L]$$
$$\text{rate} = -d/dt\,[Ru^{III}(EDTA)L]^{z-} = k_{aq}[Ru^{III}(EDTA)L]^{z-}$$

Kinetic data:

L	k_L (M^{-1} s^{-1})	k_{aq} (s^{-1})
pyrazine (N‿N)	2.0×10^4	2.0
pyridine–CONH$_2$	8.3×10^4	0.7
pyridine	6.3×10^3	0.061
imidazole (N‿NH)	1.9×10^3	-
-NCS$^-$	2.7×10^2	0.5
CH$_3$CN	3.0×10^1	3.2

10^3-fold variation of k_L values with entering group. Hence, it is an associative mechanism; probably Ia.

10-fold variation of k_{aq} values with leaving group; mechanism less clear.

Effect of rate on

mechanism	leaving group X	entering group Y
A*	very small	large
Ia†	minor	major
D*	large	none
Id†	major	minor

*Intermediate. †Transition state

(c) Aquation of Co(III) complexes (leaving group varied):

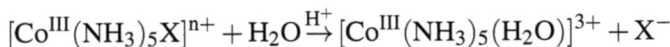

$$[Co^{III}(NH_3)_5X]^{n+} + H_2O \xrightarrow{H^+} [Co^{III}(NH_3)_5(H_2O)]^{3+} + X^-$$
$$\text{rate} = -d/dt\,[Co^{III}(NH_3)_5X]^{n+} = k[Co^{III}(NH_3)_5X]^{n+}$$

Kinetic data:

X^-	$k \ (s^{-1}), 298 \ K$
$(CH_3O)_3PO$	2.5×10^{-4}
NO_3^-	2.7×10^{-5}
I^-	8.3×10^{-6}
Cl^-	1.7×10^{-6}
SO_4^{2-}	1.2×10^{-6}
F^-	8.6×10^{-8}
N_3^-	2.1×10^{-9}
$-NCS^-$	5.0×10^{-10}

Rates depend strongly on leaving group and vary over 10^6!

Rate constants sensitive to bond being broken. Therefore, it is a dissociative mechanism.

Slow (rate-determining) step is Co–X bond breaking.

> Identification of a mechanism by the (lack of) dependence of rate constants on nature of entering or leaving group is entirely qualitative but usually reliable. However, better demonstrations of mechanisms require, *inter alia*, satisfactory fit of rate data to a rate equation for a particular mechanism. It is nearly impossible to "prove" a mechanism, one reason being that certain different mechanisms have functionally similar rate laws.
>
> The "best" or accepted mechanism is that with the least evidence against it.

In octahedral complexes the common inert metal ions are Cr(III) and ls-Co(III). Many kinetic studies have been done with their complexes because substitution rates are conveniently slow to measure. For nearly all octahedral complexes substitution reactions are dissociative (Fig. 7.5).

It should be remembered that σ-bonding effects dominate d-orbital energetics.

From simple considerations of d-orbital energetics it has been established that in d^3 Cr(III), hs-Co(III), hs-d^5, d^8, d^9, and d^{10} (and other)

SP
first intermediate

TBP

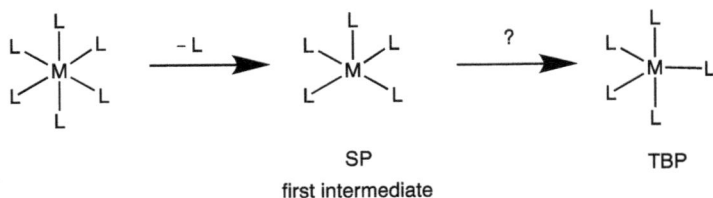

Fig. 7.5 Dissociative mechanism for substitution reactions of octahedral complexes.

cases less energy is required to generate SP intermediate from ideal octahedral geometry; therefore, greater probability of faster dissociative reactions. This is in agreement with experiment, as reactions of d^3 Cr(III), ls-d^6 (Co(III), Fe(II), Rh(III), Ru(II)) (common cases) are slow.

(2) Planar complexes

The great majority of planar complexes are found with the most common d^8 metal ions (Rh(I), Ir(I); Ni(II), Pd(II), Pt(II); Au(III)). From simple considerations of d-orbital energetics it has been established that in d^8 cases planar is more stable than tetrahedral geometry.

    Generalized substitution reaction:

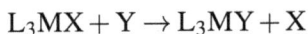

$$L_3MX + Y \rightarrow L_3MY + X$$

For nearly all such reactions the experimental rate law has the form

$$\text{rate} = -d/dt[ML_3X] = (k_s + k_y[Y])\,[ML_3X]$$

Note: (i) presence of two rate constants indicates a dual reaction pathway, (ii) the product $[Y][ML_3X]$ suggests that the k_y pathway is associative (A, I_a).

    Rate effects of entering and leaving ligands:

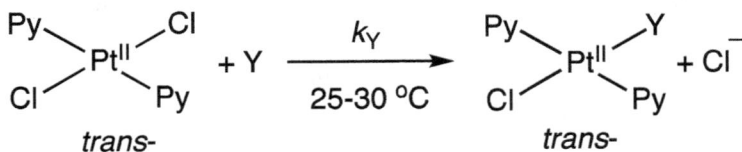

Kinetic data:

Y	k_Y (M^{-1} s^{-1})
CH$_3$OH	2.7×10^{-7} (k_s)
Cl$^-$	4.5×10^{-4}
NH$_3$	4.7×10^{-4}
N$_3^-$	1.6×10^{-3}
I$^-$	1.1×10^{-1}
CN$^-$	4.0
PPh$_3$	2.5×10^2

Identity of Y results in 10^9 range in k_y!
Hence, associative reactions.

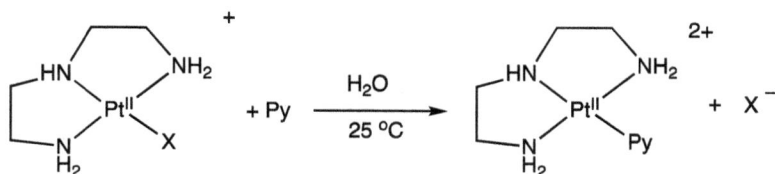

Kinetic data:

X	k_{obs} (M^{-1} s^{-1})
CN$^-$	1.7×10^{-8}
SCN$^-$	3.0×10^{-7}
I$^-$	1.0×10^{-6}
Cl$^-$	3.5×10^{-5}
H$_2$O	1.9×10^{-3}
NO$_3^-$	very fast

Identity of X results in 10^5 range in k_{obs}!
Hence, associate (I_a) reactions

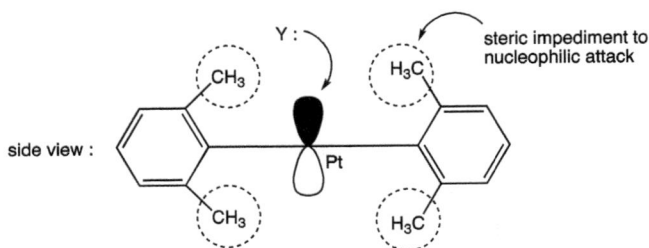

Fig. 7.6   Depiction of steric effects.

Steric effects on reaction rates

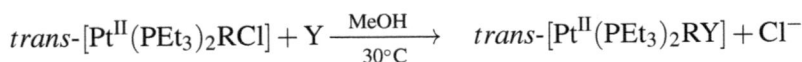

$$\textit{trans-}[\text{Pt}^{\text{II}}(\text{PEt}_3)_2\text{RCl}] + \text{Y} \xrightarrow[30°C]{\text{MeOH}} \textit{trans-}[\text{Pt}^{\text{II}}(\text{PEt}_3)_2\text{RY}] + \text{Cl}^-$$

Kinetic data:

| | k_Y (M^{-1} s^{-1}) | |
R	Y = CN$^-$	SC(NH$_2$)$_2$
	3.61	6.30
	0.234	0.652
	8.49 x 10^{-3}	4.94 x 10^{-2}

Reactions are often studied in the presence of a large excess of entering ligand Y, so that [Y] ≈ constant during reaction (Fig. 7.7).

Rate $= k_{obs}$ [ML$_3$X]

Straight line plots show that

$$k_{obs} = k_s + k_2\,[\text{Y}],$$

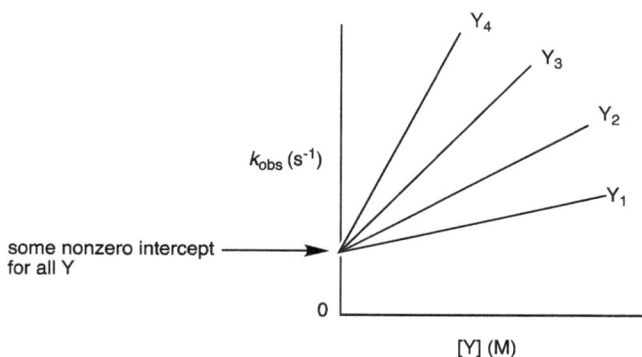

Fig. 7.7 Variation of rates as a function of [Y].

establishing the form of the rate law

$$\text{rate} = (k_s + k_Y [Y]) [ML_3X]$$

k_s pathway = independent of Y, implying attack by the same species in all reactions. This must be solvent (s).

k_Y pathway: attack by Y

Leading experimental observations of some generality:

(i) Substitutions are stereospecific
   trans → *trans*   *cis* → *cis*

(ii) Rate = $(k_s + k_Y [Y]) [ML_3X]$
   Two reaction pathways are k_s and k_Y; k_Y pathway is presumably associative.

(iii) Rates are sensitive to both bonds made (M–Y) and bonds broken (M–X)

Mechanism of substitution, k_Y pathway, I_a is displayed in Fig. 7.8.

Similar steric course for cis-complexes; square planar substitution reactions are associative processes, often considered to be I_a.

The trans effect

The discovery of the trans effect is attributed to I. I. Chernyaev (Russian and Soviet chemist: 1893–1966). It is the effect of a coordinated ligand on the rate of substitution of a ligand opposite (*trans*) to it. The trans effect can operate in a molecule with mutually *trans* ligand, but is most prevalent in

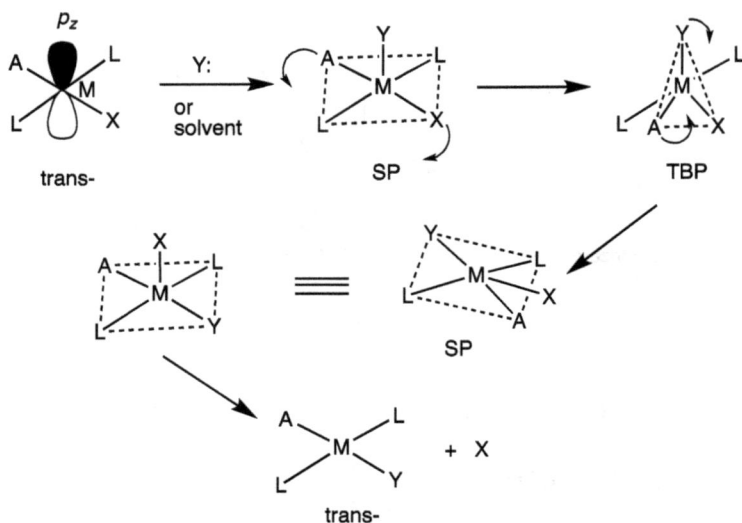

Fig. 7.8   Mechanism of substitution reaction.

and has been most extensively studied in planar d^8 complexes, especially those of Pt(II).

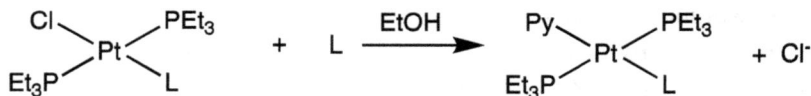

Kinetic data for the effect of the trans ligand of trans-$[Pt^{II}Cl(PEt_3)_2L]$:

L	$k_s\ (s^{-1})$	$k_{py}\ (M^{-1}\ s^{-1})$	T (°C)
H^-	1.8×10^{-2}	4.2	0
CH_3^-	2×10^{-4}	7×10^{-2}	25
$C_6H_5^-$	2×10^{-5}	2×10^{-2}	25
Cl^-	1.0×10^{-6}	4×10^{-4}	25

Trans ligands change reaction speed by $\sim 10^6$! (allowing for temperature difference).

Average (not completely general) *trans effect* order:

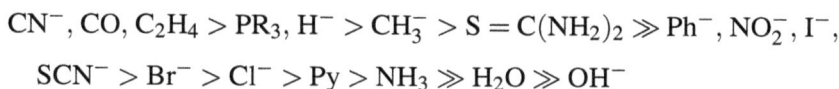

$$CN^-, CO, C_2H_4 > PR_3, H^- > CH_3^- > S = C(NH_2)_2 \gg Ph^-, NO_2^-, I^-,$$

$$SCN^- > Br^- > Cl^- > Py > NH_3 \gg H_2O \gg OH^-$$

7.3 Electron transfer (redox) reactions

Electron transfer reactions are defined as due to transfer of one or more electrons between complexes with no change in bound ligands (outer-sphere) and with change in coordinated ligands (inner-sphere). Redox reactions can occur by the direct transfer of electrons (see Chapter 4). They may occur by the transfer of atoms (atom transfer reactions) and ions. Because redox reactions involve both an oxidizing and a reducing agent, the electron transfer reactions are usually bimolecular. The only exceptions are disproportionation reactions (see Chapter 4).

In the 1950s, in homogeneous solutions, Henry Taube (Chapter 4) recognized two mechanisms of electron transfer (redox) reactions to explain redox reactions for coordination complexes: the *inner-sphere* and *outer-sphere* mechanisms.

In an *inner-sphere mechanism*, the electron transfer proceeds through usually a binuclear ligand-bridged complex between the two species involved.

(a) Inner-sphere (atom transfer) reactions

Classic example:

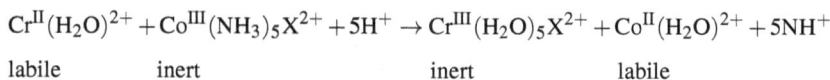

$$Cr^{II}(H_2O)^{2+} + Co^{III}(NH_3)_5X^{2+} + 5H^+ \rightarrow Cr^{III}(H_2O)_5X^{2+} + Co^{II}(H_2O)^{2+} + 5NH^+$$

    labile   inert   inert   labile

Transfer of $X = Cl^-, Br^-, I^-, OH^-, H_2O$, etc. from Co to Cr is quantitative.

reaction intermediate:

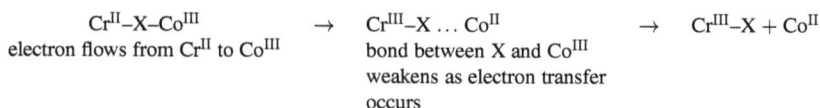

$Cr^{II}-X-Co^{III}$	$\rightarrow$	$Cr^{III}-X \dots Co^{II}$	$\rightarrow$	$Cr^{III}-X + Co^{II}$
electron flows from Cr^{II} to Co^{III}		bond between X and Co^{III} weakens as electron transfer occurs		

$$\text{Co}^{III}(NH_3)_5X:^{2+} \longrightarrow \overset{\overset{OH_2}{\vdots}}{Cr}^{II}(H_2O)_5^{2+} \longrightarrow (NH_3)_5Co\text{-}X\text{-}Cr(H_2O)_5^{4+} + H_2O$$

intermediate

$$\text{Co}(NH_3)_5(H_2O)^{2+} + Cr(H_2O)_5X^{2+} \longleftarrow (NH_3)_5Co^{II}...X\text{-}Cr^{III}(H_2O)_5^{4+}$$

$$:\overset{..}{O}H_2$$

$$\text{Co}(NH_3)_4(H_2O)_2^{2+} + NH_3 \quad \overset{H^+}{\underset{NH_4^+}{\curvearrowright}}$$

$$\text{Co}(H_2O)_6^{2+} + 5NH_4^+$$

$$Cr(H_2O)_5X^{2+} + {}^*Cr(H_2O)_6^{2+} \rightleftharpoons Cr(H_2O)_6^{2+} + {}^*Cr(H_2O)_5X^{2+}$$

inert   labile (^{51}Cr; γ emitter)

$$K_{eq} = 1$$

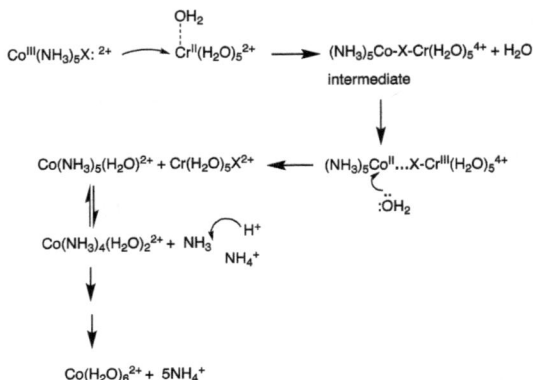

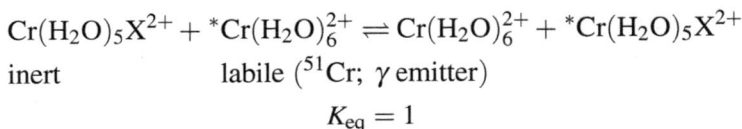

all *Cr introduced in reaction as Cr(II); one-half of this amount is recovered as Cr(III).

Prototype inner-sphere reaction:

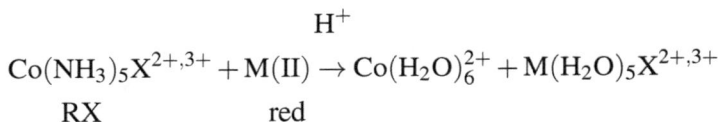

$$H^+$$

$$Co(NH_3)_5X^{2+,3+} + M(II) \rightarrow Co(H_2O)_6^{2+} + M(H_2O)_5X^{2+,3+}$$

RX   red

rate $= -d/dt\,[RX] = k\,[RX][[red]\ (red = Cr(H_2O)_6^{2+})$

Kinetic data:

X	$k\ (M^{-1}\ s^{-1})$	
NH_3	9×10^{-5}	
H_2O	5×10^{-1}	cannot bridge, not inner sphere; slow because of 2+/3+ reactant charge
$CH_3CO_2^-$	2×10^{-1}	
CN^-	6×10^1	
F^-	3×10^5	generally good bridging ligands; 2+/2+ reactant charge
OH^-	2×10^6	
I^-	3×10^6	
Cl^-	6×10^5	

Common strong reductants used in aqueous solution: Cr^{2+}, V^{2+}, Eu^{2+}

(b) Outer-sphere reactions

In an outer-sphere mechanism, the electron transfer is accomplished with the primary coordination spheres remaining invariant.

Examples:

$$[Fe^{II}(CN)_6]^{4-} + [Mo^V(CN)_8]^{3-} \rightarrow [Fe^{III}(CN)_6]^{3-} + [Mo^{IV}(CN)_8]^{4-}$$

inert   inert

$$[Ru^{II}(NH_3)_6]^{2+} + [Co^{III}(bpy)_3]^{3+} \rightarrow [Ru^{III}(NH_3)_6]^{3+}$$

inert   inert

$$+ [Co^{II}(bpy)_3]^{2+} \ (bpy = 2,2'\text{-bipyridine})$$

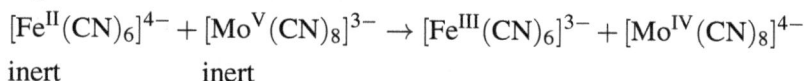

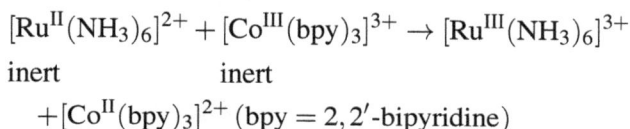

Outer sphere reactions are also possible when only one reactant is inert but lacks bridging ligands:

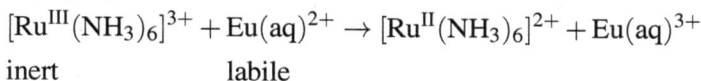

$$[Ru^{III}(NH_3)_6]^{3+} + Eu(aq)^{2+} \rightarrow [Ru^{II}(NH_3)_6]^{2+} + Eu(aq)^{3+}$$

inert   labile

Types of outer sphere reactions:

• Electron self-exchange

$[Ru(NH_3)_6]^{3+,2+}$ $[Fe(CN)_6]^{3-,4-}$

• Cross transfer

Characteristics of both types:

2nd order kinetics

rate $= k\,[Ox][Red]$

> Electron is transferred under Franc-Condon conditions (rate of electron transfer is fast compared to nuclear motion in the transition state).

To understand how the simplest type of outer-sphere redox reactions occur, i.e., *electron-exchange reactions*, let us consider the following self-exchange reaction:

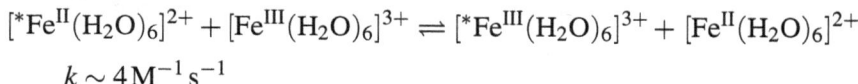

$$[^*Fe^{II}(H_2O)_6]^{2+} + [Fe^{III}(H_2O)_6]^{3+} \rightleftharpoons [^*Fe^{III}(H_2O)_6]^{3+} + [Fe^{II}(H_2O)_6]^{2+}$$

$$k \sim 4\,M^{-1}s^{-1}$$

As mentioned earlier, the rates of such reactions [R. A. Marcus (1923–), a Canadian-born American theoretical chemist (Nobel Prize in Chemistry: 1992) provided the theory of electron transfer reactions in chemical systems.] can be monitored by using isotopic labeling. For this reaction the equilibrium constant $K_{eq} = 1$ and $\Delta G° = 0$. The activation energy for this reaction is ~ 32 kJ/mol, as the reactants have different structures. In fact, the average Fe^{III}–OH_2 distance is shorter than the average Fe^{II}–OH_2 distance by ~ 0.12 Å. Therefore, electron transfer from one ion to another cannot occur without expenditure of some energy. The rate constant k at $25°C$ is $\simeq 4$ M^{-1} s^{-1}.

The d-electron distribution for these two grossly octahedral complexes $[Fe^{II}(H_2O)_6]^{2+}$ and $[Fe^{III}(H_2O)_6]^{3+}$ are $(t_{2g})^4 (e_g)^2$ $S = 2$ and $(t_{2g})^3 (e_g)^2$ $S = 5/2$ (as H_2O is a weak field ligand), respectively.

Two major contributors to the activation energy required for the electron transfer to take place are reorganization of metal–ligands bonds in $[Fe^{II}(H_2O)_6]^{2+}$ and $[Fe^{III}(H_2O)_6]^{3+}$, and the solvation shell to accommodate the new dimensions of the reactants. Thus, before the electron transfer event occurs the Fe–OH_2 bond lengths of the two participating species must change from their equilibrium positions. The metal–ligand bonds in $[Fe^{III}(H_2O)_6]^{3+}$ species expand to a distance halfway between the distances in $[Fe^{II}(H_2O)_6]^{2+}$ and $[Fe^{III}(H_2O)_6]^{3+}$ species. The metal–ligand bonds in $[Fe^{II}(H_2O)_6]^{3+}$ species contract to the same distance. Only then can electron transfer take place and the products relax back to their equilibrium geometries. This model for the outer-sphere electron transfer process is in accord with the *Franc-Condon principle*, which states that electron transfer is much faster than nuclear motion. In other words, as far as the electrons are concerned, the nuclear positions are frozen during the time period required for electron transfer. Notice that this reorganization energy (Franc-Condon energy) must be expended. Otherwise the initial products of self-exchange would be $[*Fe^{III}(H_2O)_6]^{3+}$ in $[*Fe^{II}(H_2O)_6]^{2+}$ and $[Fe^{II}(H_2O)_6]^{2+}$ in $[Fe^{III}(H_2O)_6]^{3+}$ ligand environment. The energy (heat) released when these species relax back to the stable geometry would be created from nothing. This violates the principle of conservation of energy.

Let us consider generalized outer-sphere reaction,

$$A + B \rightleftharpoons A^- + B^+$$

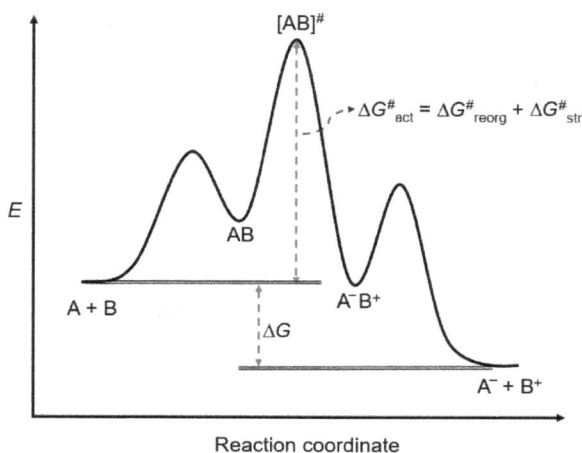

Fig. 7.9 Energy profile of an outer-sphere reaction.

which proceeds spontaneously to the right. The energy profile is shown below (Fig. 7.9).

Transition state ($\ddagger$) must involve structural change in reactants. Each adjust its M–L bonds to a "compromise" length. If the M–L bond is a harmonic oscillator, then the energy,

$$E = 1/2\, k\,(\Delta r)^2;\ v(\text{cm}^{-1}) = (1/2\pi c)\sqrt{(k/\mu)}$$

where k, Δr, c, and μ are the force constant, displacement from equilibrium position, velocity of light, and reduced mass of the molecule, respectively.

The contribution from geometrical change to the free energy of activation is

$$\Delta G^{\ddagger}_{\text{reorg}} \cong \Delta H^{\ddagger}_{\text{reorg}} = (nk_1/2)\,(r_1 - r_{\ddagger})^2 + (nk_2/2)\,(r_2 - r_{\ddagger})^2$$

where r_1, r_2 are ground state (equilibrium) M–L distance of the reactants, n is the number of bonds, and $r_{\ddagger}$ is the M–L bond distance changes in the transition state.

Minimizing M–L bond distance changes in the transition state,

$$d/dr_{\ddagger}\,(\Delta H^{\ddagger}_{\text{reorg}})_{r1,r2} = -n\,r_{\ddagger}(k_1 + k_2) - n\,(k_1 r_1 + k_2 r_2) = 0$$

$$\text{Therefore, } r_{\ddagger} = (k_1 r_1 + k_2 r_2)/(k_1 + k_2)$$

Let us now apply these ideas to the following self-exchange reactions:

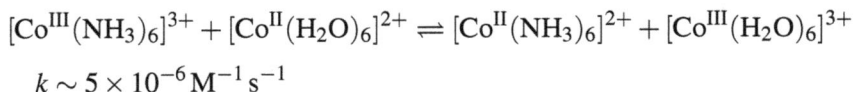

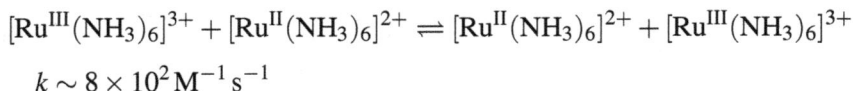

$$[Co^{III}(NH_3)_6]^{3+} + [Co^{II}(H_2O)_6]^{2+} \rightleftharpoons [Co^{II}(NH_3)_6]^{2+} + [Co^{III}(H_2O)_6]^{3+}$$

$$k \sim 5 \times 10^{-6} M^{-1} s^{-1}$$

$$[Ru^{III}(NH_3)_6]^{3+} + [Ru^{II}(NH_3)_6]^{2+} \rightleftharpoons [Ru^{II}(NH_3)_6]^{2+} + [Ru^{III}(NH_3)_6]^{3+}$$

$$k \sim 8 \times 10^2 M^{-1} s^{-1}$$

		bond distance, Å		
reactants	state change	Av M(II)–L	Av M(III)–L	Δ, Å
$[Ru(NH_3)_6]^{2+,3+}$	$^1A_{1g} \rightarrow {}^2T_{2g}$ $(t_{2g})^6 \rightarrow (t_{2g})^5$ $S=0 \quad S=1/2$	2.144	2.104	0.040
$[Co(NH_3)_6]^{2+,3+}$	$^4T_{1g} \rightarrow {}^1A_{1g}$ $(t_{2g})^5 (e_g)^2 \rightarrow (t_{2g})^6$ $S=3/2 \quad S=0$	2.114	1.936	0.178

Considerable elongation of the Co(III)–N bond and compression of the Co(II)–N bond is necessary before electron transfer can occur. The much faster rate for the exchange for the ruthenium system is consistent with a small bond length adjustment prior to electron transfer. The two systems are not entirely analogous, since cobalt goes from a ls d^6 Co(III) complex to a hs d^7 Co(II) complex while ruthenium remains ls in both the oxidized and reduced forms.

The values of totally symmetric stretching frequencies:

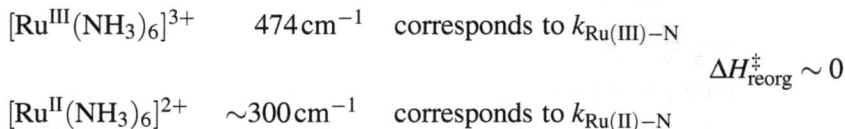

$[Ru^{III}(NH_3)_6]^{3+}$ $474 \, cm^{-1}$ corresponds to $k_{Ru(III)-N}$

$$\Delta H^{\ddagger}_{reorg} \sim 0$$

$[Ru^{II}(NH_3)_6]^{2+}$ $\sim 300 \, cm^{-1}$ corresponds to $k_{Ru(II)-N}$

From r and k data, $r_{\ddagger}$ (Ru) $= 2.115$ Å: Ru(II)–N bonds contract by $2.144-2.115 = 0.029$ Å and Ru(III)–N bonds expand by $2.115-2.104 = 0.011$ Å. The change in Ru–N bonds to reach the transition state, $\lesssim 0.029$ Å, is very small ($\Delta H^{\ddagger}_{reorg} \lesssim 3$ kcal/mol) and can be readily achieved by out-of-phase totally symmetric Ru–N stretches in the transition state. The electron transfer can occur when both complexes achieve $r_{\ddagger}$.

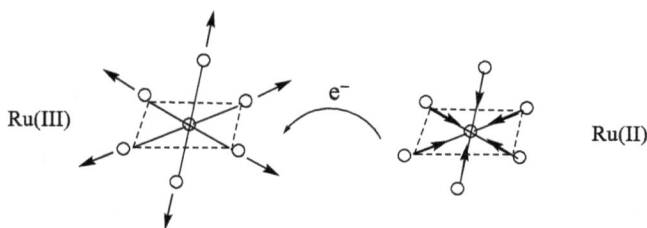

For the Co complexes similar data were calculated, $r_{\ddagger}$ (Co) = 1.989 Å: Co(II)–N bonds contract by 2.114–1.989 = 0.125 Å and Co(III)–N bonds expand by 1.989–1.936 = 0.053 Å. The change in Co–N bonds to reach the transition state, $\gtrsim 0.053$ Å, is not very small ($\Delta H^{\ddagger}_{\text{reorg}} \cong$ 8–9 kcal/mol).

This factor alone will make Co system to exchange rate slower but there is another contributing factor. Because no one-step process can effect the change in the spin state (see above); several discrete steps are required for the Co system but not for the Ru system.

Most likely scheme of electron self-exchange for Co system:

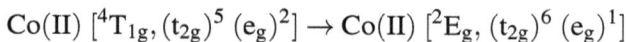

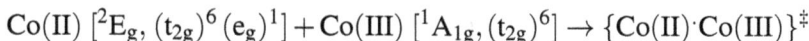

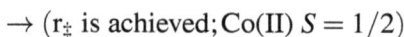

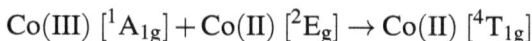

$$\text{Co(II)} \, [^4T_{1g}, (t_{2g})^5 \, (e_g)^2] \rightarrow \text{Co(II)} \, [^2E_g, (t_{2g})^6 \, (e_g)^1]$$

$$\text{Co(II)} \, [^2E_g, (t_{2g})^6 \, (e_g)^1] + \text{Co(III)} \, [^1A_{1g}, (t_{2g})^6] \rightarrow \{\text{Co(II)} \cdot \text{Co(III)}\}^{\ddagger}$$

$$\rightarrow (r_{\ddagger} \text{ is achieved}; \text{Co(II)} \, S = 1/2)$$

$$\text{Co(III)} \, [^1A_{1g}] + \text{Co(II)} \, [^2E_g] \rightarrow \text{Co(II)} \, [^4T_{1g}]$$

Thus,

for the Ru system $\Delta H^{\ddagger}_{\text{reorg}}$ is main contributor to $\Delta G^{\ddagger}_{\text{activation}}$

for Co system both $\Delta H^{\ddagger}_{\text{reorg}}$ and $\Delta H^{\ddagger}$ ($S = 3/2 \rightarrow S = 1/2$) are contrib-

utors to $\Delta G^{\ddagger}_{\text{activation}}$.

For Co system both $\Delta H^{\ddagger}_{\text{reorg}}$ and $\Delta H^{\ddagger}$ ($S = 3/2 \rightarrow S = 1/2$) are larger than that of Ru system, accounting for much slower reaction rate.

Other examples of outer-sphere self-exchange reactions:

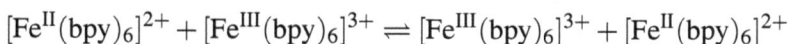

$$[\text{Fe}^{II}(\text{bpy})_6]^{2+} + [\text{Fe}^{III}(\text{bpy})_6]^{3+} \rightleftharpoons [\text{Fe}^{III}(\text{bpy})_6]^{3+} + [\text{Fe}^{II}(\text{bpy})_6]^{2+}$$

$$k > 10^6 \, \text{M}^{-1}\text{s}^{-1}$$

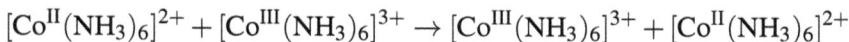

$$[\text{Co}^{II}(\text{NH}_3)_6]^{2+} + [\text{Co}^{III}(\text{NH}_3)_6]^{3+} \rightarrow [\text{Co}^{III}(\text{NH}_3)_6]^{3+} + [\text{Co}^{II}(\text{NH}_3)_6]^{2+}$$

$$k \sim 10^{-6} \, \text{M}^{-1}\text{s}^{-1}$$

In the absence of any bridging ligand these self-exchange electron transfer reactions occur via the outer sphere mechanism.

These electron transfer reactions involve M^{II} (d^6 Fe^{II} or d^7 Co^{II}) complexes of bpy (bidentate 2,2'-bipyridine) or NH_3 ligands, as the reductant and M^{III} (d^5 Fe^{III} or d^6 Co^{III}) complexes of bpy or NH_3 ligands, as the oxidant. As discussed earlier, in each case, the M^{III}–L bonds will be shorter than M^{II}–L bonds due to higher charge density on the M^{3+} than the M^{2+} ion. Further differences can occur due to the differences in the d^n configurations – occupation of the higher energy e_g orbital(s) (considering CF theory) or antibonding e_g^* orbital(s) (considering MO theory) will cause a lengthening of the M–L bonds.

Both bpy and NH_3 are strong field ligands. Hence, with bpy both Fe(II) and Fe(III) complexes and with NH_3 both Co(II) and Co(III) complexes are ls. For iron–bpy system both the Fe(II) and Fe(III) states the electrons occupy only the t_{2g} set and for cobalt–NH_3 system in the Co(II) state the odd electron enters the e_g set but for the Co(III) state the electrons occupy only the t_{2g} set. The electronic distribution in t_{2g} and e_g set of orbitals (CF theory) in M(II) and M(III) oxidation states are as follows:

With bpy ligand, $Fe(II)(t_{2g})^6$, $S = 0$ and $Fe^{III}(t_{2g})^5$, $S = 1/2$

With NH_3 ligand, $Co(II) (t_{2g})^6 (e_g)^1$, $S = 1/2$ and Co^{III} with $(t_{2g})^6$, $S = 0$

Outer-sphere electron transfer mechanisms are faster, when $\Delta H^{\ddagger}_{\text{reorg}} \sim 0$ ($r_{ox} \cong r_{red}$). It is closely achieved in octahedral complexes when only t_{2g} electrons are involved.

$|S_{ox} - S_{red}| = 1/2$; only one elementary electron transfer step; no spin "rearrangement".

Further reading

C. E. Housecroft and A. G. Sharpe, *Inorganic chemistry*, 2nd ed., Person Education Ltd. (2005).

D. F. Shriver and P. W. Atkins, *Inorganic Chemistry*, 3rd ed., Oxford University Press (1999).

J. E. Huheey, E. A. Keiter, and R. L. Keiter, *Inorganic Chemistry: Principles of Structure and Reactivity*, 4th edition, Addison-Wesley Publishing Company (1993).

W. L. Jolly, Modern Inorganic Chemistry, 2nd edition, McGraw-Hill Inc., McGraw-Hill International Editions Chemistry Series (1991).

F. A. Cotton and G. Wilkinson, Basic Inorganic Chemistry, Wiley Eastern Limited (1988).

K. F. Purcell and J. C. Kotz, Inorganic Chemistry, Saunders Golden Sunburst Series, W. B. Saunders Company, Holt-Saunders Japan (1985).

B. E. Douglas, D. H. McDaniel, and J. J. Alexander, Concepts and Models of Inorganic Chemistry, 2nd edition, John Wiley & Sons, Inc. (1983).

Exercises

7.1 Consider Taube's classical experiment in acidic aqueous solution,

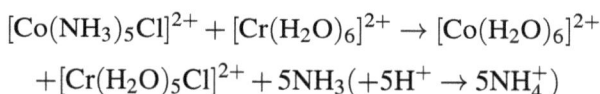

$$[Co(NH_3)_5Cl]^{2+} + [Cr(H_2O)_6]^{2+} \rightarrow [Co(H_2O)_6]^{2+}$$
$$+ [Cr(H_2O)_5Cl]^{2+} + 5NH_3(+5H^+ \rightarrow 5NH_4^+)$$

How do we know Cl^- does not fall off before or after electron transfer? What is the effect of the nature of X^- in $[Co(NH_3)_5X]^{2+}$ likely to be if,

a) step 1 is rate limiting?

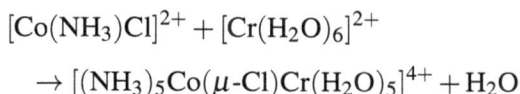

$$[Co(NH_3)Cl]^{2+} + [Cr(H_2O)_6]^{2+}$$
$$\rightarrow [(NH_3)_5Co(\mu\text{-}Cl)Cr(H_2O)_5]^{4+} + H_2O$$

b) step 2 is rate limiting?

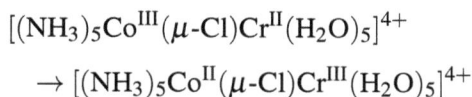

$$[(NH_3)_5Co^{III}(\mu\text{-}Cl)Cr^{II}(H_2O)_5]^{4+}$$
$$\rightarrow [(NH_3)_5Co^{II}(\mu\text{-}Cl)Cr^{III}(H_2O)_5]^{4+}$$

7.2 While the electron self-exchange rate $(M^{-1} s^{-1})$ for $[Co(NH_3)_6]^{2+/3+}$ is 10^{-6}, the same for $[Fe(bipy)_3]^{2+/3+}$ (bipy $= 2,2'$-bipyridine) is $>10^6$. Comment on this observation.

Chapter 8

Organometallic Chemistry of *d*-Block Elements

The understanding of metal-ligand bonding in metal-carbonyls and metal-olefins, and related compounds containing *d*-block transition metals in low formal oxidation states is very important. The chemistry associated with such compounds is very diverse and is of paramount importance in chemical industry.[1,2] However, because of the strong tendency for the metals in these compounds to achieve a closed-shell configuration, it is possible to make very convincing analogies between the chemistry of many of these compounds and that of corresponding non-metallic main group compounds in which there is a similar tendency for the constituent elements to achieve a closed-shell configuration. This domain of chemistry, interfacing inorganic with organic chemistry, is called *organometallic chemistry*. Organometallic chemistry is the study of the synthesis, structure, and reactivity of chemical compounds that usually but not necessarily contain metal-carbon bonds. It is the chemistry of formally low-valent transition metals, not necessarily with only metal-carbon bonds. The metal-cyanide complexes (e.g. ferrocyanide, ferricyanide) with Fe–C bonds do not fall under organometallic category. Notably, many well-known organometallic molecules contain metal-phosphine bonds. The simplest organometallic compound contains the M−C single bond of metal alkyls. These compounds are often used as homogeneous catalysts.

The period just after the second world war encompassed a virtual explosion in our knowledge of organometallic chemistry and in the use

---

[1]J. S. Thayer, *Adv. Organometallic Chem.* **1975**, *13*, 1.
[2]G. W. Parshall and R. E. Putscher, *J. Chem. Educ.* **1986**, *63*, 183.

of this chemistry in catalytic processes (See Chapter 9). A whole new technology of organometallic catalysis, especially for olefin polymerization, blossomed. Nobel prizes were awarded to K. Ziegler (German chemist: 1898–1973) and G. Natta (Italian chemical engineer: 1903–1979) in 1963, and to G. Wilkinson (English chemist: 1921–1996) and E. O. Fischer (German chemist: 1918–2007) in 1973 for their contributions to this area of research.

8.1 Types of ligand

Most ligands form the M−L σ-bond by using a lone pair (a pair of electrons that are nonbonding in the free ligand). For ligands such as PR_3 or pyridine, these lone pairs are often the HOMO and the most basic electrons in the molecule. Classical Werner coordination complexes (Chapter 6) always involve lone-pair donor ligands.

There are two other types of ligand found in organometallic compounds, σ and π, of which H_2 and $H_2C=CH_2$, respectively, are typical examples.

8.2 Metal-carbonyl bonding

Carbon monoxide is the most common π-accepting ligand in organometallic chemistry. The molecular orbital (MO) diagram for CO (10 valence electron) is discussed in Chapter 3. The HOMO of CO has σ symmetry (precisely σ^* orbital), which is largely nonbonding lone pair on C atom (Fig. 8.1). When CO acts as a ligand, this σ orbital serves as a very poor ligand by donating a pair of electrons from this level to appropriate metal orbitals, forming a M−CO σ-bond. The LUMO of CO is the two low-lying doubly degenerate pair of antibonding π orbitals (π^* orbitals). These two orbitals play a crucial role in M−CO bonding. These π^* orbitals can overlap with filled metal d orbitals (as the metals are in low formal oxidation state) of same symmetry (e.g. t_{2g} orbitals in an O_h complex). The π interaction facilitates transfer of metal t_{2g} electrons to CO π^* orbital(s) leading to M−CO back-bonding. Thus the M−CO bonding gives rise to a partial triple bond in C−O. This in effect increases the strength of M−CO bonding

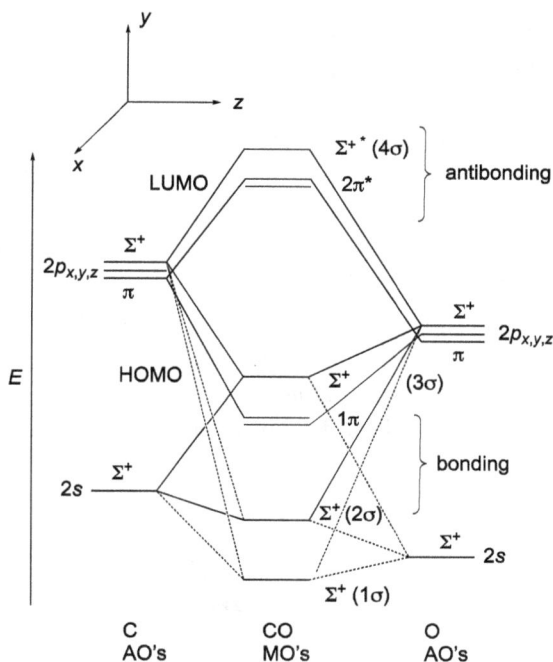

Fig. 8.1 Qualitative MO energy-level diagram for CO.

but in turn the strength of C–O bond strength decreases, as electrons are pushed to the antibonding orbital(s). Carbon monoxide bonds to transition metals using synergic π^*-back-bonding. The π-bonding has the effect of weakening the C–O bond compared with free CO. This leads to the delocalization of electrons from metal to ligand orbitals. Thus, CO acts also as a π acceptor.

Infrared spectra (IR) have been widely used in the study of metal carbonyls since the C–O stretching frequencies give diagnostic strong sharp bands well separated from other vibrational modes of any other ligands. The above-mentioned metal-ligand bonding can explain the observed trend in C–O stretching frequencies obtained from IR spectra of metal-carbonyl complexes.

Let us consider the compounds $[V(CO)_6]^-$ (1860 cm^{-1}), $[Cr(CO)_6]$ (2000 cm^{-1}), and $[Mn(CO)_6]^+$ (2090 cm^{-1}) with their C–O IR stretching vibrations in parentheses, along with the value for free $CO(g)$ 2143 cm^{-1}. All three metal-haxacarbonyl compounds have 18 electrons in their valence shell (see below), with six electrons in the t_{2g} set of orbitals. The only difference is the formal charge on the complexes – mononegative, neutral, and monopositive, respectively. An increase in the negative charge on the metal atom will enhance the flow of electrons from metal t_{2g} to the π^* orbitals. In the chosen complexes the extent of flow of electrons is expected to follow the order: $[V(CO)_6]^- > [Cr(CO)_6] > [Mn(CO)_6]^+$. Consequently, the lowest IR stretching frequency is expected for $[V(CO)_6]^-$.

Notably, C–O stretching frequencies follows the order: terminal CO (such as in $Mn_2(CO)_{10}$; see below) > two bridging CO (such as in $Co_2(CO)_8$; see below) > three bridging CO (such as in $Fe_2(CO)_9$; see below).

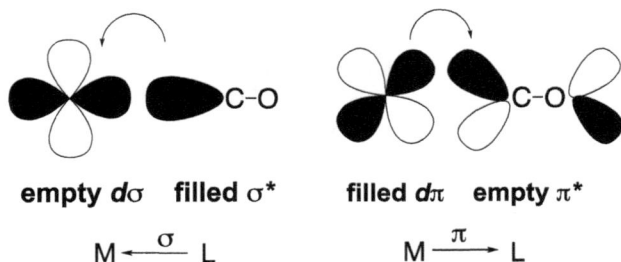

empty $d\sigma$ filled σ^* filled $d\pi$ empty π^*

$$M \xleftarrow{\ \sigma\ } L \qquad\qquad M \xrightarrow{\ \pi\ } L$$

8.3 Metal-olefin bonding

Ethylene has no lone pair, yet it binds strongly to low-valent metals. The C=C π bond in $H_2C=CH_2$ can act as a ligand by donating a pair of electrons to a metal atom in the formation of an organometallic compound such as Zeise's salt $K[Pt^{II}Cl_3(\eta^2\text{-}C_2H_4)]\cdot H_2O$ (Danish organic chemist: W. C. Zeise: 1789–1847) in 1830. In this case the HOMO is the C=C π bond and hence the term π-complex. Since the C=C π bond lies both above and below the molecular plane, the metal has to bind out of the $H_2C=CH_2$ plane, where the electrons are. This type of binding is represented as (η^2-C_2H_4) (pronounced "eta–two ethylene") where η represents the *hapticity* of the ligand, defined as the number of atoms in the ligand bonded to the metal. Simple alkene ligands are dihapto two-electron donors to suitably oriented metal orbitals, utilizing its filled π orbital, and thus form a $M–C_2H_4\,\pi$ bond. As in the case of metal-carbonyls, the π^* orbital of $H_2C=CH_2$ can accept electron density from filled metal orbitals of the appropriate symmetry. This metal-ethene bonding interaction comprising electron donation from filled π orbital of $H_2C=CH_2$ to metal orbital and simultaneous electron acceptance into the empty π^* orbital of $H_2C=CH_2$ from the filled metal orbital is called the Dewar-Chatt-Duncanson (American theoretical chemist, M. J. S. Dewar: 1918–1997; British chemist, J. Chatt: 1914–1994; L. A. Duncanson) model presented in early fifties. Both of these effects tend to reduce the C–C bond order, leading to an elongated C–C distance and a lowering its vibrational frequency.

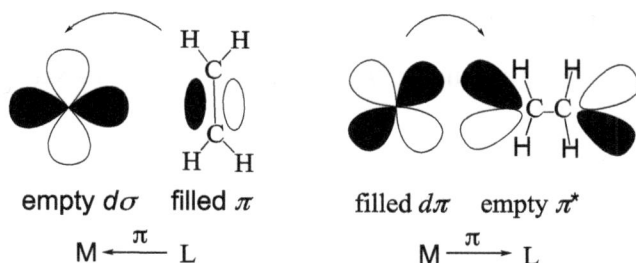

empty $d\sigma$ filled π filled $d\pi$ empty π^*

$$M \xleftarrow{\ \pi\ } L \qquad\qquad M \xrightarrow{\ \pi\ } L$$

Electron donating and accepting properties appear to be fairly evenly balanced in most metal-ethene complexes. Notably, the degree of donation and back-donation can be altered by substituents. Tetracyanoethene $(NC)_2C=C(CN)_2$ is an abnormally strong electron acceptor ligand on account of its electron-withdrawing CN groups. Its π acceptor character dominates over its role as a donor. When the degree of donation of electron density from the metal atom to the alkene ligand is small, substituents on the ligand are bent only slightly away from the metal, and the C–C bond length is only slightly greater than in the free ethene. With electron-rich metals or electron-withdrawing substituents on the alkene, the back-donation is greater. Substituents on the alkene are then bent away from the metal, and the C–C bond length approaches that of a single C–C bond.

These differences led some chemists to depict the first group as simple π-complexes and the second as metallocycles with M–C single bonds.

Example 8.1 Pinpoint two important differences between metal-carbonyls and metal-alkene complexes in terms of (i) donation and back-donation and (ii) structural aspects.

Answer

(i) CO coordinates using filled σ^* orbital to an empty dx^2-y^2 orbital of metal to form a M–C σ bond and for back-donation it uses its empty π^* orbitals. Alkenes use its π orbital to donate to the metal to form a M-alkene π bond and for back-donation it uses its empty π^* orbitals.
(ii) The C–C bond axis in the coordinated alkene is perpendicular to the molecular plane of the metal, whereas in metal carbonyls the CO is in the same plane as that of metal.

The planar alkene molecule becomes nonplanar on complexation with the metal.

Example 8.2 Consider $[V(CO)_6]^-$, $[Cr(CO)_6]$, $[Mn(CO)_6]^+$. Comment on the extent of back-bonding.

Answer

In all three compounds the metal is in d^6 configuration. Negatively charged metal ion will participate in enhanced back-bonding with CO, followed by neutral, and then by positively charged ion. Hence, extent of back-bonding: $[V(CO)_6]^- > [Cr(CO)_6] > [Mn(CO)_6]^+$.

Example 8.3 For $[Ti(CO)_6]^{2-}$, $[V(CO)_6]^-$, $[Cr(CO)_6]$, and $[Mn(CO)_6]^+$, v_{CO} / cm^{-1} values are 1750, 1860, 2000, and 2090, respectively. Explain.

Answer

Extent of back-bonding increases means the bond order of metal-coordinated C–O decreases. It is due to addition of electrons to the empty π^* antibonding orbitals.

Example 8.4 The V–C bond length in $[V(CO)_6]^-$ and $[V(CO)_6]$ are 1.93 and 2.00 Å, respectively. Explain the reason.

Answer

Because of additional negative charge the back-bonding is more in $[V(CO)_6]^-$ than in $[V(CO)_6]$. Hence, C–O bond becomes weaker and V–C bond becomes stronger resulting in a shortening of bond length.

8.4 Counting of valence electrons (18-electron rule)

The rule states that most reactions of diamagnetic organometallic transition metal complexes, including those involved in homogeneous catalysis, proceed by elementary reactions which take the metal from 18 to 16, or 16 to 18 valence electron, or leave the electron count unchanged.

The 18-electron rule is a way to help us decide whether a given d-block transition metal organometallic complex is likely to be thermodynamically stable.[3] In combining six metal orbitals with six ligand orbitals, we make a bonding set of six (the $M-L$ σ bonds) that are stabilized, and an antibonding set of six (the $M-L$ $\sigma*$ levels) that are destabilized when six L groups approach to bonding distance. The remaining three d orbitals, the $d\pi$ set, do not overlap with the ligand orbitals, and remain nonbonding. In a d^6 ion, we have $6e^-$ (six electrons) from Co(0) and $12e^-$ from six CO ligands, giving $18e^-$ in all (example: $Co(CO)_6$). This means that all the levels up to and including the $d\pi$ set are filled, and the $M-L$ $\sigma*$ levels remain unfilled.

In the case of four-coordinate square planar complexes ($Pt^{II}L_4{}^{2-}$; L $=$ Cl$^-$) the energy of one of the d orbitals (dx^2-y^2) is raised very high. Therefore, effectively eight orbitals take part in accommodating a bonding set of four (the $M-L$ σ bonds) and four d orbitals. In a d^8 ion, we have $8e^-$ from Pt(II) and $8e^-$ from four ligands, giving $16e^-$ in all.

Electron counts for common ligands and hapticity:

$[Mn(CO)_5]$: $5 \times 2e^-$ (CO) $+ 7e^-$ (Mn) $= 17e^-$, actually it is a dimer $[Mn_2(CO)_{10}]$ (structure I)

At each Mn: $5 \times 2e^- + 7e^- + 1e^-$ (Mn–Mn bond) $= 18e^-$

---

[3]P. R. Mitchell and R. V. Parish, *J. Chem. Educ.* **1969**, *46*, 811.

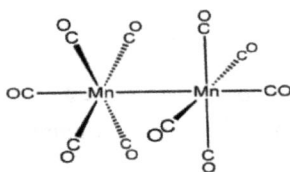

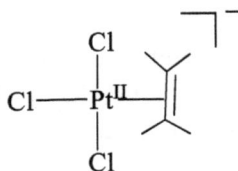

I   II

Zeise's salt $K[Pt^{II}(C_2H_4)Cl_3]\cdot H_2O$ was one of the first organometallic compounds, containing a transition metal ion, to be prepared.

$[Pt^{II}(C_2H_4)Cl_3]^-$: $2e^-$ $(C_2H_4) + 3 \times 2e^-$ $(3\,Cl^-) + 8e^-$ $(Pt^{II}) = 16e^-$ (structure II)

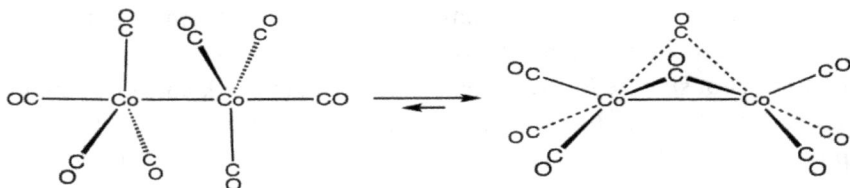

III

$[Co(CO)_4]$: $4 \times 2e^- + 9e^- = 17e^-$, actually it is a dimer $[Co_2(CO)_8]$.

Left of the equilibrium (structure in solution): No. of valence electron at each Co: $4 \times 2e^-$ (4 terminal CO) $+ 9e^-$ (Co) $+ 1e^-$ (Co–Co bond) $= 18e^-$ (structure III)

Right of the equilibrium (X-ray structure): No. of valence electron at each Co: $3 \times 2e^-$ (3 terminal CO) $+ 9e^-$ (Co) $+ 2 \times 1e^-$ (1 bridging CO) $+ 1e^-$ (Co–Co bond) $= 18e^-$ (structure III)

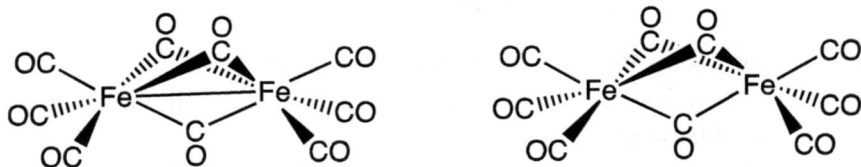

IV

$[Fe_2(CO)_9]$: No. of valence electron at each Fe (left): $3 \times 2e^-$ (3 terminal CO) $+ 8e^-$ (Fe) $+ 3 \times 1e^-$ (3 bridging CO) $+ 1e^-$ (Fe–Fe bond) $= 18e^-$ (old bonding description). Theoretical analyses have consistently indicated the absence of a direct Fe–Fe bond. This bonding model proposes an Fe–C–Fe three-center-two-electron *banana bond* (cf. B_2H_6) for one of the bridging carbonyls (recent bonding description; no. of valence electron at each Fe (right): $3 \times 2e^-$ (3 terminal CO) $+ 8e^-$ (Fe) $+ 2 \times 1e^-$ (2 bridging CO) $+ 2e^-$ (Fe–CO–Fe 3c-2e$^-$ bond $= 18e^-$) (structure IV).

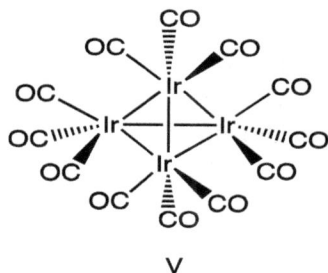

V

$[Ir(CO)_3]_4$: No. of valence electron at each Ir $= 3 \times 2e^-$ (3 terminal CO) $+ 9e^-$ (Ir) $+ 3 \times 1e^-$ (3 Ir–Ir bond) $= 18e^-$ (structure V).

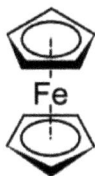

$[Fe(\eta^5\text{-}C_5H_5)_2]$ VI $[Cr(\eta^6\text{-}C_6H_6)_2]$ VII

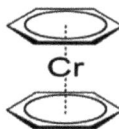

No. of valence electron at Fe $= 2 \times 6e^-$ (2 $C_5H_5^-$) $+ 6e^-$ (Fe(II)) $= 18e^-$ (structure VI). Cyclopentadienide ion $C_5H_5^-$ is a six-electron donor.

No. of valence electron at Cr $= 2 \times 6e^-$ (2 C_6H_6) $+ 6e^-$ (Cr(0)) $= 18e^-$ (structure VII). C_6H_6 is a six-electron donor.

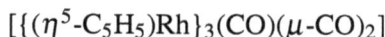

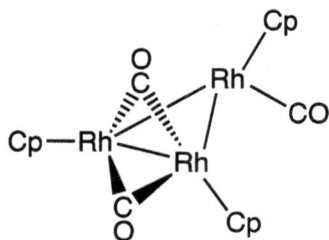

$[\{(\eta^5\text{-}C_5H_5)Rh\}_3(CO)(\mu\text{-}CO)_2]$

VIII

No. of valence electron at two Rh = $6e^-$ (1 Cp^- or 1 $C_5H_5^-$) + 2 × $1e^-$ (2 bridging CO) + $8e^-$ (Rh(I)) + 2 × $1e^-$ (2 Rh–Rh bond) = $18e^-$ (structure VI). Cp^-: $C_5H_5^-$ is a six-electron donor.

No. of valence electron at one Rh = $6e^-$ (1 Cp^-) + 1 × $2e^-$ (1 terminal CO) + $8e^-$ (Rh(I)) + 2 × $1e^-$ (2 Rh–Rh bond) = $18e^-$ (structure VIII).

8.5 Organometallic reactions: oxidative addition, reductive elimination, insertion (migratory insertion), and β-hydrogen (β-hydride) elimination

Oxidative addition is often a step in catalytic cycles, in conjunction with its reverse reaction, reductive elimination.

Oxidative addition

The term *oxidative addition* is used when an X−Y bond has been broken by the insertion of a metal fragment L_nM into the X−Y bond. X−Y can be any one of a large number of groups: H–H, Cl–Cl, Me–I etc. Thus, oxidative addition is a process that increases both the oxidation state and coordination number of a metal center.

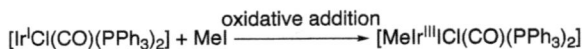

$$L_nM + \underset{Y}{\overset{X}{|}} \longrightarrow L_nM \overset{X}{\underset{Y}{<}}$$

$$[Ir^I Cl(CO)(PPh_3)_2] + MeI \xrightarrow{\text{oxidative addition}} [MeIr^{III}ICl(CO)(PPh_3)_2]$$

Valence electron count:

2 × 2 (2 PPh_3) + 1 × 2 (1 CO) + 1 × 2 (1 Cl^-)   2 × 2 (2 PPh_3) + 1 × 2 (1 CO) + 1 × 2 (1 Cl^-) +
+ 8 (Ir(I)) = $16e^-$ 1 × 2 (1 I^-) + 1 × 2 (1 Me^-) + 6 (Ir(III)) = $18e^-$

The formal oxidation state of Ir in the product is +3. Thus, due to oxidative addition reaction, the formal oxidation state of Ir changes from +1 to +3.

$$[Rh^I(PPh_3)_3Cl] + H_2 \text{ (oxidative addition)} \rightarrow cis\text{-}[Rh^{III}(PPh_3)_3(H)_2Cl]$$

Valence electron count:

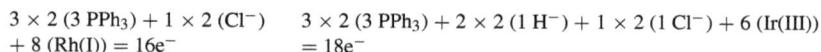

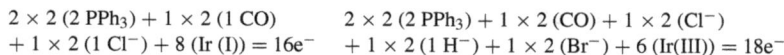

3×2 (3 PPh$_3$) $+ 1 \times 2$ (Cl$^-$) $\qquad$ 3×2 (3 PPh$_3$) $+ 2 \times 2$ (1 H$^-$) $+ 1 \times 2$ (1 Cl$^-$) $+ 6$ (Ir(III))
$+ 8$ (Rh(I)) $= 16e^- \qquad\qquad\quad = 18e^-$

The formal oxidation state of Rh in the product is $+3$. Thus, due to oxidative addition reaction, the formal oxidation state of Rh changes from $+1$ to $+3$.

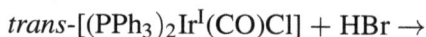

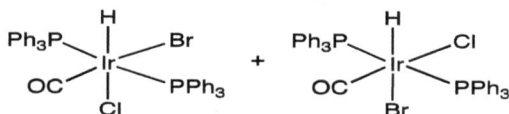

$\quad$ *trans*-[(PPh$_3$)$_2$Ir$^{\rm I}$(CO)Cl] + HBr $\rightarrow$

Valence electron count:

2×2 (2 PPh$_3$) $+ 1 \times 2$ (1 CO) $\qquad$ 2×2 (2 PPh$_3$) $+ 1 \times 2$ (CO) $+ 1 \times 2$ (Cl$^-$)
$+ 1 \times 2$ (1 Cl$^-$) $+ 8$ (Ir (I)) $= 16e^- \quad + 1 \times 2$ (1 H$^-$) $+ 1 \times 2$ (Br$^-$) $+ 6$ (Ir(III)) $= 18e^-$

The formal oxidation state of Ir in the products is $+3$. As a consequence of oxidative addition reaction, the formal oxidation state of Ir changes from $+1$ to $+3$.

A special case of oxidative addition is *cyclometalation*, in which a C$-$H bond in a ligand oxidatively adds to a metal to give a ring. Because of this ring formation, the reaction can be highly selective, for example, only one of the nine distinct CH bonds in benzoquinoline is cleaved when cyclometalation occurs.

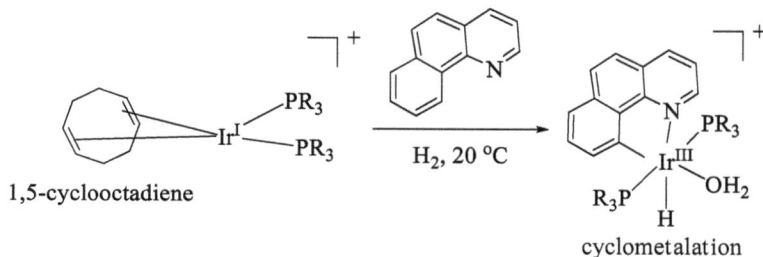

1,5-cyclooctadiene

cyclometalation

Valence electron count:

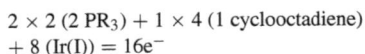

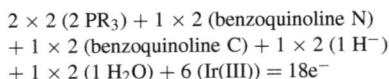

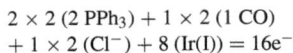

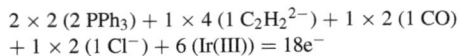

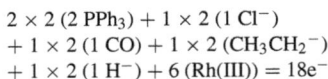

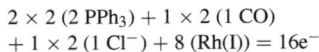

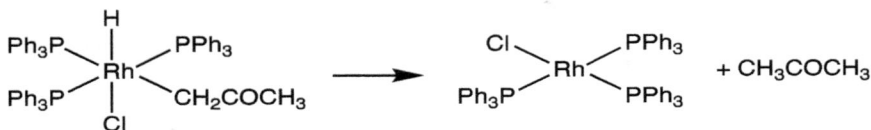

2×2 (2 PR$_3$) + 1×4 (1 cyclooctadiene)
+ 8 (Ir(I)) = 16e$^-$

2×2 (2 PR$_3$) + 1×2 (benzoquinoline N)
+ 1×2 (benzoquinoline C) + 1×2 (1 H$^-$)
+ 1×2 (1 H$_2$O) + 6 (Ir(III)) = 18e$^-$

trans-[(PPh$_3$)$_2$Ir^I(CO)Cl] + acetylene (oxidative addition) $\rightarrow$

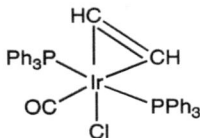

Valence electron count:

2×2 (2 PPh$_3$) + 1×2 (1 CO)
+ 1×2 (Cl$^-$) + 8 (Ir(I)) = 16e$^-$

2×2 (2 PPh$_3$) + 1×4 (1 C$_2$H$_2{}^{2-}$) + 1×2 (1 CO)
+ 1×2 (1 Cl$^-$) + 6 (Ir(III)) = 18e$^-$

The formal oxidation state of Ir in the product is $+3$. Thus, the oxidative addition reaction brings about a change in Ir formal oxidation state from $+1$ (one negative charge from Cl$^-$) to $+3$ (one negative charge from Cl$^-$ and two negative charges from $^-$HC=CH$^-$.

Reductive elimination

A very common decomposition pathway for metal-alkyls is *reductive elimination*. This leads to a decrease by two units in the electron count, the formal oxidation state, and the number of bonds.

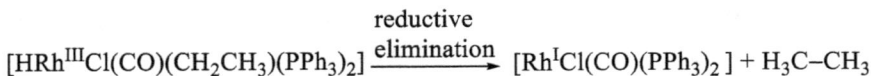

[HRh^{III}Cl(CO)(CH$_2$CH$_3$)(PPh$_3$)$_2$] $\xrightarrow{\text{reductive elimination}}$ [Rh^{I}Cl(CO)(PPh$_3$)$_2$] + H$_3$C–CH$_3$

Valence electron count:

2×2 (2 PPh$_3$) + 1×2 (1 Cl$^-$)
+ 1×2 (1 CO) + 1×2 (CH$_3$CH$_2{}^-$)
+ 1×2 (1 H$^-$) + 6 (Rh(III)) = 18e$^-$

2×2 (2 PPh$_3$) + 1×2 (1 CO)
+ 1×2 (1 Cl$^-$) + 8 (Rh(I)) = 16e$^-$

Valence electron count:

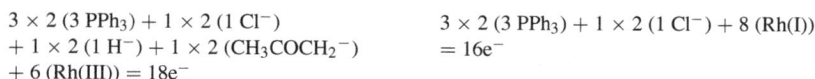

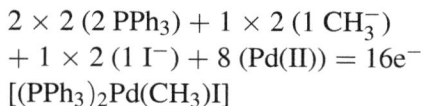

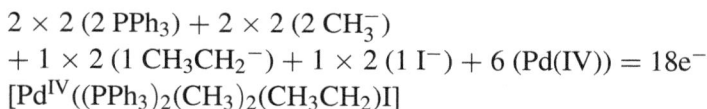

3×2 (3 PPh$_3$) $+ 1 \times 2$ (1 Cl$^-$)
$+ 1 \times 2$ (1 H$^-$) $+ 1 \times 2$ (CH$_3$COCH$_2{}^-$)
$+ 6$ (Rh(III)) $= 18e^-$

3×2 (3 PPh$_3$) $+ 1 \times 2$ (1 Cl$^-$) $+ 8$ (Rh(I))
$= 16e^-$

The formal oxidation state of Rh product is $+1$. Thus, reductive elim-ination reaction brings about a change in Rh formal oxidation state from $+3$ (three negative charges from H$^-$, Cl$^-$, and CH$_3$COCH$_2{}^-$) to $+1$ (one negative charge from Cl$^-$).

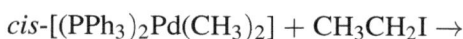

cis-[(PPh$_3$)$_2$Pd(CH$_3$)$_2$] $+$ CH$_3$CH$_2$I $\rightarrow$

Ph$_3$P, I
    Pd $+$ CH$_3$CH$_2$CH$_3$ $+$
H$_3$C, PPh$_3$

Ph$_3$P, I
    Pd $+$ CH$_3$CH$_3$
H$_3$CH$_2$C, PPh$_3$

Valence electron count:

2×2 (2 PPh$_3$) $+ 2 \times 2$ (2 CH$_3{}^-$)
$+ 8$ (Pd(II)) $= 16e^-$
[(PPh$_3$)$_2$Pd(CH$_3$)$_2$]

2×2 (2 PPh$_3$) $+ 1 \times 2$ (1 I$^-$) $+$
1×2 (1 CH$_3$CH$_2{}^-$) $+ 8$ (Pd(II)) $= 16e^-$

2×2 (2 PPh$_3$) $+ 1 \times 2$ (1 CH$_3{}^-$)
$+ 1 \times 2$ (1 I$^-$) $+ 8$ (Pd(II)) $= 16e^-$
[(PPh$_3$)$_2$Pd(CH$_3$)I]

2×2 (2 PPh$_3$) $+ 2 \times 2$ (2 CH$_3{}^-$)
$+ 1 \times 2$ (1 CH$_3$CH$_2{}^-$) $+ 1 \times 2$ (1 I$^-$) $+ 6$ (Pd(IV)) $= 18e^-$
[Pd^{IV}((PPh$_3$)$_2$(CH$_3$)$_2$(CH$_3$CH$_2$)I]

A four-coordinate complex [Pd^{II}(PPh$_3$)$_2$(CH$_3$)I] becomes a six-coordinate complex [Pd^{IV}((PPh$_3$)$_2$(CH$_3$)$_2$(CH$_3$CH$_2$)I] due to oxidative addition reaction. Then the six-coordinate complex is transformed to a four-coordinate complex [Pd^{II}(PPh$_3$)$_2$(CH$_3$CH$_2$)I] due to reductive elim-ination with the release of C$_2$H$_6$. Thus it is a case of an oxidative addition followed by a reductive elimination.

Insertion (or migratory insertion)

The *insertion* is particularly important because it allows to make a metal-alkyl from a metal-alkene and a metal-hydride. We shall see

(see below) how this sequence can lead to a whole series of catalytic trans-formations of alkenes, such as hydrogenation of alkenes with H_2 to give alkanes, hydroformylation with H_2 and CO to give aldehydes, and hydro-cyanation with HCN to give nitriles. Such catalytic reactions are among the most important applications of organometallic chemistry. Olefin insertion is the reverse of the β-hydrogen (β-hydride) elimination reaction.

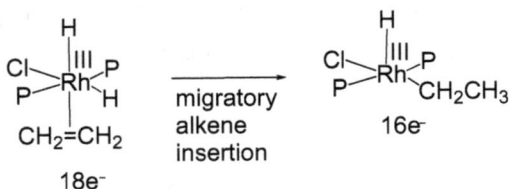

migratory alkene insertion

18e⁻ → 16e⁻

$2 \times 2 + 1 \times 2 + 2 \times 2 + 1 \times 2 + 6 = 18e^-$ $2 \times 2 + 1 \times 2 + 1 \times 2 + 1 \times 2 + 6 = 16e^-$

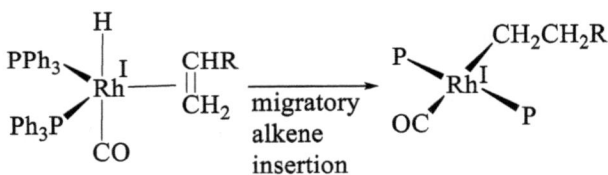

migratory alkene insertion

$2 \times 2 + 1 \times 2 + 1 \times 2 + 1 \times 2 + 8$
$= 18e^-$ $2 \times 2 + 1 \times 2 + 1 \times 2 + 8$
     $= 16e^-$

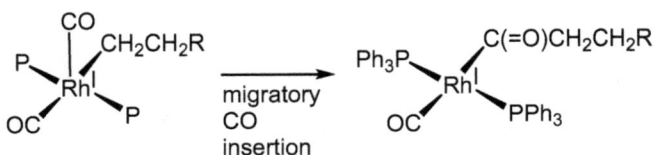

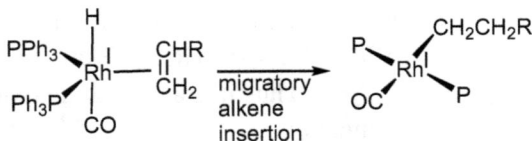

migratory CO insertion

$2 \times 2 + 2 \times 2 + 1 \times 2 + 8$
$= 18e^-$ $2 \times 2 + 1 \times 2 + 1 \times 2 + 8$
     $= 16e^-$

migratory alkene insertion

$2 \times 2 + 1 \times 2 + 1 \times 2 + 1 \times 2 + 8$
$= 18e^-$ $2 \times 2 + 1 \times 2 + 1 \times 2 + 8$
     $= 16e^-$

$P = PPh_3$

β-Hydrogen (β-hydride) elimination

The major decomposition pathway for metal-alkyls is *β-elimination,* which converts a metal-alkyl into a hydridometal-alkene complex.

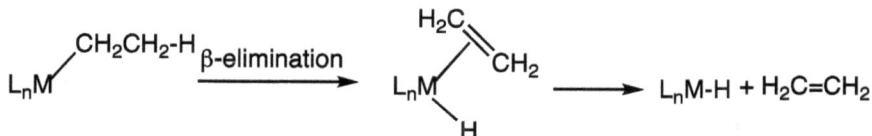

In this reaction, an elimination of H^- takes place from a β-hydrogen of metal-coordinated $CH_3CH_2^-$ and ends up getting coordinated to the metal. In the course of this reaction, a metal-coordinated alkane complex is transformed to a metal-coordinated alkene complex. The formal oxidation state of the metal remains invariant at $+3$ (reactant: Cp^-, $CH_3CH_2^-$, and CH_3^-; product: Cp^-, CH_3^-, and H^-).

Example 8.5 Using the $18e^-$ rule as a guide, indicate the probable number of carbonyl ligands in (i) $[(\eta^5\text{-}C_5H_5)Rh(CO)_n]$ and (ii) $[(\eta^6\text{-}C_6H_6)W(CO)_n]$.

Which one of the two complexes $[W(CO)_6]$ and $[(PPh_3)_2Ir(CO)Cl]$ should undergo faster exchange with ^{13}CO? Justify the answer.

Answer

(i) $[(\eta^5\text{-}C_5H_5)Rh^I(CO)_n]$: No. of valence electron $= 1 \times 6e^- + 8e^- + 2 \times 2e^- = 18e^-$

(ii) $[(\eta^6\text{-}C_6H_6)W^0(CO)_n]$: No. of valence electron $= 1 \times 6e^- + 6e^- + 3 \times 2e^- = 18e^-$

$[(\eta^5\text{-}C_5H_5)Rh^I(CO)_2]$ and $[(\eta^6\text{-}C_6H_6)W^0(CO)_3]$.

$[W^0(CO)_6]$: No. of valence electron $= 6 \times 2e^- + 6e^- = 18e^-$

$[(PPh_3)_2Ir^I(CO)Cl]$: No. of valence electron $= 2 \times 2e^-$ (2 PPh$_3$) $+ 8e^-$ (Ir(I)) $+ 1 \times 2e^-$ (1 CO) $+ 1 \times 2e^-$ (1 Cl$^-$) $= 16e^-$

Ir complex will undergo faster exchange with ^{13}CO, as it is coordinatively unsaturated.

Example 8.6 Verify whether $[(CH_3)Mn(CO)_5]$ is thermodynamically stable.

Answer

No. of valence electron $= 5 \times 2e^-$ (5 CO) $+ 6e^-$ (Mn(I)) $+ 1 \times 2e^-$
$(1\ CH_3^-) = 18e^-$
  It is thermodynamically stable.

Example 8.7 Write the structure of the products (A and B) in the following reactions:

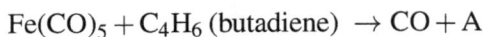

$$Fe(CO)_5 + C_4H_6\ \text{(butadiene)} \rightarrow CO + A$$

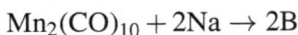

$$Mn_2(CO)_{10} + 2Na \rightarrow 2B$$

Answer

A: $4 \times 2e^-$ (4 CO) $+ 8e^-$ (Fe(0)) $+ 1 \times 2e^-$ (1 C$_2$H$_4$) $= 18e^-$

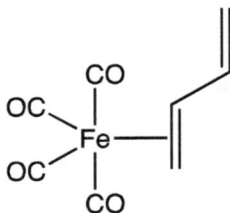

B: Na$^+$[Mn(CO)$_5$]$^-$: $5 \times 2e^-$ (5 CO) $+ 8e^-$ (Mn($-$I)) $= 18e^-$

Example 8.8 When a sample of [(η^5-C$_5$H$_5$)Mo(CO)$_3$]$_2$ is heated a product is formed [(η^5-C$_5$H$_5$)Mo(CO)$_2$]$_2$ is formed. Both the complexes show in their IR spectra only one type of C–O stretching frequency: the former at 1970 cm^{-1} and the latter at 1840 cm^{-1}. Draw the structure of the reactant and the product, keeping the 18e$^-$ rule in mind.

Answer

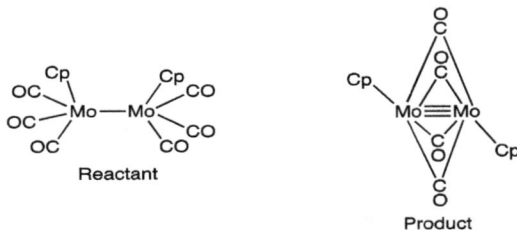

Total no. of valence electron at each Mo
$$= 1 \times 6e^- \ (1 \ Cp^-) + 3 \times 2e^- \ (3 \ \text{terminal CO}) + 5e^- \ (Mo(I)) + 1 \times 1e^-$$
(1 Mo–Mo bond) $= 18e^-$
(reactant)
$$= 1 \times 6e^- \ (1 \ Cp^-) + 4 \times 1e^- \ (4 \ \text{bridging CO}) + 5e^- \ (Mo(I)) + 3 \times 1e^-$$
(3 Mo–Mo bonds) $= 18e^-$
(product)

8.6 Noninnocent ligands in organometallic chemistry

Transition metal-catalyzed chemical transformations in general and specifically homogeneous organometallic catalysis involve formally transfer of two electrons from or to the metal pertaining to oxidative addition and reductive elimination steps (see above). First-row transition metal ions undergo one electron transformations and thus can mediate radical-type reactions. This limits their potential applicability in organic reactions that require two-electron transfer. However, this limitation can be overcome by exploiting 'ligand redox noninnocence' or in other words using noninnocent ligands in stoichiometric or catalytic reactions. These ligands offer an intriguing way to approach multielectron reactivity and play a vital role in catalysis.

A combination of both metal and ligand redox in a metal-ligand system is expected to lead to new catalytic reactions and improvements in existing processes. This concept has been slowly and steadily entering the area of catalysis. Ligand redox noninnocence is clearly becoming a useful synthetic tool to enhance the reactivity and control selectivity. Redox-active ligands can participate in catalytic transformations by storing and providing electrons, modifying the Lewis acidity of the metal, and assisting the formation/breaking of substrate covalent-bond electron-transfer events.

Ligand redox-based oxidative addition and reductive elimination
An elegant example of metal–carbon bond formation by oxidative additions without a change of the metal *d*-electron configuration, by virtue of electron transfer from redox-active ligands to the metal, is discussed here. The metal-ligand system chosen is an unusual paramagnetic square planar complex $[Co^{III}\{(L^{AP})^{2-}\}_2]^-$, which contains two redox-active

2-amidophenolate ligands $(L^{AP})^{2-}$. The 2-aminophenolate ligands are easily oxidized in a one-electron process to form stable 2-iminosemiquinonate radical anion $(L^{ISQ})^{\cdot -}$ and then to neutral iminobenzoquinone $(L^{IBQ})^{0}$ redox level and thus behave as robust, "redox non-innocent ligands" (Scheme 8.1). The Co(III) system adopts an unusual intermediate spin $(dxy)^{2}(dz^{2})^{2}(dxz)^{1}(dyz)^{1}$ electron configuration $(S = 1)$, wherein the doubly occupied dz^{2} orbital must be the highest occupied molecular orbital (HOMO) of the complex, thus explaining the nucleophilic character of the metal (Fig. 8.3). Reaction of the anionic complex $[Co^{III}\{(L^{AP})^{2-}\}_2]^{-}$ with electrophilic alkyl halogenides R–X leads to formation of Co–C bonds by oxidative addition, which shows the characteristics of S_N2-type reactions (i.e., reactivity order $I^{-} > Br^{-} > Cl^{-}$ and strong influence of steric factors in R). The observed nucleophilic reactivity is atypical for d^{6} Co(III) ions, and is normally reserved for low-valent, low spin d^{8} M(I) species (M = Co, Rh, Ir).[4,5]

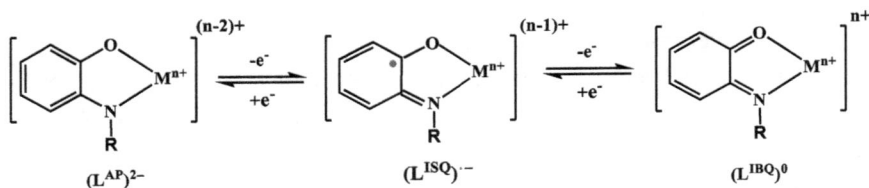

Scheme 8.1   The ligand-centered redox processes of metal-coordinated 2-aminophenolates.

Oxidative addition at the Co(III) center would formally lead to a Co(V) species, but the actual process occurs without a change of the true oxidation state of the metal. The two electrons required to make the Co–C bond are actually obtained from the two redox-active $(L^{AP})^{2-}$ ligands, which are each oxidized by one-electron to form a neutral diamagnetic Co(III) complex with two antiferromagnetically coupled $(L^{ISQ})^{\cdot -}$ ligand radicals (Fig. 8.3). The anionic starting complex $[Co^{III}\{(L^{AP})^{2-}\}_2]^{-}$ and some of

[4]W. I. Dzik, J. I. van der Vlugt, J. N. H. Reek, and B. de Bruin, *Angew. Chem. Int. Ed.* **2011**, *50*, 3356.
[5](a) A. L. Smith, K. I. Hardcastle, and J. D. Soper, *J. Am. Chem. Soc.* **2010**, *108*, 14358.
(b) A. L. Smith, L. A. Clapp, K. I. Hardcastle, and J. D. Soper, *Polyhedron* **2010**, *29*, 164.

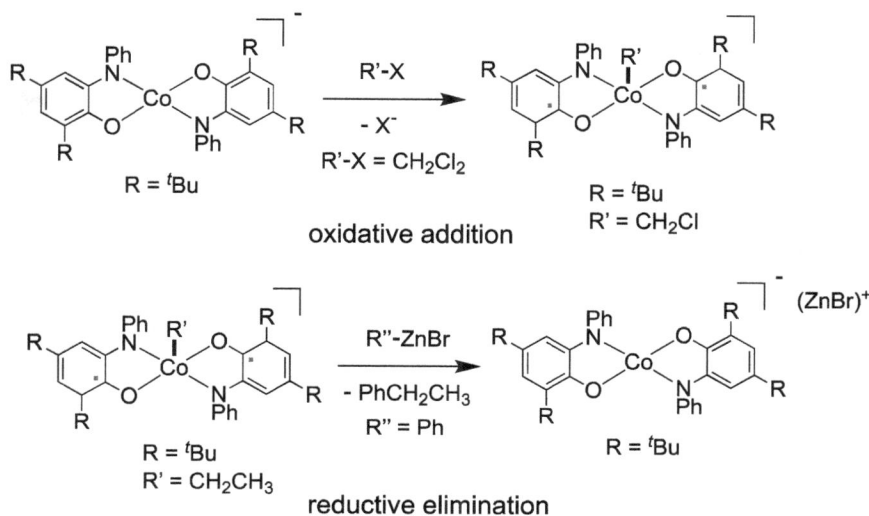

Fig. 8.3 Oxidative addition and reductive elimination reactions on a Co(III) complex of a bidentate redox-active ligand.

the neutral oxidative addition products (R = CH$_2$Cl and Et) were characterized by X-ray diffraction. The changes of the bond-lengths within the ligand framework clearly confirm that the actual redox processes occur at the ligands, and hence provide solid proof for the involvement of redox-active ligands as electron reservoirs to achieve two-electron reactions with cheap, readily available transition metal ions.

Even more noteworthy is that the same system allows for C–C bond-forming reductive elimination reactions, again without a change of the metal oxidation state. Treatment of the [Co^{III}(R$^-$){(L^{ISQ})$^{\cdot-}$}$_2$] species with organozinc reagents led to reductive C–C bond formation with regeneration of the complex [Co^{III}{(L^{AP})$^{2-}$}$_2$]$^-$. Together, these reactions comprise a complete, well-defined stoichiometric cycle for Co(III)-mediated Negishi-type cross-coupling (formation of R$'$–R$''$) reactions.[4,5]

The abovementioned remarkable discoveries can clearly expand the development of cheap catalysts based on abundant first-row transition metals. The elegance of this approach relies on the use of redox-active ligands to circumvent some of the limitations of first-row transition metal ions.

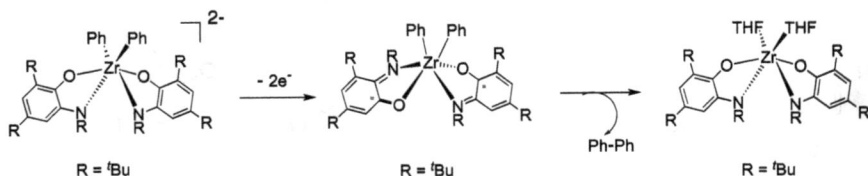

Fig. 8.4 Reductive elimination reaction on a Zr(IV) complex of a tetradentate redox-active ligand.

Ligand-centered redox processes in C–C bond-forming reductive eliminations without a change in the metal oxidation state were also demonstrated for a Zr(IV) complex (Fig. 8.4).[6] Treatment of a THF solution of structurally characterized $[Zr^{IV}\{(L^{AP})^{2-}\}_2(Ph)_2]^{2-}$ with $[Fe(\eta^5\text{-}C_5H_5)_2](PF_6)$ at $-78°C$ generated $[Zr^{IV}\{(L^{ISQ})^{\cdot-}\}_2(Ph)_2]$.

8.7 The σ complexes

Molecular hydrogen has neither a lone pair nor a π bond, yet it also binds to metals. The only available electron pair is the H–H σ bond, and this becomes the donor. Back donation in this case is accepted by the σ* orbital of H_2. The metal binds in the side-on mode to H_2 to maximize σ–dσ overlap. The electron donation from the filled H–H σ bond to the empty d_σ orbital on the metal, and the back donation from the filled $M(d_\pi)$ orbital to the empty H–H σ*.

For σ donors such as H_2, or an alkane, forming the M–L σ bond partially depletes the H–H σ bond because electrons that were fully engaged in keeping the two H atoms together in free H_2 are now also delocalized over the metal (hence the name *two-electron, three-center bond* for this interaction). Back bonding into the H–H σ* causes additional weakening or even breaking of the H–H σ bond because the σ* is antibonding with respect to H–H.

---

[6] M. R. Haneline and A. F. Heyduk, *J. Am. Chem. Soc.* **2006**, *128*, 8410.

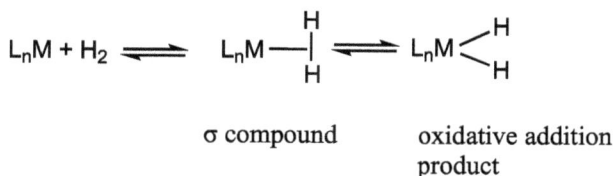

$$L_nM + H_2 \rightleftharpoons L_nM-\begin{array}{c} H \\ | \\ H \end{array} \rightleftharpoons L_nM\begin{array}{c} H \\ \diagup \\ \diagdown H \end{array}$$

σ compound   oxidative addition
                                     product

Fig. 8.2   Activation of H_2 and formation of metal hydride bonds.

Free H_2 has an H$-$H distance of 0.74 Å, but the H$-$H distances in H_2 complexes go all the way from 0.82 to 1.5 Å. Eventually the H$-$H bond breaks and a dihydride is formed (Fig. 8.2). This is the *oxidative addition reaction*. Formation of a σ complex can be thought of as an incomplete oxidative addition.

In general, the basicity of electron pairs decreases in the following order: lone pairs $>$ π-bonding pairs $>$ σ-bonding pairs, because being part of a bond stabilizes electrons. The usual order of binding ability is therefore as follows: lone-pair donor $>$ π donor $>$ σ donor.

8.8 Bis(cyclopentadienyl)M(II) complexes

The bonding in the normal metallocenes will be discussed here only briefly.[7] In a D_{5d} geometry (e.g. ferrocene, Cp_2Fe) the orbitals of two parallel $C_5H_5^-$ ligands give rise to three sets of approximately degenerate orbitals: a low-lying filled pair of a_{1g} and a_{2u} symmetry, a set of filled orbitals, e_{1g} and e_{1u}, and a high-lying empty set of antibonding orbitals of symmetry e_{2g} and e_{2u}. These interact with the orbitals of the metal as shown in Fig. 8.5. There is a strong interaction with the metal s and p orbitals and also a strong bonding interaction with the e_{1g} (dxz, dyz) set. The remaining three d orbitals of the metal, the $a_{1g}(dz^2)$ and the $e_{2g}(dx^2-y^2, dxy)$ set, remain essentially nonbonding. Thus the d-level splitting is

$$e_{2g} \leq a_{1g} < e_{1g}^* \quad \text{or} \quad (dx^2-y^2, dxy) \leq (dz^2) < (dxz, dyz).$$

[7] J. W. Lauher and R. Hoffmann, *J. Am. Chem. Soc.* **1976**, *98*, 1729.

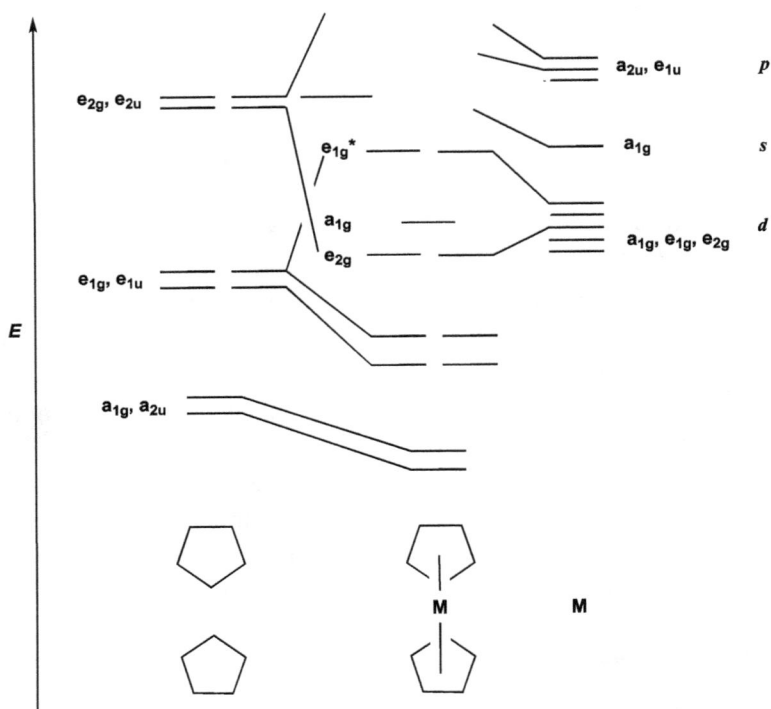

Fig. 8.5   Orbital interaction diagram for a D_{5d} metallocene.

Further reading

N. N. Greenwood and A. Earnshaw, Chemistry of the Elements, 2nd edition, Elsevier Butterworth-Heinemann (2010)

C. E. Housecroft and A. G. Sharpe, *Inorganic Chemistry*, 2nd edition, Pearson Education Limited (2005)

R. H. Crabtree, *The Organometallic Chemistry of the Transition Metals*, 4th edition, Wiley Interscience (2005)

D. F. Shriver and P. W. Atkins, *Inorganic Chemistry*, 3rd edition, Oxford University Press (1999)

F. A. Cotton, G. Wilkinson, C. A. Murillo, and M. Bochmann, *Advanced Inorganic Chemistry*, 6th edition, John Wiley & Sons, Inc. (1999)

J. E. Huheey, E. A. Keiter, and R. L. Keiter, *Inorganic Chemistry: Principles of Structure and Reactivity*, 4th edition, Addison-Wesley Publishing Company (1993)

F. A. Cotton and G. Wilkinson, *Advanced Inorganic Chemistry*, 5th edition, John Wiley & Sons (1988)

F. A. Cotton and G. Wilkinson, Basic Inorganic Chemistry, Wiley Eastern Limited (1988)

N. N. Greenwood and A. Earnshaw, Chemistry of the Elements, 1st edition, Pergamon Press (1984)

Exercises

8.1 Draw the plausible structure and show the electron count of $[\text{Co}(\eta^4\text{-C}_4\text{H}_4)(\eta^5\text{-C}_5\text{H}_5)]$.

8.2 Complete and identify the reaction:

$$[\text{Ir}(\text{PPh}_3)_2(\text{CO})\text{Cl})] + \text{F}_2\text{C}{=}\text{CF}_2 \rightarrow$$

8.3 Nickelocene, Cp_2Ni, has a structure analogous to that of Cp_2Fe, but the Ni–C distance is about 0.16 Å longer than the Fe–C distance. Rationalize the fact.

Chapter 9

Organometallic Catalysis

The organometallic field has provided a series of important conceptual insights, unusual structures, and useful catalysts both for industrial processes and for organic synthesis. Many catalysts are capable of very high levels of asymmetric induction in preferentially forming one enantiomer of a chiral product.

Homogeneous and heterogeneous catalysis both began to be used in the chemical industry about 1910, but heterogeneous catalysis grew steadily while the use of soluble catalysts started in the 1950's. A major difference was that many large-scale hydrogenation reactions were found to proceed efficiently with solid catalysts. These reactions included the synthesis of methanol, ammonia, and aniline. In contrast, the applications of homogeneous catalysis were largely limited to reactions of acetylene, an expensive starting material. In spite of its cost, however, acetylene was used to produce organic chemicals such as acetaldehyde, acrylonitrile, and vinyl monomers.

While the largest components are based on heterogeneous (solid) catalysis, the value of products derived from organometallic catalysis and soluble (homogeneous) catalysts is also very impressive. Traditionally, heterogeneous catalysts have been used for production of large-scale commodity chemicals such as methanol and ammonia, and in the production of high-octane gasoline from petroleum. Homogeneous catalysts, which are soluble in the reaction medium, are commonly used in the production of high-purity, high-value chemicals. The catalysts used for production of polypropylene and high-density polyethylene are often hybrid species in which an organometallic functional group is bound to a solid support. The distinctions between these classes of catalysts are rather arbitrary because similar chemical reactions often take place in the coordination sphere of a catalytic metal whether it is present in solution or adsorbed on a solid surface.

9.1 Why 4d metal ions?

Often the compounds of all metals in a group exhibit catalytic activity in a particular reaction, but the $4d$-metal compounds are often superior as catalysts to their lighter and heavier congeners. In some cases the difference may be associated with balanced substitutional lability and inertness of $4d$ organometallic compounds in comparison with their $3d$ and $5d$ analogs.

9.2 Hydrogenation of alkene

The importance of reddish-violet compound $[Rh^I Cl(PPh_3)_3]$, the well-known *Wilkinson's catalyst*, arises from its effectiveness as a catalyst for highly selective hydrogenations of organic molecules which are of great importance in the pharmaceutical industry. Using this catalyst rapid homogeneous hydrogenation of olefins occurs at 298 K and 1 bar H_2 pressure:

$$RCH=CH_2 + H_2 \rightarrow RCH_2CH_3$$

Fig. 9.1 shows a simplified but reasonable *catalytic cycle*. A catalyzed reaction pathway is usually represented by a catalytic cycle. In benzene/ethanol $[Rh^I Cl(PPh_3)_3]$ dissociates to some extent to release a PPh_3 molecule and a solvent molecule (solv) fills the site generating $[Rh^I Cl(PPh_3)_2(solv)]$ ($[Rh^I Cl(PPh_3)_3] + solv \rightleftharpoons [Rh^I Cl(PPh_3)_2(solv)] + PPh_3; K \sim 10^{-4}$).

The essential steps in this catalytic cycle (Fig. 9.1) are the oxidative addition of H_2 (activation of H_2 by $[Rh^I Cl(PPh_3)_3]$) brings about oxidation of Rh(I) by two electrons to Rh(III) and concomitant two-electron reduction of H_2 to generate two H^- ions and formation of two additional bonds to $[Rh^I Cl(PPh_3)_3]$ takes place (a four-coordinate compound becomes a six-coordinate with two H^- ions), the formation of a Rh(III)-coordinated alkene complex (π complex), alkene insertion (migratory insertion of a H^- ion) triggers a change in the nature of Rh–C bond (π-complex to

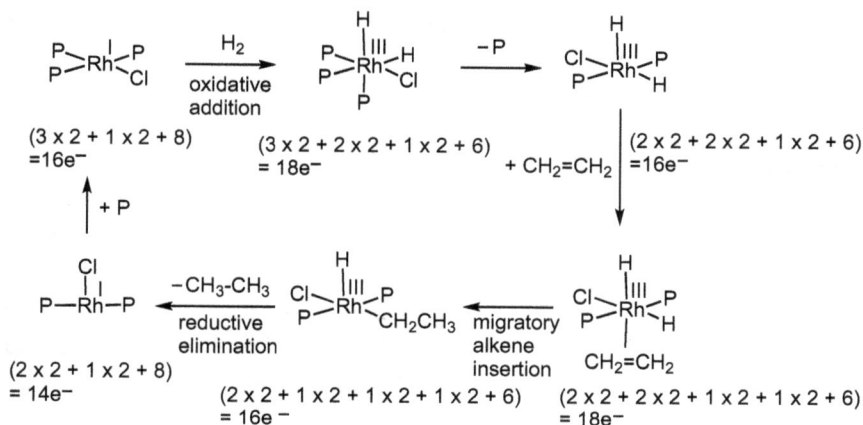

Fig. 9.1 The catalytic cycle for the hydrogenation of an alkene (shown as ethylene) in benzene; possible solvent coordination is not shown. P stands for PPh$_3$. For each species, the oxidation state of Rh and total number of valence electron at Rh are indicated.

σ-complex), and finally the reductive elimination of the alkane (two-electron reduction of Rh(III) to Rh(I) and concomitant breakage of two bonds: one Rh–H and insertion of a H$^-$ ion to Rh(III)–CH$_2$CH$_2$R to release RCH$_2$CH$_3$); R=H in Fig. 9.1).

For effective catalysis, the size of the alkene is important. The rate of hydrogenation is hindered by sterically demanding alkenes. It should be noted that the reaction events in the catalytic cycle considers an interplay between electronic and steric effects. The importance of both steric and electronic effects control equilibria, rates, and product distributions in organometallic reactions. Electronic effect facilitates binding of the substrate(s) to a 14e$^-$/16e$^-$ species to attain an 18e$^-$ species and steric effect facilitates removal of bulky group(s).

Generally, electronic effects, in which ligands change the electron density on the metal by pushing or pulling electrons through chemical bonds, can also be important, particularly if a reaction involves a change in the metal's formal oxidation state. The importance of steric effects was realized in studies of competition equilibria for binding of different ligands to metal. It became clear that steric effects are important in a broad range of organometallic reactions.

The *catalyst turnover number* is the number of moles of product formed per mole of catalyst used.

It should be mentioned here that the PEt_3 analog of $[Rh(PPh_3)_3Cl]$ reacts with H_2 to give a stable and catalytically inactive dihydride $[RhH_2Cl(PEt_3)_3]$.

Another rhodium-phosphine compound $[Rh^IH(CO)(PPh_3)_3]$ with catalytic activity was also synthesized in Wilkinson's laboratory. It was found that, for steric reasons, it selectively catalyzes the hydrogenation of alk-1-enes (i.e. terminal olefins) rather than alk-2-enes and it has been used in the hydroformylation of alkenes, (i.e. the addition of H and the formyl group -CHO) also known as the *oxo process* because it introduces oxygen into the hydrocarbons (see below).

9.3 Hydroformylation reaction (oxo process)

The hydroformylation reaction (or the oxo process) is the conversion of alkenes to aldehydes:

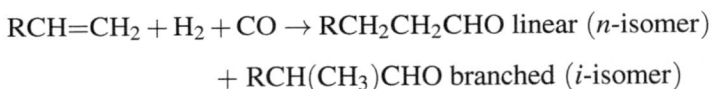

$$RCH{=}CH_2 + H_2 + CO \rightarrow RCH_2CH_2CHO \text{ linear } (n\text{-isomer})$$

$$+ RCH(CH_3)CHO \text{ branched } (i\text{-isomer})$$

Two kinds of products linear (n-isomer) and branched (i-isomer) are observed. This is a process of enormous industrial importance, being used to convert alk-1-enes into aldehydes. Aldehydes can then be converted to alcohols for the production of polyvinylchloride (PVC) and polyalkenes and, in the case of the long-chain alcohols, in the production of detergents. A simplified catalytic cycle (reaction scheme) is shown in Fig. 9.2. This catalyst $[Rh^IH(CO)(PPh_3)_3]$ has the great advantage over the conventional cobalt carbonyl catalyst that it operates efficiently at much lower temperatures and pressures and produces straight-chain as opposed to branched-chain products.

In this mechanism, the first step is dissociation of a PPh_3 from $[Rh^IH(CO)(PPh_3)_3]$ (total number of valence electron = 18). The coordinatively unsaturated compound thus formed $[Rh^IH(CO)(PPh_3)_2]$ (valence electron: 16) promotes hydroformylation. The basic principles (interplay between electronic and steric effects) are quite similar to that of hydrogenation of olefin catalyzed by $[Rh^ICl(PPh_3)_3]$ (Fig. 9.1).

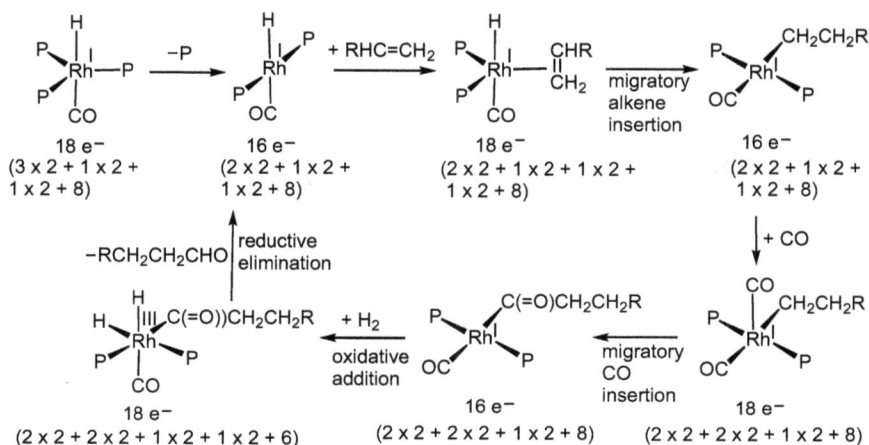

Fig. 9.2 The catalytic cycle for the hydroformylation of an alkene catalyzed by *trans*-[Rh^{I}H(CO)(PPh$_3$)$_3$] (temp. 373 K, press. ~10-15 bar). P stands for PPh$_3$. For each species, the oxidation state of Rh and total number of valence electron at Rh are indicated.

The cobalt-based catalysts were the first to be employed, and still most widely used process depends on the use of [Co$_2$(CO)$_8$] as catalyst. We now discuss the catalytic cycle of cobalt carbonyl catalyst (Fig. 9.3), which typically requires 423 K and ~250 bar pressure. This is used in industry. The reaction events start with loss of CO from [HCo(CO)$_4$] rather than PPh$_3$ from rhodium catalyst [Rh^{I}H(CO)(PPh$_3$)$_3$].

In this mechanism (catalytic cycle), a pre-equilibrium is established in which [Co$_2$(CO)$_8$] combines with H$_2$ at high pressure to generate the mononuclear compound [HCo(CO)$_4$]. Addition of the alkene is the first step. At this point, the reaction scheme splits into two routes depending on which carbon atom is involved in Co–C bond formation. The cobalt-alkene bond formation is followed by CO addition and migration insertion of H into the cobalt-alkene bond (π-complex) and formation of a cobalt-alkane bond (σ-complex). It should be noted here that a significant portion of branched aldehyde is also formed in the [Co$_2$(CO)$_8$] catalyzed hydroformylation. This may result from isomerization to a 2-alkylcobalt intermediate followed by the insertion of CO. The two pathways are shown as the inner and outer cycles in Fig. 9.3. In each, the next step is migratory insertion of CO to cobalt-alkyl bond forming acyl compounds. The

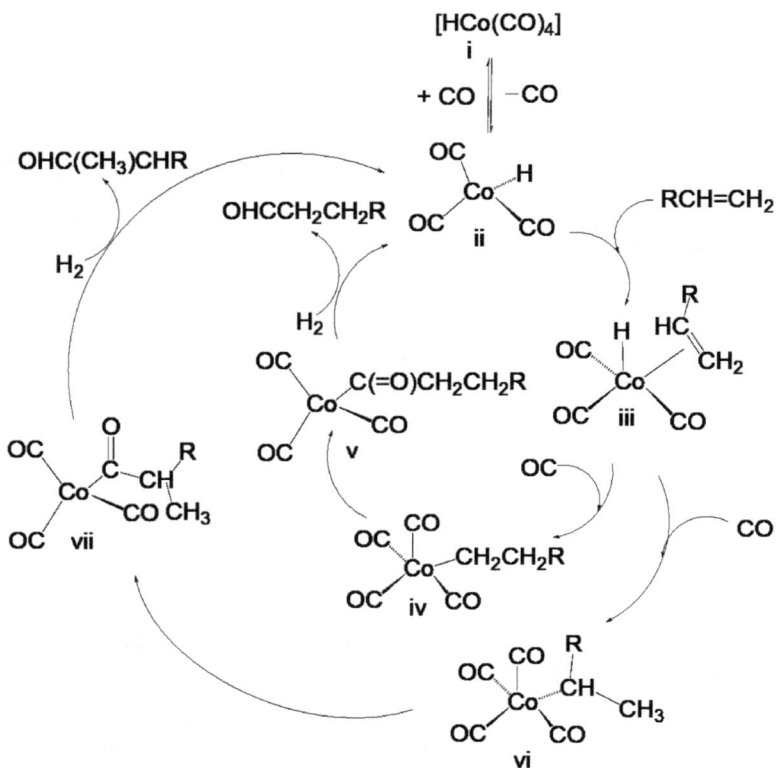

Fig. 9.3 The catalytic cycle for the hydroformylation of an alkene catalyzed by $[Co_2(CO)_8]$ ($[Co_2(CO)_8] + H_2 \rightarrow [2HCo(CO)_4]$).

last step is oxidative addition of H_2, implying the formation of Co(III)-dihydrido species, and the transfer of one H atom to the $-C(=O)CH_2CH_2R$ or $-C(=O)CH(CH_3)R$ group to afford elimination of the aldehyde. The inner cycle eliminates a linear aldehyde, while the outer cycle produces a branched isomer.

Total number of valence electron for

i $[HCo^I(CO)_4]$: $4 \times 2 + 1 \times 2 + 8 = 18$
ii $[CoH(CO)_3]$: $3 \times 2 + 1 \times 2 + 8 = 16$
iii $[Co^IH((CO)_3(RCH=CH_2)]$: $3 \times 2 + 1 \times 2 + 1 \times 2 + 8 = 18$
iv $[Co^I((CO)_4(CH_2CH_2R)]$: $4 \times 2 + 1 \times 2 + 8 = 18$
v $[Co^I(CO)_3(C(=O)CH_2CH_2R)]$: $3 \times 2 + 1 \times 2 + 8 = 16$

vi $[Co^I(CO)_4(CH(R)CH_3)]$: $4 \times 2 + 1 \times 2 + 8 = 18$
vii $[Co^I(CO)_3(COCH(R)CH_3)]$: $3 \times 2 + 1 \times 2 + 8 = 16$

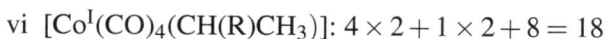

Two major complications in the process are the hydrogenation of aldehydes to alcohols, and alkene isomerization (which is also catalyzed by $HCo(CO)_3$). The first of these problems (see hydrogenation of alkene) can be controlled by using H_2:CO ratios greater than 1:1 (e.g. 1.5:1). The isomerization problem (regioselectivity) can be addressed by using other catalysts (see above) or can be turned to advantage by purposely preparing mixtures of isomers for separation at a later stage.

9.4 Oxidation of alkenes (Wacker process)

The Wacker process is primarily used to produce CH_3CHO from $H_2C=CH_2$.

$$H_2C=CH_2 \text{ (g)} + O_2 \text{ (g)} \longrightarrow CH_3CHO \text{ (g)}$$

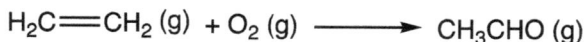

It was invented in the late 1950s (the then West Germany). Although this is no longer a major industrial process, it has some interesting mechanistic features that are worth noting. The Pd metal is oxidized to Pd(II) by $Cu^{II}Cl_2$, with the $[Cu^ICl_2]^{1-}$ being regenerated by atmospheric oxidation of Cu(I): $4Cu^I + O_2 + 4H^+ \rightarrow 4Cu^{II} + 2H_2O$.

$$H_2C=CH_2 + Pd^{II}Cl_2 + H_2O \longrightarrow CH_3CHO + Pd(0) + 2HCl$$

$$Pd(0) + 2Cu^{II}Cl_2 + 4Cl^- \longrightarrow Pd^{II}Cl_4^{2-} + 2[Cu^ICl_2]^{1-}$$

$$2[Cu^ICl_2]^{1-} + 1/2O_2 + 2HCl \longrightarrow 2[Cu^{II}Cl_2] + 2Cl^- + H_2O$$

The first step of the catalytic cycle (Fig. 9.4) involves substitution of a Cl^- ion from $[Pd^{II}Cl_4]^{2-}$ by $H_2C=CH_2$ to form $[Pd^{II}Cl_3(\eta^2\text{-}C_2H_4)]^{1-}$. The next step involves nucleophilic attack by H_2O with loss of H^+. In the third step, β-elimination occurs and formation of the Pd–H bond results with loss of a Cl^- ion. This is followed by attack by Cl^- with migratory insertion of an H atom to give a σ-bonded '$Pd^{II}CH(OH)CH_3$' unit. Elimination of CH_3CHO, H^+, and Cl^- with reduction of Pd(II) to Pd(0) is the last step. To keep the catalytic cycle running, Pd(0) is then oxidized by

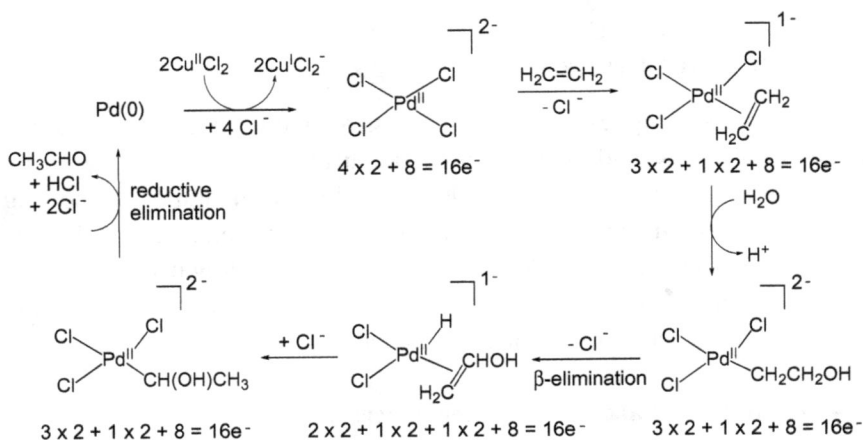

Fig. 9.4   The catalytic cycle for the Wacker process.

Cu(II). The Cu(I) ion, thus formed, is oxidized by O_2 in the presence of H^+. The notable feature of this catalytic cycle is the use of a co-catalyst $(Cu^{II}Cl_2)$.

9.5   Olefin hydrocyanation

During the 1960s Du Pont (USA) developed a successful new process using homogeneous catalysis to add two moles of HCN to butadiene to give adiponitrile (ADN). The ADN is then hydrogenated (with a hetero-geneous catalyst) to hexamethylenediamine. The reaction of the diamine with adipic acid gives the polyamide known as Nylon-6,6-the major form of nylon used in the United States.[1]

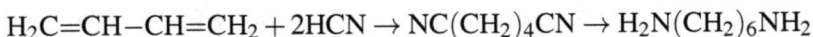

$$H_2C=CH-CH=CH_2 + 2HCN \rightarrow NC(CH_2)_4CN \rightarrow H_2N(CH_2)_6NH_2$$

$$H_2N(CH_2)_6NH_2 + HO_2C(CH_2)_4CO_2H \rightarrow Nylon\text{-}6,6$$

Butadiene hydrocyanation is carried out in two steps using an NiL_4 catalyst (L is a triarylphosphite):

Step I: Addition of the first HCN, gives the desired linear 3-pentenenitrile (3PN) and the undesired branched 2-methyl-3-butenenitrile (2M3BN);

---

[1]C. A. Tolman, *J. Chem. Educ.* **1986**, *63*, 199.

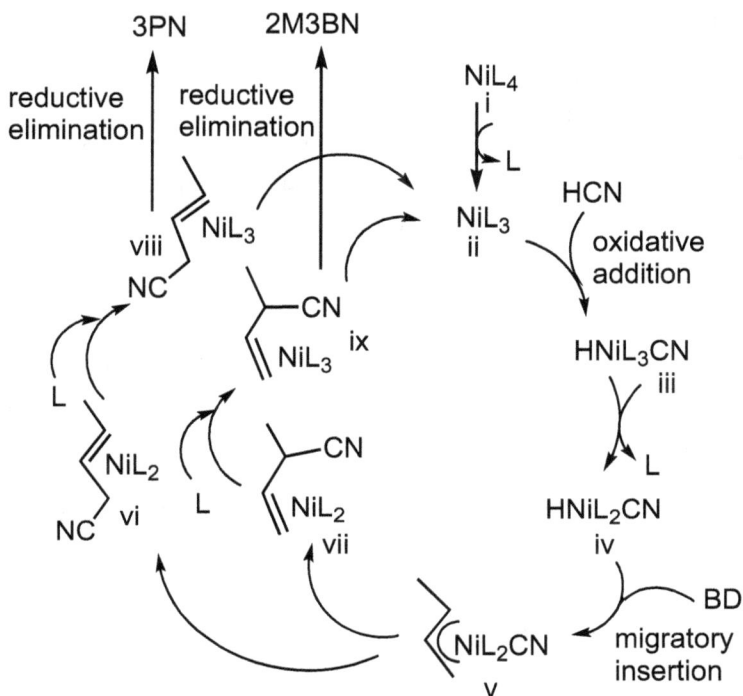

Fig. 9.5 The mechanism of butadiene (BD) hydrocyanation. L represents P(OEt)$_3$.

these may he separated by distillation and the branched product isomerized to 3PN with a similar catalyst.

Step II: Addition of the second HCN, involves both a double-bond migration to 4-pentenenitrile (4PN) and anti-Markovnikov addition of HCN to the terminal double bond. Thus, olefin isomerization is an important part of the chemistry in Step II.

Fig. 9.5 shows the catalytic cycle involved in Step I of the hydrocyanation process. Most of the steps involve ligand dissociation or association, which are sensitive to ligand steric effects, but there are also oxidative addition and reductive elimination steps in which the formal oxidation state of the metal changes and which are particularly sensitive to electronic effects.

Total number of valence electron for

  i $[Ni^0L_4]$: $4 \times 2(4L) + 10(Ni^0) = 18$

  ii $[Ni^0L_3]$: $3 \times 2(3L) + 10 = 16$

 iii $[HNi^{II}L_3CN]$: $3 \times 2(3L) + 1 \times 2(CN^-) + 1 \times 2(H^-) + 8(Ni^{II}) = 18$

 iv $[HNi^{II}L_2CN]$: $2 \times 2(2L) + 1 \times 2(CN^-) + 1 \times 2(H^-) + 8(Ni^{II}) = 16$

  v $[(CH_3(CH^-)CH{=}CH_2)Ni^{II}L_2CN]$: $2 \times 2(2L) + 1 \times 2(CN^-) + 1 \times 4\{(CH_3(CH^-)CH{=}CH_2)\} + 8(Ni^{II}) = 18$

 vi $[(CH_3CH{=}CHCH_2CN)Ni^0L_2]$: $2 \times 2(2L) + 1 \times 2\{(CH_3CH{=}CHCH_2CN)\} + 10(Ni^0) = 16$

 vii $[(CH_3CH(CN)CH{=}CH_2)Ni^0L_2]$: $2 \times 2(2L) + 1 \times 2 \{(CH_3CH(CN)CH{=}CH_2)\} + 10(Ni^0) = 16$

viii $[(CH_3CH{=}CHCH_2CN)Ni^0L_3]$: $3 \times 2(3L) + 1 \times 2 \{(CH_3CH{=}CHCH_2CN)\} + 10(Ni^0) = 18$

 ix $[(CH_3CH(CN)CH{=}CH_2)Ni^0L_3]$: $3 \times 2(3L) + 1 \times 2\{(CH_3CH(CN)CH{=}CH_2)\} + 10(Ni^0) = 18$

When $L = P(OEt)_3$, the $HNi^{II}L_3CN$ and π-$(C_4H_7{}^-)Ni^{II}L_2CN$ (v) complexes are stable enough to he observed spectroscopically in solution by NMR or IR.

In Fig. 9.5 the ratio of the rate of formation of vi : vii $= 2.5$, which determines the kinetically controlled ratio of linear (3PN) / branched (2M3BN) mononitrile products. The first step in getting the catalytic reaction underway is the dissociation of a $P(OEt)_3$ L from Ni^0L_4. Both rates and equilibrium constants of this reaction are strongly dependent on the steric size of L, as is the enthalpy of formation of NiL_4 from $Ni(COD)_2$ (COD = 1,5-cyclooctadiene).

2M3BN coordinates more strongly than 4PN, which is why it is important not to feed 2M3BN to Step II. Fortunately, 3PN binds much less strongly than does 4PN, so that ADN is the major Step I product, even though 3PN is the major cyanoolefin in solution. The 2PN's bind rather strongly and (like acrylonitrile) form relatively stable α-cyanoalkylnickel cyanide complexes which do not undergo carbon-carbon coupling to regenerate Ni(0). *It is one of the great strokes of good fortune in this system that 3PN is isomerized to 4PN much faster than to 2PN, even though 2PN is the thermodynamically favored isomer.*

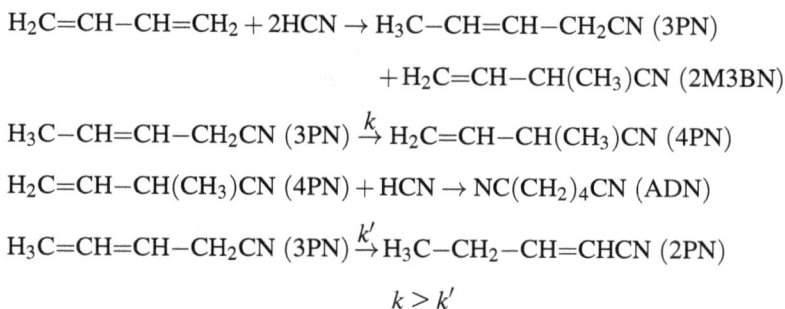

$$H_2C=CH-CH=CH_2 + 2HCN \rightarrow H_3C-CH=CH-CH_2CN \ (3PN)$$

$$+ H_2C=CH-CH(CH_3)CN \ (2M3BN)$$

$$H_3C-CH=CH-CH_2CN \ (3PN) \xrightarrow{k} H_2C=CH-CH(CH_3)CN \ (4PN)$$

$$H_2C=CH-CH(CH_3)CN \ (4PN) + HCN \rightarrow NC(CH_2)_4CN \ (ADN)$$

$$H_3C=CH=CH-CH_2CN \ (3PN) \xrightarrow{k'} H_3C-CH_2-CH=CHCN \ (2PN)$$

$$k \geq k'$$

9.6 Homogeneous/heterogeneous catalysis: Propylene polymerization

Since its discovery in the mid-1950's, polypropylene has grown to be one of today's most important commodity polymers, often prepared by use of organometallic catalysts, either in solution (homogeneous) or supported on a solid surface (heterogeneous).

In the early 1950's when Ziegler in Germany, investigating his new triethylaluminium-catalyzed synthesis of olefins (known as the "Aufbau Reaction"), serendipitously discovered the "nickel effect", which was caused by a colloidal nickel contaminant. Later on, the catalyst (TiCl$_4$ + AlEt$_3$) was subsequently developed for the manufacture of high-density polyethylene.[2]

Although Ziegler's invention formed the cornerstone for the subsequent development and commercialization of the majority of present-day poly-olefin manufacturing processes, the full potential of the discovery was realized by Natta's group in Italy. In early 1954 Natta first succeeded in isolating crystalline polypropylene using Ziegler's TiCl$_4$ + AlEt$_3$ catalyst. Natta determined that crystalline polypropylene (which he designated "isotactic") comprises extended sequences of monomer units with the same

---

[2]B. L. Goodall, *J. Chem. Educ.* **1986**, *63*, 191.

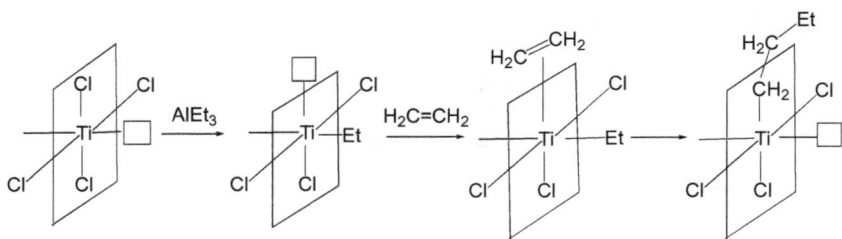

Fig. 9.6 Mechanism for olefin polymerization.

configuration; the amorphous (or random) analogs were dubbed "atactic". Natta also first recognized the importance of the catalyst crystal structure and particularly the crystal surface in determining polymer isotacticity and found that when certain solid titanium chlorides were used instead of $TiCl_4$ highly isotactic polypropylene was produced. Natta and coworkers demonstrated that the $TiCl_3$ catalyst lattice structure determined the stereoselectivity of the catalyst and hence polymer isotacticity; brown (β)-$TiCl_3$ gave low isotacticity, while purple (α, γ, δ)-$TiCl_3$ gave high isotacticity. The elegant work of Arlman and Cossee[3] presented a rational and still widely accepted explanation of these facts. An essential feature of the Cossee mechanism is that the catalytically active centers are those Ti ions which possess chloride vacancies – such vacancies on the (α, γ, δ)-$TiCl_3$ surface being necessary to ensure the electroneutrality of the crystal.

The full details of the reaction events (mechanism) of 'Zieglar-Natta catalysis' are still uncertain. The 'Cosee-Arlman mechanism' is regarded as most plausible (Fig. 9.6). In the first step, the alkylaluminium ($AlEt_3$) alkylates a solid surface-bound $TiCl_4$ and then an alkene molecule (C_2H_4) coordinates to the neighboring vacant site. In the propagation steps for the polymerization to occur the Ti-coordinated alkene undergoes a migratory insertion reaction (metal-alkene to metal-alkane switchover). This triggers opening up of another neighboring vacant site and the reaction continues for the polymer chain to grow. The release of the polymer from Ti center occurs due to β-hydrogen (β-hydride) elimination and the chain

---

[3]E. J. Arlman and P. Cossee, *J. Cat.* **1964**, 3, 99.

is terminated. The stereospecific nature of the polymerization reflects the constraints on the orientation in which the monomer can attach to the Ti atom, next to the growing chain.

Example 9.1 Take a closer look at the reactants and the products of the following organometallic reaction sequence in a catalytic cycle. Justify the formal oxidation state of the metal and the number of valence electron indicated/presented in the parentheses.

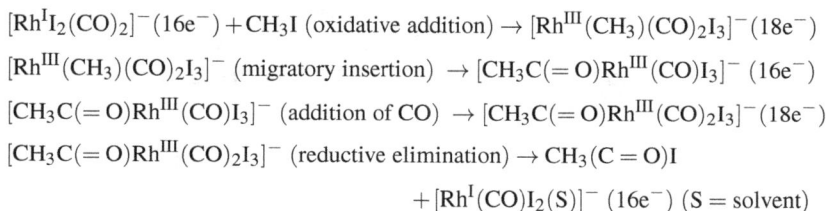

$[Rh^I I_2(CO)_2]^- (16e^-) + CH_3I$ (oxidative addition) $\rightarrow [Rh^{III}(CH_3)(CO)_2I_3]^- (18e^-)$

$[Rh^{III}(CH_3)(CO)_2I_3]^-$ (migratory insertion) $\rightarrow [CH_3C(=O)Rh^{III}(CO)I_3]^- (16e^-)$

$[CH_3C(=O)Rh^{III}(CO)I_3]^-$ (addition of CO) $\rightarrow [CH_3C(=O)Rh^{III}(CO)_2I_3]^- (18e^-)$

$[CH_3C(=O)Rh^{III}(CO)_2I_3]^-$ (reductive elimination) $\rightarrow CH_3(C=O)I$

$$+ [Rh^I(CO)I_2(S)]^- (16e^-) (S = solvent)$$

Example 9.2 Write the structure of A, B, and C in the following reaction sequence:

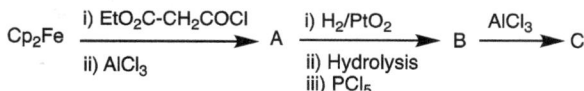

$$Cp_2Fe \xrightarrow[\text{ii) AlCl}_3]{\text{i) EtO}_2\text{C-CH}_2\text{COCl}} A \xrightarrow[\substack{\text{ii) Hydrolysis}\\\text{iii) PCl}_5}]{\text{i) H}_2/\text{PtO}_2} B \xrightarrow{\text{AlCl}_3} C$$

Answer

A   B   C

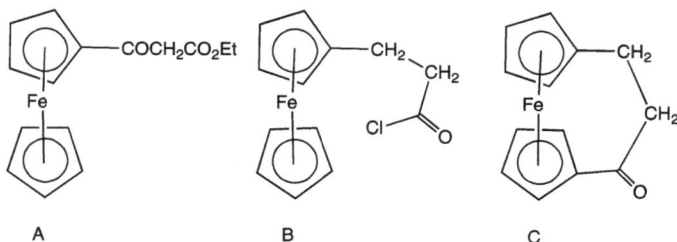

Example 9.3 Identify the product of the following reaction:

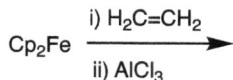

$$Cp_2Fe \xrightarrow[\text{ii) AlCl}_3]{\text{i) H}_2\text{C=CH}_2}$$

Answer

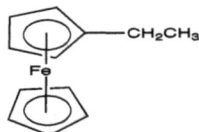

Further reading

N. N. Greenwood and A. Earnshaw, Chemistry of the Elements, 2nd edition, Elsevier Butterworth-Heinemann (2010)

C. E. Housecroft and A. G. Sharpe, *Inorganic Chemistry*, 2nd edition, Pearson Education Limited (2005)

R. H. Crabtree, *The Organometallic Chemistry of the Transition Metals*, 4th edition, Wiley Interscience (2005)

D. F. Shriver and P. W. Atkins, *Inorganic Chemistry*, 3rd edition, Oxford University Press (1999)

F. A. Cotton, G. Wilkinson, C. A. Murillo, and M. Bochmann, *Advanced Inorganic Chemistry*, 6th edition, John Wiley & Sons, Inc. (1999)

J. E. Huheey, E. A. Keiter, and R. L. Keiter, *Inorganic Chemistry: Principles of Structure and Reactivity*, 4th edition, Addison-Wesley Publishing Company (1993)

W. L. Jolly, Modern Inorganic Chemistry, 2nd edition, McGraw-Hill Inc., McGraw-Hill International Editions Chemistry Series (1991)

F. A. Cotton and G. Wilkinson, *Advanced Inorganic Chemistry*, 5th edition, John Wiley & Sons (1988)

F. A. Cotton and G. Wilkinson, Basic Inorganic Chemistry, Wiley Eastern Limited (1988)

K. F. Purcell and J. C. Kotz, Inorganic Chemistry, Saunders Golden Sunburst Series, W. B. Saunders Company, Holt-Saunders Japan (1985)

N. N. Greenwood and A. Earnshaw, Chemistry of the Elements, 1st edition, Pergamon Press (1984)

Exercises

9.1 Identify A and B of the following reaction:

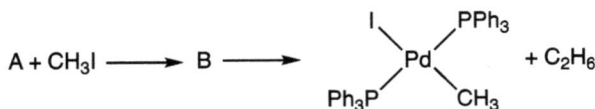

9.2 Identify the products:

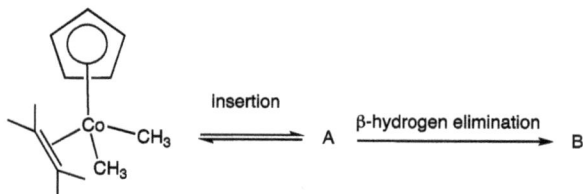

9.3 Show the termination reaction in a Ziegler-Natta polymerization of ethylene.

Chapter 10

Bioinorganic Chemistry

Bioinorganic chemistry is an interdisciplinary field that relies on a solid understanding of both biological and inorganic chemistry and employs wide-ranging synthetic, analytical, and physical techniques from both chemistry and biology.

Nature uses metal ions for functional as well as structural roles. The subject of biological inorganic chemistry/bioinorganic chemistry is one of the most intellectually attractive and experimentally demanding frontiers in modern chemical science. Bioinorganic chemistry largely focuses on the role of metal ions in biology.[1]

It has been known for decades that metal ions play an essential role in living systems. In fact. the role of iron as an essential metal has been known since the 18th century, whereas the role of other elements, such as cobalt, copper, manganese, and zinc has been known for near about a century. It has also been known for a long time that excesses of these elements can be very dangerous. In fact, a narrow concentration window exists for most of the so-called trace elements. Many elements of the periodic table are nowadays known and accepted to be essential or beneficial for life on earth. However, on the molecular level the role of these elements is understood for only a few of them, and even, not fully.

[1](a) E.-I. Ochiai, *J. Chem. Educ.* **1978**, *55*, 631. (b) R. H. Holm, P. Kennepohl, and E. I. Solomon, *Chem. Rev.* **1996**, *96*, 2239.

Table 10.1   Some important elements in biology: occurrence, concentrations, and biological roles

Element	Concentrations in some specie			Daily required [mg]	Role in biology
	human [mg]	sea water [ppm]	earth crust [ppm]		
Iron	4500	10	50×10^3	10	Many enzymes, respiratory proteins
Copper	100	3	55	2	Many enzymes, dioxygen transport
Zinc	2000	10	70	12	Hydrolytic enzymes, nucleic acid synthesis
Molybdenum	5	10	1.5	0.2	Many redox enzymes, nitrogenase (plants)
Manganese	20	2	950	3	Enzyme activation, photosynthesis
Cobalt	1	0.1	25	0.3	Vitamin B12
Selenium	0.5	0.4	0.05	0.1	Glutathione peroxidase, antioxidant
Magnesium	4×10^4	10^6	20×10^3	350	Photosynthesis, Nucleic acid processes
Calcium	10^6	4×10^5	35×10^3	800	Bones, teeth; muscle activation

In addition, many other elements of the periodic table influence the quality of life either as a toxic pollutant[2] or as a drug to cure a specific disease.[3] Examples of these are well known from general literature, but again little is understood about the details of their mode of action on the molecular level. The major challenge of modern bioinorganic chemistry is to understand the molecular basis of all these interactions, and to apply this knowledge in medicine, biology, environmental sciences, catalysis, and technology. Table 10.1 lists several elements, together with some statistical

---

[2] E.-I. Ochiai, *J. Chem. Educ.* **1995**, *72*, 479.

[3] (a) S. J. Lippard, *Science* **1993**, *261*, 699. (b) M. J. Abrams and B. A. Murrer, *Science* **1993**, *261*, 725. (c) T. C. Pinkerton, C. P. Desilets, D. J. Hoch, M. V. Mikelsons, and G. M. Wilson, *J. Chem. Educ.* **1985**, *62*, 965. (d) J. R. Dilworth and S. J. Parrott, *Chem. Soc. Rev.* **1998**, *27*, 43. (e) X. Wang, X. Wang, S. Jin, N. Muhammad, and Z. Guo, *Chem. Rev.* **2019**, *119*, 1138.

information and a few comments about their biological roles.[4] The compilation is limited and restricted to some of the most important transition elements, the non-metal Se and to Ca and Mg.

The metallobiomolecules comprising proteins and enzymes are highly elaborated coordination complexes whose metal-containing sites (coordination units), comprising one or more metal atoms and their ligands, are usually the loci of electron transfer, binding of exogenous molecules, and catalysis and are termed 'active sites'.[5] Enzymes are responsible for many different types of catalytic activity. Understanding the mechanisms through which a catalytic activity is favored over the others remains a significant challenge. The behavior of metal ions in proteins is not very different from the fundamental chemistry of a particular metal ion outside the protein environment. Thus, the study of synthetic analogues of the metalloenzyme active sites has turned out to be of great significance. A synergistic approach to the study of metallo-proteins/-enzymes has yielded crucial information because synthetic analogues have provided insights into the effects of systematic variations in metal-coordination geometry, nature of donor sites and subtle other factors, with high-level of precision.[5]

Metal-ligand complexes have been developed as metallodrugs with both therapeutic and diagnostic purposes for various diseases. The major advantages of metallodrugs over small organic molecules and natural compounds of biological origin, are their biological and chemical diversities.[3] Moreover, luminescent metal complexes can exert diagnostic potential with excellent optical characteristics; diagnostic metallodrugs of magnetic and radioactive metals have thus been extensively used as contrast agents.[3]

10.1 Role of metal ions

Many inorganic elements and their compounds are now known to be essential to organisms. Organic compounds are of course essential, because they provide organisms with such essential compounds as proteins, nucleotides, carbohydrates, vitamins, and so forth. Inorganic compounds (particularly

---

[4]J. Reedijk, *Naturwissenschaften*, **1987**, *74*, 71.

[5](a) J. A. Ibers and R. H. Holm, *Science* **1980**, *209*, 223. (b) K. D. Karlin, *Science* **1993**, *261*, 701. (c) E. I. Solomon and M. D. Lowery, *Science* **1993**, *259*, 1575.

metal ions and their complexes) are essential cofactors in a variety of enzymes and proteins. They conceivably provide essential services which cannot be or can only poorly be rendered by organic compounds.

The biochemical roles played by essential inorganic elements and compounds are[1a]:

i) structural
ii) carrying and transporting electrons and oxygen
iii) catalytic roles in oxidation-reduction reactions
iv) catalytic roles in acid-base and other reactions

Let us lay out some basic principles regarding the type of question we are interested in here is why a certain inorganic element (compound) is specifically required for a certain function in biological systems.[1a] They are:

1) Rule of abundance

When a function can he accomplished by two or more entities, organisms would utilize the more abundant, readily available one. The four elements Fe, Cu, Zn, and Mo are the most frequently found in the catalytic sites of enzymes. The use of the most abundant alkali metals, sodium and potassium, in controlling ion balance[6] and enzyme activities is also in accord with this rule.

Most organisms utilize calcium compounds such as carbonate and phosphate as protective and skeletal material. Undoubtedly this is due to the insolubility of calcium carbonate and phosphate. However, the corresponding strontium compounds are equally insoluble and could substitute calcium compounds. It is obvious that calcium is much more abundant than strontium.

Zinc in zinc-enzymes can be in most cases replaced by cobalt in vitro without losing catalytic activity. If organisms are grown in cobalt-rich media, they can produce enzymes in which $Zn(II)$ is replaced by $Co(II)$. Thus zinc and cobalt seem to be interchangeable, but organisms selected zinc because zinc is much more abundant both in sea and earth's upper crust.

---

[6]B. Dietrich, *J. Chem. Educ.* **1985**, *62*, 954.

2) Rule of efficiency

The rule of efficiency asserts that organisms would choose the more efficient entity as long as it is readily available. Flavin adenine dinucleotide (FAD) is a redox-active coenzyme associated with various proteins, which is involved with several enzymatic reactions in metabolism. A flavoprotein is a protein that contains a flavin group, which may be in the form of FAD or flavin mononucleotide (FMN). Flavodoxins are small, soluble electron-transfer proteins (Fig. 10.1).

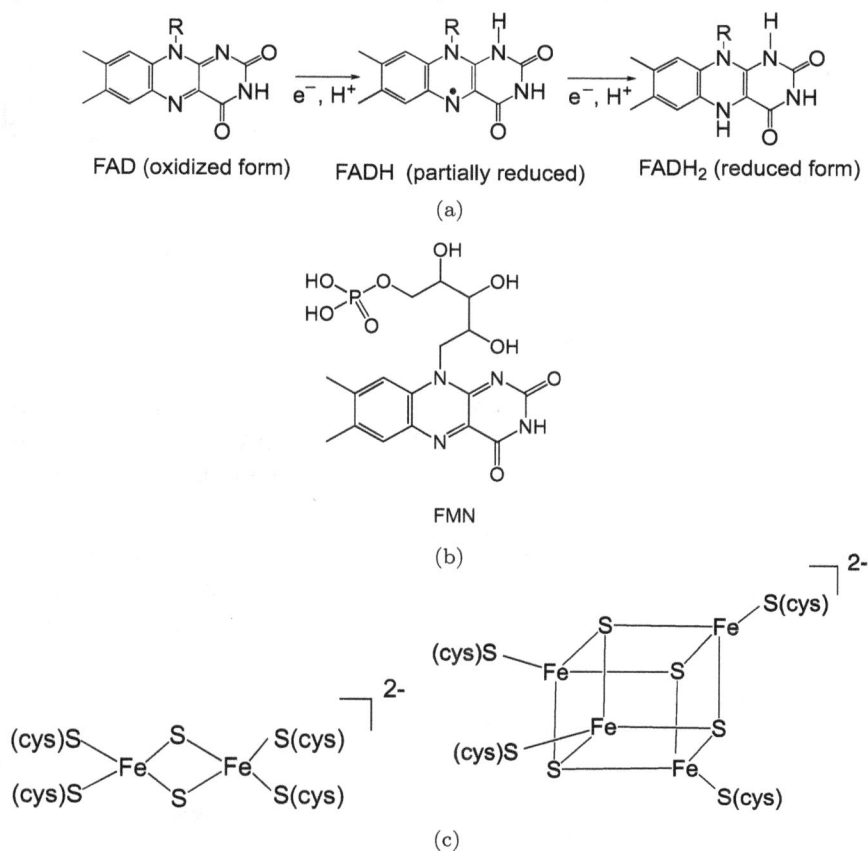

Fig. 10.1 (a) Electron transfer reactions of FAD, (b) structure of FMN, and (c) structures of plant and bacterial ferredoxin, respectively.

Flavodoxins and ferredoxins (Fig. 10.1) function as electron-carriers in very similar ways, being interchangeable in most cases. However their compositions are entirely different. Flavodoxin contains flavin mononucleotide (FMN) as the prosthetic group, whereas the functional units in ferredoxins are iron-sulfur clusters. Flavodoxin in general is less efficient than ferredoxin and the synthesis of flavodoxin occurs only during growth in iron-poor media in many microorganisms.[1,7]

3) Rule of basic fitness

An inorganic element (generally a metal) to be selected should have a basic ability or potential to carry out the desired function. That is, a certain element (or elements) would inherently fit to a particular function.

The reduction potentials of O_2^-/H_2O_2 and O_2/O_2^- are $+0.96$ V and -0.45 V (at pH 7), respectively. Suppose that the mechanism of superoxide dismutase (SOD) reaction[1,8] is simply as follows:

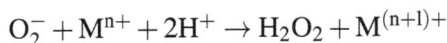

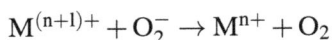

$$O_2^- + M^{n+} + 2H^+ \rightarrow H_2O_2 + M^{(n+1)+}$$

$$M^{(n+1)+} + O_2^- \rightarrow M^{n+} + O_2$$

$$\text{Adding, } 2O_2^- + 2H^+ \rightarrow O_2 + H_2O_2$$

Then it would be inferred that a redox system whose potential lies somewhere in the middle of the range $+0.96$ V to -0.45 V would function as a catalyst for the SOD reaction. A closer look at the redox potential values indicates that such systems are aqua Fe(III)/Fe(II), Cu(II)/Cu(I), and Mn(III)/Mn(II) couples. In fact, SODs from different sources contain Cu, Fe, or Mn. A Cu,Zn-containing SOD is found to have a reduction potential value of 0.42 V. As far as the reduction potential is concerned, the aqua V(IV)/V(III) system is also fit for SOD function. One of the functions of hemovanadin, which is a vanadium-containing protein found in a marine invertebrate, ascidian, might well be a SOD.[1]

---

[7](a) E.-I. Ochiai, *J. Chem. Educ.* **1993**, *70*, 128. (b) E.-I. Ochiai, *J. Chem. Educ.* **1997**, *74*, 348. (c) S. C. Lee, W. Lo, and R. H. Holm, *Chem. Rev.* **2014**, *114*, 3579.

[8](a) J. S. Valentine and D. M. de Freitas, *J. Chem. Educ.* **1985**, *62*, 990. (b) Y. Sheng, I. A. Abreu, D. E. Cabelli, M. J. Maroney, A.-F. Miller, M. Teixeira, and J. S. Valentine, *Chem. Rev.* **2014**, *114*, 3854.

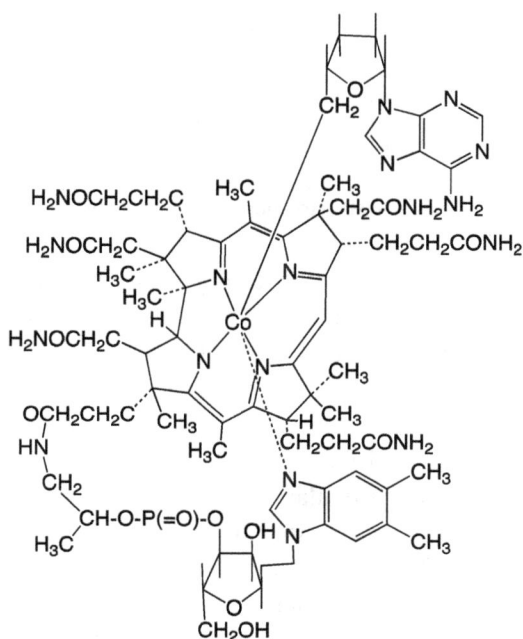

Fig. 10.2   The structure of adenosyl cobalamin (vitamin B$_{12}$ coenzyme).

The important requirements for the candidates for the job of vitamin B$_{12}$ cobalamin, which has a Co–C bond (Fig. 10.2), appear to be rather well defined. They are: (1) the metal ion should readily take three consecutive oxidation states (I, II, and III) in aqueous media, (2) the lowest oxidation state should be highly nucleophilic, and (3) the middle oxidation state should perhaps have one unpaired electron. The requirement (2) implies that the lowest oxidation state of the catalytic metal ion should have d^8 or d^{10} configuration. Only a Co(III)/Co(II)/Co(I) system satisfies all these requirements. This explains why cobalt (cobalamin) uniquely fits the job.

4) Evolutionary improvement of efficiency and specificity

The rule of basic fitness dictates which element(s) is suitable for a specific enzymatic reaction only in terms of thermodynamics. The element or its simpler compounds selected may be able to do the required function but may not be very efficient in the senses of kinetics and specificity.

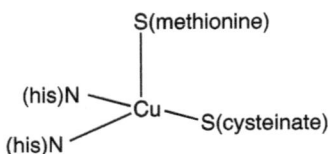

Fig. 10.3   The active site structure (ChemDraw) of plastocyanin.

The substrate specificity and the efficiency seem to be controlled mainly by the protein portions of metalloproteins or metalloenzymes. What we are concerned with here is the efficiency of many enzymes. Their rates are usually very much higher than those of simple model compounds. Vallee (American biochemist B. L. Vallee: 1919–2010) and Williams (English chemist R. J. P. Williams: 1926–2015) proposed that the structure of the active site of an enzyme, particularly a metalloenzyme, is rather distorted or strained and that it is responsible for its high catalytic activity. They called it *entatic effect.*[1a]

Blue copper proteins such as plastocyanin (Fig. 10.3) and azurin function as elrctron-carriers, in which copper oscillates between Cu(II) and Cu(I) states. The favorable coordination structure of Cu(II) is square-based and that of Cu(I) is tetrahedral. The reduction of a square-based Cu(II), therefore, would require a significant rearrangement in the structure about copper ion. One way to reduce this cost is to start with a compound whose structure is between the regular square-based and the tetrahedral one. X-ray crystal structure has revealed that the coordination structure about copper ion in the blue proteins is indeed distorted tetrahedral.[9]

Other examples are zinc enzymes.[10] The structural studies by X-ray crystallography and by spectroscopy of the cobalt(II)-substituted enzymes have established that the coordination structures about zinc ion in carbonic anhydrase, carboxypeptidase A, and alkaline phosphatase are distorted tetrahedral, being between a regular tetrahedron and a five-coordinate trigonal bipyamidal. The function of these enzymes is to hydrate the substrates. Not only the substrate but also the water molecule would have to be

---

[9]D. R. McMillin, *J. Chem. Educ.* **1985**, *62*, 997.

[10](a) I. Bertini, C. Luchinat, and R. Monnanni, *J. Chem. Educ.* **1985**, *62*, 924. (b) W. N. Lipscomb and N. Sträter, *Chem. Rev.* **1996**, *96*, 2375. (c) G. Parkin, *Chem. Rev.* **2004**, *104*, 699.

activated. The distorted tetrahedral structures of the enzymes usually have one water molecule and three amino acid residues coordinated about the zinc ion. If the coordination structure about the zinc atom is distorted from tetrahedron in such a way that the substrate can readily approach the zinc ion (forming a five-coordinate trigonal bipyramidal type of structure), the whole reaction would be facilitated. In a regular tetrahedron, this binding of an additional ligand would not be very easy.

10.2   The importance of iron

The importance of iron in our civilization is perhaps best summarized by the quotation from Rudyard Kipling cited by Philip Aisen in his article on physicochemical aspects of iron metabolism[11]:

> *"Gold is for the mistress - silver, for the maid -*
> *Copper for the craftsman cunning at his trade!*
> *'Good!' said the Baron, sitting in his hall,*
> *'But iron' - Cold Iron - is master of them all".*

Iron, an essential element in living organisms, has two properties at physiological pH and in aqueous medium, which make it at the same time interesting for biological systems in terms of catalysis and transport, and biologically inaccessible. Iron can exist in aqueous solution in two oxidation states: Fe(II) and Fe(III). Secondly, the solution chemistry of iron is essentially dominated by the hydrolysis and polymerization of aqueous Fe(III) to insoluble and potentially biologically inaccessible ferric hydroxides and oxyhydroxides. For living organisms to be able to obtain iron from their environment they firstly evolved a series of molecules (essentially derived from catechols or from a series of different hydroxamic acids) designated as *siderophores*.[12] These molecules are powerful chelators of Fe(III) and are secreted into the extracellular medium by a large number of unicellular organisms. The ferri-siderophores (the Fe(III)-chelate complexes) are then assimilated by the cells via a receptor specific for the

---

[11] R. Crichton and M. Charloteaux-Wauters, *Eur. J. Biochem.* **1987**, *164*, 485.
[12] (a) H. Boukhalfa and A. L. Crumbliss, *BioMetals* **2002**, *15*, 325. (b) R. C. Hider and X. Kong, *Nat. Prod. Rep.* **2010**, *27*, 637.

complex and once inside the cell the iron is released either by reduction of the Fe(III) to Fe(II) (for which the siderophore has little affinity) or by hydrolysis of the ferri-siderophore accompanied by release of its Fe(III). Iron is commonly used in the Fe(II) oxidation state, but in our oxidizing atmosphere Fe(III) is the more prevalent oxidation state. It is involved in a great many biological functions.

(a) Iron-transport protein: transferrin

For multicellular organisms, the problem of the transport of iron to the different cell types of the organism and its resorption from the dietary sources of iron available to it (in general such organisms do not possess a system of siderophores[12] with their appropriate receptors, and must therefore find the iron that they require in their alimentation) also demands a solution. This role is assured by a serum globulin, *transferrin*, which binds iron through complexation, transports it in the circulation and, via specific receptors, is taken up by the cells of the organism. The release of iron is pH-dependent and the availability of proton pumps within specific intracellular compartments assures the intracellular release of iron. The iron-free protein is recycled from the cell and released into the circulation for re-utilization.

Structurally, transferrin consists of a single polypeptide chain with N-terminal and C-terminal lobes. Each lobe contains two domains that are joined by a short peptide to create a hydrophobic metal binding site. Transferrin reversibly binds two Fe(III) ions with high affinity ($K_a = \sim 10^{20}$ L/mol). The iron coordination in each lobe includes two tyrosine, an aspartate, and a histidine residue, along with a bidentate carbonate anion (CO_3^{2-}) (Fig. 10.4).[13]

Conformational changes occur in transferrin that are associated with iron binding or release. When iron is bound, each domain moves to enclose iron in the metal binding site. Similarly, upon release of iron, the metal-binding domains move apart; these "closed" and "open" conformations have been observed through crystallography. Once iron is bound to transferrin it is internalized by the cell through a process known as receptor-mediated endocytosis; iron is then released and either used or stored in ferritin.[14]

---

[13]M. Hémadi, N.-T. Ha-Duong and J.-M. El Hage Chahine, *J. Mol. Biol.* **2006**, *358*, 1125.
[14]M. J. Donlin, R. F. Frey, C. Putnam, J. K. Proctor, and J. K. Bashkin, *J. Chem. Educ.* **1998**, *75*, 437.

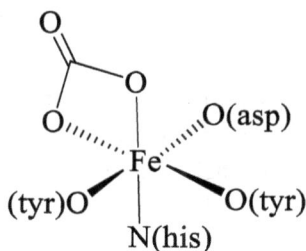

Fig. 10.4 Metal coordination environment of transferrin.

Once the iron is released within the cell, the same problems that we have evoked above again present their necessary fidelity to the rules of the solution chemistry of iron: either the iron which has been liberated from its soluble siderophore complex will be complexed by an appropriate intracellular chelator, or else it will follow its inexorable hydrolysis and polymerization towards biological inaccessibility.

(b) Iron-storage protein: ferretin

The second development in the evolution of the capacity of living organisms to assimilate and utilize iron: the appearance of intracellular iron storage compounds, which allow the cell to dispose of a pool of bioavailable iron in a form which is soluble in physiological conditions, and non-toxic. The archetype of these compounds is the iron storage protein, *ferritin*.[14]

The ability to store and release iron in a controlled fashion is essential. Cells have solved this problem of iron storage by developing ferritins, a family of iron-storage proteins that sequester iron inside a protein coat as a hydrous ferric oxide–phosphate mineral similar in structure to the mineral ferrihydrite. The protein is a spherical shell comprising 24 subunits with a combined molecular weight of 474 000 g/mol. The walls of the ferritin shell are approximately 10 Å thick and surround a spherical space approximately 80 Å in diameter. This spherical space can contain a maximum of 4500 iron atoms.[14]

The crystal structures of two ferritin proteins have been solved. Among the important structural features of ferritin are the two types of channels that form at the protein-subunit interfaces. Iron is probably transported through the 3-fold channels, which are lined with hydrophilic side chains of the amino acid residues, aspartate and glutamate. A second channel, the 4-fold channel, is formed where 4 subunits meet. It is lined with hydrophobic side chains of the amino acid leucine. Molecular oxygen, reducing agents, and other small molecules may enter the ferritin cavity through these hydrophobic channels.

The growth of the mineral core and the release of iron from the core in ferritins are redox-switched processes that depend on the different thermodynamic stability (solubility) and kinetic lability of aqueous Fe(II) and Fe(III). At the physiological pH of 7, Fe(II) is soluble whereas Fe(III) is not; therefore, iron is transported into and out of the protein in the Fe(II) oxidation state. Exactly how iron is released from ferritin in vivo has not been established. However, numerous studies on the in vitro release of iron from ferritin have demonstrated that a variety of reducing agents and chelators can be used to trigger iron release.

10.3   Dioxygen and hemeproteins

Dioxygen is a biradical species, but it is not especially reactive by itself. However, it can readily react with transition metal ions that have proper reduction potentials.[7a,b,15] A few examples of such reactions are given below.

Transport, activation, and metabolism of O_2 are very important processes in most living organisms. These functions are often realized by metalloproteins containing iron or copper. Iron-containing heme proteins include the best-known examples of this type, e.g., hemoglobin and myoglobin (O_2 transport), cytochrome c oxidase (activation and reduction of O_2), and peroxidase (elimination of radical intermediates). Other metalloproteins with an iron or copper-containing dinuclear metal center serve as biological alternatives. Examples of this type of protein are the two dioxygen transport proteins, hemerythrin (iron) and hemocyanin (copper). The dinuclear metal centers in these proteins are closely similar to the

---

[15]E.-I. Ochiai, *J. Chem. Educ.* **1996**, *73*, 130.

active sites of other proteins, whose function is not dioxygen transport but enzymatic catalysis (see below).

The following reaction represents the autooxidation of Fe(II) by dioxygen.

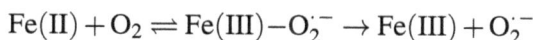

$$\text{Fe(II)} + O_2 \rightleftharpoons \text{Fe(III)} - O_2^{\cdot -} \rightarrow \text{Fe(III)} + O_2^{\cdot -}$$

However, the autooxidation of Fe(II) is actually much more complicated than this suggests.

(a) Hemoglobin and myoglobin

In the animal kingdom, dioxygen is transported by three metal-containing proteins known as *hemoglobin*, *hemerythrin*, and *hemocyanin* (for latter two: see below). Although the heme prefix appears in all three names, only hemoglobin has a heme group (see below). Hemerythrin and hemocyanin are nonheme compounds in which the metal is bound directly to the protein (see below).

Hemoglobin is the most widely distributed of the three dioxygen carriers.[16] It consists of a heme and a single polypeptide chain, about 150 amino acids long, wrapped around the heme. Such monomeric hemoglobins, when found in muscle cells, are called *myoglobin*. They store O_2 and help promote its transfer from blood to mitochondria. In vertebrate blood, hemoglobin is present as a tetrameric molecule made of four myoglobin-like subunits.

The structure of horse hemoglobin was determined by M. F. Perutz and his associates in late 1959 and published in 1960, culminating a study that began in 1936 as a PhD research. John Kendrew and co-workers also published the structure of sperm whale myoglobin in 1960. The two molecules turned out to be remarkably similar.

The heme group is made of a porphyrin ring with iron in the center. Four of the six coordination sites around the metal ion are occupied by the nitrogens of the macrocycle porphyrin. In the fifth site is the imidazole nitrogen of a histidine residue. This linkage is found in all hemoglobins, and it is the only covalent bond between the heme and the protein (Fig. 10.5).

---

[16]N. M. Senozan and R. L. Hunt, *J. Chem. Educ.* **1982**, *59*, 173.

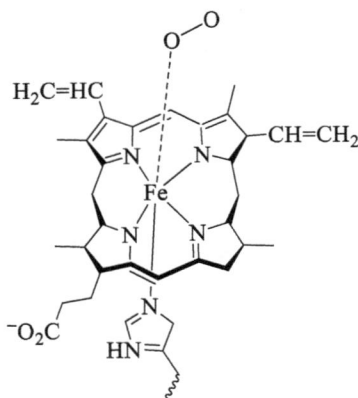

Fig. 10.5 The ChemDraw structure of the heme prosthetic group, with axial histidine coordination.

In deoxy-Hb, the sixth coordination space is empty; upon exposure to air, an O_2 molecule enters here. In oxygen-free Hb, the iron is in the 2+ oxidation state and has a hs (high spin) configuration with four unpaired electrons. The hs-Fe(II) ion is too large to fit in the porphyrin ring and hence sits ~0.6 Å out of the heme plane toward the imidazole nitrogen. With the entry of O_2, the metal reverts to ls (low spin) state, shrinks, and moves toward the plane of the porphyrin ring. This brief motion is believed to trigger a whole series of subtle changes in the arrangement of the subunits and to be responsible for many remarkable aspects of oxygen-hemoglobin equilibrium. The state of iron in oxy-Hb is a ls-ferric-superoxide complex, $Fe(III)-O_2^{\cdot-}$, with one electron transferred to the O_2. The two electrons (one in ls-Fe(III) and the other one in $O_2^{\cdot-}$) are antiferromagnetically coupled giving rise to a diamagnetic $(S = 0)$ state.

The obvious function of the protein chain globin is to prevent the iron from being irreversibly oxidized to the Fe(III) state. An unprotected heme cannot interact with O_2 without getting irreversibly oxidized to the Fe(III) state. Kinetic studies on ferrous compounds suggest that a crucial step in the irreversible oxidation of the heme is the approach of two irons to form an Fe–O–Fe dimer. The protein chain prevents the irons from coming into close contact and allows reversible combination with oxygen. The protein chain is also responsible for the solubility of Hb. While the hydrophobic groups that is inside of the globin provide a proper "oily" environment for the heme group, hydrophilic side chains that point outward render the protein water soluble. A third and subtle function of the polypeptide chain is to reduce the heme's affinity for carbon monoxide.

The equilibrium between a respiratory molecule and oxygen is best discussed in terms of an oxygenation curve in which fractional saturation, $\bar{y}$, defined as[16]

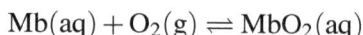

$$Mb(aq) + O_2(g) \rightleftharpoons MbO_2(aq)$$

$$K = [MbO_2]/[Mb] \times PO_2$$

where [Mb] represents the concentration of oxygen-free monomeric Mb, [MbO$_2$] the concentration of the oxygenated form, PO_2 the partial pressure of oxygen in equilibrium with the respiratory molecule, and K the equilibrium constant between the respiratory molecule and oxygen.

$$KPO_2 = [MbO_2]/[Mb]$$

$$1 + KPO_2 = \{[MbO_2] + [Mb]\}/[Mb]$$

$$KPO_2/\{1 + KPO_2\} = \bar{y} = [MbO_2]/\{[MbO_2] + [Mb]\}$$

= (number of moles of O$_2$ actually bound)/(maximum number of moles of O$_2$ that can be bound)

The low $P_{1/2}$ values indicate that various myoglobins are good storage molecules but quite useless as transporters of oxygen. An efficient carrier should exhibit a pronounced difference in the degree of saturation between the pressures at which oxygen is absorbed and the pressures at which it is released. Most animals absorb oxygen from an environment where the PO_2 values lie between 20 to 150 mm Hg (the latter in open air or well-aerated waters) and release it to the tissues where the pressures are typically one-third to one-half as large. Within such a pressure range, Mb's remain almost completely saturated and are incapable of discharging any oxygen.

The presence of several subunits alters the nature of the oxygenation curves entirely. Due to a delicate interplay between the subunits, which is still only partially understood, the binding of oxygen is now cooperative. The affinity of a given heme for oxygen increases as the other hemes in the molecule are oxygenated. Consequently, the degree of saturation at first does not respond much to the pressure, then begins to rise abruptly.

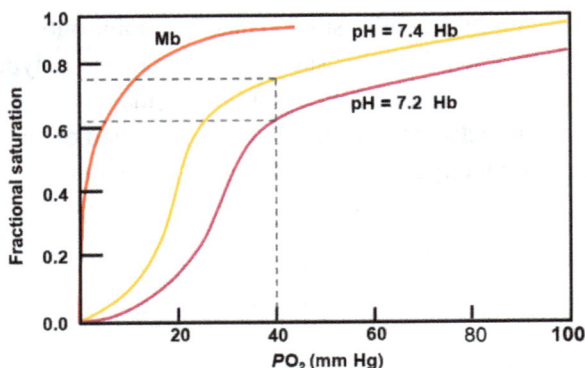

Fig. 10.6 Oxygenation curves for human Mb and Hb. Mb curve (Mb is at 40°C and pH 7.45. Hb curves (Hb) are at 37°C and pH 7.2 and 7.4. The pressure in the lungs is about 100 mm Hg. In the tissues it is about 40 mm Hg. At pH 7.4 the saturation of Hb between these pressures drops from 0.98 to 0.75. Thus, it can deliver 23% of its maximum binding capacity, or $4 \times 0.23 = 0.92$ molecules of O_2 per molecule of Hb. The actual efficiency is somewhat higher because of the lower pH of the respiring tissues. Assuming a pH of 7.4 in the lungs and 7.2 in the tissues, the fractional saturation drops from 0.98 to 0.62, and the amount of O_2 delivered by each Hb molecule rises to $4 \times 0.36 = 1.44$, an increase of almost 50%.

As the pressure continues to increase, the curve levels off and approaches asymptotically to 1, the full saturation value. The result is an *S*-shaped or sigmoidal oxygenation curve shown in Fig. 10.6. The advantage of a sigmoidal curve is apparent. If the steep portion falls between the ambient and tissue pressures, Hb delivers a good portion of the oxygen it is capable of binding. In the case represented in Fig. 10.6, each tetrameric molecule at pH 7.4 delivers on the average 0.92 molecules of O_2, that is 23% of its maximum binding capacity.

The Bohr effect, CO_2 transport

The influence of H^+ ions on the oxygen affinity of a respiratory molecule is called the *Bohr Effect* in honor of Danish physician Christian Bohr (1855–1911), father of Niels Bohr (Danish physicist (1885–1962), Nobel Prize in Physics, 1922), who in 1904 with his co-workers, discovered that dissolving CO_2 in blood decreased its affinity for oxygen. The Bohr Effect is of considerable physiological importance. It increases respiratory efficiency by making Hb release a greater percentage of its O_2 at the slightly decreased pH of the respiring tissues. A pH difference of only 0.2 units causes human Hb to discharge 50% more oxygen.

The Bohr Effect helps carbon dioxide transport too. If a decrease in pH shifts the oxygen-Hb equilibrium in favor of deoxy-Hb,

$$\text{deoxy-Hb} + O_2 \rightleftharpoons \text{oxy-Hb}$$

then, according to Le Chatelier's principle, oxy-Hb must be a stronger acid and deoxy-Hb a stronger base. This way, an attempt to decrease the pH of the medium, which shifts the equilibrium to the left, is somewhat compensated by the removal of some of the protons by the deoxy-Hb. Likewise, an increase in pH leads to the production of oxy-Hb which, being a stronger acid, releases protons and to some extent counteracts the rise in pH. The removal of H^+ ions by deoxy-Hb and their release by oxy-Hb facilitate the transport of CO_2 in the following way:

The CO_2 produced in respiring tissues diffuses into erythrocytes where it is converted to carbonic acid, H_2CO_3, with the help of the enzyme *carbonic anhydrase* (see below).[10a] Neither CO_2 nor H_2CO_3 is very soluble and, unless the protons from the dissociation of carbonic acid ($H_2CO_3 = HCO_3^- + H^+$) are promptly removed, only a very limited amount of CO_2 can be accommodated in blood. The generation of CO_2, however, coincides with the production of deoxy-Hb which binds the protons released by H_2CO_3 and converts it to HCO_3^-. As the HCO_3^- concentration in the erythrocyte rises, some diffuse back to the serum. Over 85% of the CO_2 in human venous blood is carried as bicarbonate ion, which is distributed between erythrocytes and serum in approximately 1 to 4 ratio. What happens in the lungs is the reverse of the situation in respiring tissues. The formation of oxy-Hb is accompanied with the release of protons. This converts the HCO_3^- to H_2CO_3 and CO_2, which diffuse out of erythrocytes first into blood plasma, then to the environment.

(b) Heme enzymes

Negatively charged axial ligands such as cysteinate (cys S^-) or tyrosinase (tyr O^-) bind more strongly to a metal in its higher oxidation states than to one in its lower oxidation states. This in turn contributes to decrease the $E^{o\prime}$ value (stabilization of higher oxidation states). A neutral ligand such as histidine (imidazole N), on the other hand, does not significantly affect the $E^{o'}$ value. The low reduction potential of cytochrome P450 (cyt P450) is due to its fifth ligand being the S^- of a cysteine residue. This can be compared with cytochrome c (cyt c), which has a neutral S (methionine) as the fifth ligand and, as a result, has a rather high $E^{o'}$ value. The sixth ligand is a histidine. The extremely low potential of catalase is due to its fifth ligand being a negatively charged tyrosinate. The low reduction potential would stabilize oxidation of the iron, probably to as high as Fe(IV) and Fe(V) (Fe(IV)-cation radical). These oxidation states are assumed to occur in the operation of cyt P450-dependent monooxygenases, catalase, and peroxidases (see below).

Cyrtochrome 450 exhibits monoxygenase activity (activation of O_2 and subsequent incorporation of one of the oxygens to the substrate and the reduction of the other oxygen to water). It hydroxylates many substrates, including drug metabolism. The resting state of the enzyme in its Fe(III) state is first reduced to Fe(II) state to trigger the subsequent steps, as shown below[7a,b,15,17]:

The overall reaction:

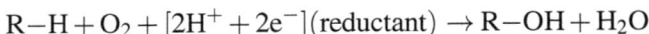

$$R-H + O_2 + [2H^+ + 2e^-](\text{reductant}) \rightarrow R-OH + H_2O$$

The molecular events:

Six-coordinate 'ls-$[(P)Fe(III)(S_{cys})(H_2O)]$' $+ e^- \rightarrow$ Five-coordinate

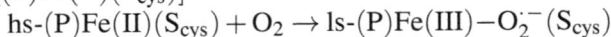

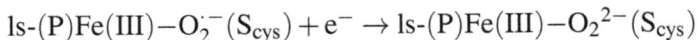

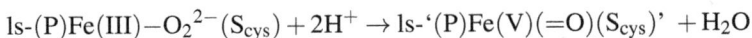

'hs-$[(P)Fe(II)(S_{cys})]$'
$$\text{hs-}(P)Fe(II)(S_{cys}) + O_2 \rightarrow \text{ls-}(P)Fe(III)-O_2^-(S_{cys})$$
$$\text{ls-}(P)Fe(III)-O_2^-(S_{cys}) + e^- \rightarrow \text{ls-}(P)Fe(III)-O_2^{2-}(S_{cys})$$
$$\text{ls-}(P)Fe(III)-O_2^{2-}(S_{cys}) + 2H^+ \rightarrow \text{ls-'}(P)Fe(V)(=O)(S_{cys})' + H_2O$$
$$\text{actually } (P^{\cdot+})Fe(IV)(=O)(S_{cys})$$

(Compound I)

$$(P^{\cdot+})Fe(IV)(=O)(S_{cys}) + R-H \rightarrow R-OH$$
$$+ \text{six-coordinate 'ls-}[(P)Fe(III)(S_{cys})(H_2O)]\text{'}$$

[17](a) I. G. Denisov, T. M. Makris, S. G. Sligar, and I. Schlichting, *Chem. Rev.* **2005**, *105*, 2253. (b) J. H. Dawson, *Science* **1988**, *240*, 433.

Many enzymes act on hydrogen peroxide (*peroxidases*) or hydroperoxides (ROOH) (*catalases*). A common mechanism for peroxidases and catalases is believed to involve the following initial reaction:

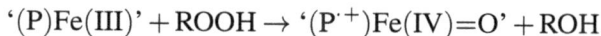

$$\text{`(P)Fe(III)'} + \text{ROOH} \rightarrow \text{`(P}^{\cdot+}\text{)Fe(IV)}{=}\text{O'} + \text{ROH}$$

where P is porphyrin and $P^{\cdot+}$ is porphyrin cation radical. The first step is a one-electron transfer from Fe(III) to the ROOH (or ROO^-), resulting in the formation of Fe(IV)=O and $RO^{\cdot}$. The radical then abstracts a H atom from the porphyrin. The axial ligand (not indicated above) is tyrosinate$(1-)$ in *catalases* and in *horseradish peroxidase* it is histidine. However, in *chloroperoxidase* (CPO catalyzes the reaction: $A{-}H + X^- + H^+ + H_2O_2 \rightarrow A{-}X + 2H_2O$) it is cysteinate$(1-)$. The reaction events (catalytic cycle) in catalase and horseradish peroxidase are as follows.

Catalase reaction steps:

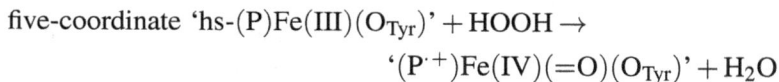

five-coordinate `hs-(P)Fe(III)(O_{Tyr})' + HOOH $\rightarrow$

$$\text{`(P}^{\cdot+}\text{)Fe(IV)}(={=}\text{O)}(O_{Tyr})\text{'} + H_2O$$

Compound I

`(P$^{\cdot+}$)Fe(IV)(=O)(O_{Tyr})' + HOOH $\rightarrow$

$$\text{five-coordinate `hs-(P)Fe(III)}(O_{Tyr})\text{'} + O_2 + H_2O$$

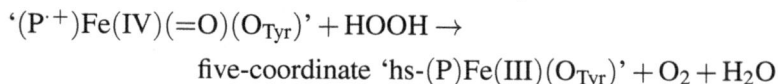

CPO reaction steps:

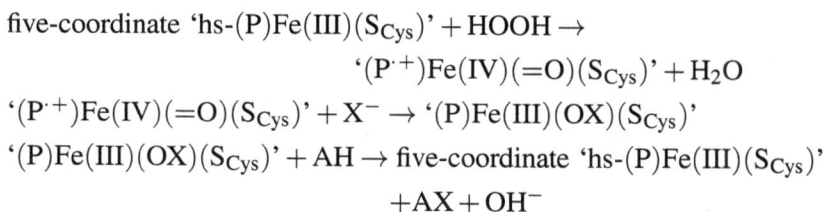

five-coordinate `hs-(P)Fe(III)(S_{Cys})' + HOOH $\rightarrow$

$$\text{`(P}^{\cdot+}\text{)Fe(IV)(=O)}(S_{Cys})\text{'} + H_2O$$

`(P$^{\cdot+}$)Fe(IV)(=O)$(S_{Cys})$' + $X^-$ $\rightarrow$ `(P)Fe(III)(OX)(S_{Cys})'

`(P)Fe(III)(OX)$(S_{Cys})$' + AH $\rightarrow$ five-coordinate `hs-(P)Fe(III)(S_{Cys})'

$$+ AX + OH^-$$

HRP reaction steps:

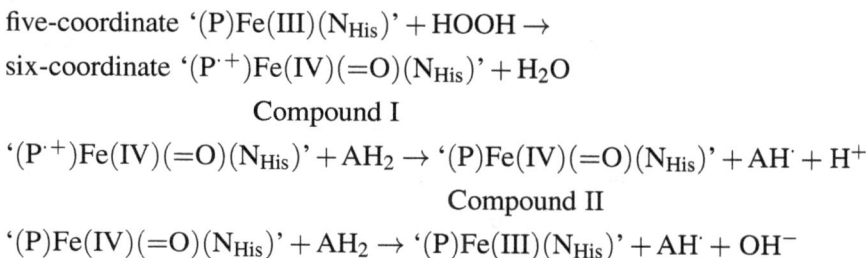

five-coordinate `(P)Fe(III)(N_{His})' + HOOH $\rightarrow$

six-coordinate `(P$^{\cdot+}$)Fe(IV)(=O)(N_{His})' + H_2O

Compound I

`(P$^{\cdot+}$)Fe(IV)(=O)$(N_{His})$' + $AH_2$ $\rightarrow$ `(P)Fe(IV)(=O)(N_{His})' + $AH^{\cdot}$ + H^+

Compound II

`(P)Fe(IV)(=O)$(N_{His})$' + $AH_2$ $\rightarrow$ `(P)Fe(III)(N_{His})' + $AH^{\cdot}$ + OH^-

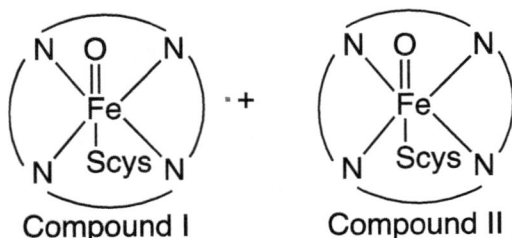

Compound I   Compound II

10.4 Aerobic metabolism

Humans are heterotrophic, highly complex organisms. They require reduced organic compounds (carbohydrates, fats, and proteins) as a food source. Stored in these molecules is a great amount of useful energy. Humans can tap the energy from these food sources and can use it to build up their own complex molecules, to move about, to sense the environment, and to keep warm. Without this input of energy, they would soon die or in thermodynamic sense, they would be at equilibrium with their environment. Fortunately, this external source of energy does grow on trees and elsewhere. The large, complex fuel molecules necessary to run the machine known as humans are directly or indirectly produced by plants.

Living things require a continual input of free energy for three major purposes: the performance of mechanical work in muscle contraction and other cellular movements, the active transport of molecules and ions, and the synthesis of macromolecules and other biomolecules from simple precursors. The free energy used in these processes, which maintain an organism in a state that is far from equilibrium, is derived from the environment. *Chemotrophs* (aerobic metabolism) obtain this energy by the oxidation of foodstuffs, whereas *phototrophs* (photosynthesis) obtain it by trapping light energy (sunlight). The free energy derived from the oxidation of foodstuffs and from light is partly transformed into a special form before it is used for motion, active transport, and biosynthesis. This special carrier of free energy is adenosine triphosphate (ATP).

Stages in the extraction of energy from foodstuffs

Hans Krebs (German-born British biologist, physician, and biochemist: 1900–1981; Nobel Prize in Physiology or Medicine in 1953) described three stages in the generation of energy from the oxidation of foodstuffs. In the first stage, large molecules in food are broken down into smaller units. Proteins are hydrolyzed to their twenty kinds of constituent amino acids, polysaccharides are hydrolyzed to simple sugars such as glucose, and fats are hydrolyzed to glycerol and fatty acids. In the second stage, these numerous small molecules are degraded to a few simple units that play a central role in *metabolism*. In fact, most of them — sugars, fatty acids, glycerol, and several amino acids — are converted into the acetyl unit of *acetyl CoA* (acetyl coenzyme A). Some ATP is generated in this stage, but the amount is small compared with that obtained from the complete oxidation of the acetyl unit of acetyl CoA. The third stage consists of the *citric acid cycle* (also known as the *Krebs cycle*) and *oxidative phosphorylation*, which are the final common pathways in the oxidation of fuel molecules. Acetyl CoA brings acetyl units into this cycle, where they are completely oxidized to CO_2. Four pairs of electrons are transferred to NAD^+ and FAD for each acetyl group that is oxidized. Then, ATP is generated as electrons flow

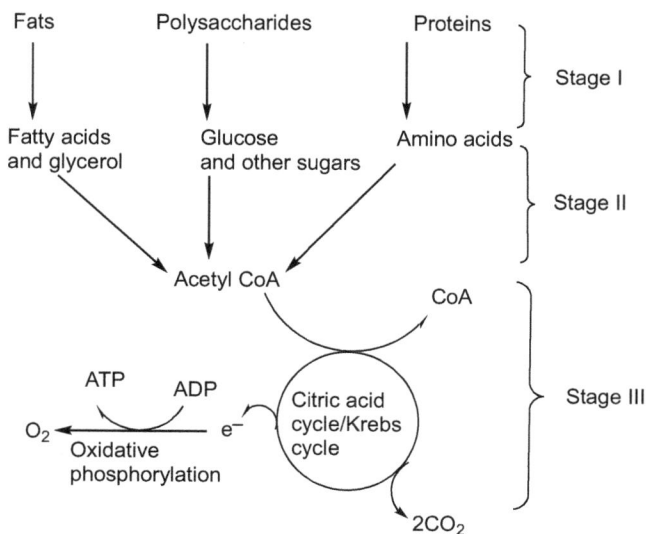

Fig. 10.7 Generation of energy from the oxidation of foodstuffs.

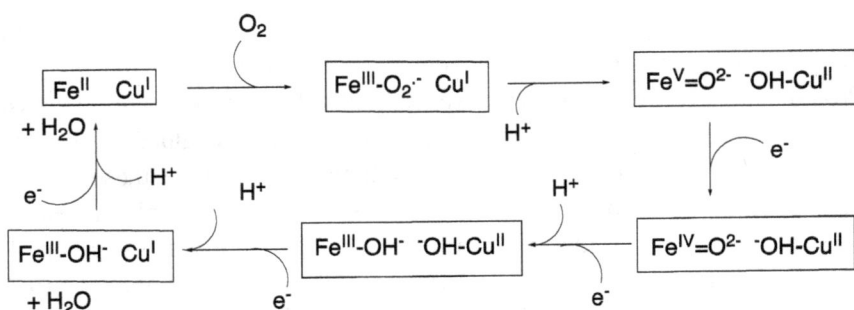

Fig. 10.8 A simplified scheme of the catalytic cycle of O_2 reduction by cytochrome c oxidase.

from the reduced forms of these carriers to O_2, a process called oxidative phosphorylation. Most of the ATP generated by the degradation of foodstuffs is formed in this third stage (Fig. 10.7).

At the end of the electron transport system, the electrons are used to reduce an oxygen molecule to O^{2-} ions. The extra electrons on the O^{2-} ions attract H^+ ions from the surrounding medium, and water is formed. The electron transport chain and the production of ATP through chemiosmosis are collectively called *oxidative phosphorylation* (Fig. 10.7).

Cytochrome *c* Oxidase (CcO) is an enzyme in mitochondria which catalyzes the transfer of four electrons to a O_2 molecule to form two molecules of H_2O. Its active site consists of four redox centers, heme *a*, heme a_3-Cu_B, Cu_A, and Cu_B. The O_2 is believed to bind to the Fe(II) and to the Cu(I) centers of heme a_3-Cu_B and O_2 is reduced to peroxide level. Further transfer of two electrons from heme *a* and the Cu_A completes the reduction of O_2 (Fig. 10.8).[15,18]

10.5 Nonheme iron proteins/enzymes

Oxygen-activating enzymes with *mononuclear* nonheme iron active sites participate in many metabolically important reactions that have environmental, pharmaceutical, and medical significance. For example, bacterial

---

[18]M. Wikström, K. Krab, and V. Sharma, *Chem. Rev.* **2018**, *118*, 2469.

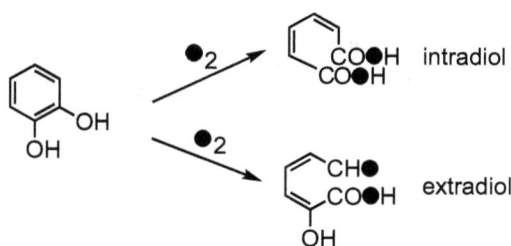

Fig. 10.9 Intradiol and extradiol cleavage reactions.

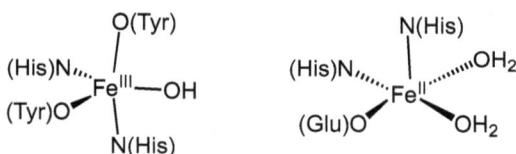

Fig. 10.10 Active site structures of protocatechuate 3,4-dioxygenase (left) and 2,3-dihydroxybiphenyl 1,2-dioxygenase (right).

catechol dioxygenases (activation of O_2 and subsequent incorporation of both oxygens into the substrate) are involved in the degradation of aromatic molecules in the environment. Depending on the position of the cleaved double bond relative to the hydroxyl groups, catechol dioxygenases can be split into two families: the intradiol-cleaving catechol dioxygenases, which cleave the carbon-carbon bond of the enediol moiety, and the extradiol-cleaving catechol dioxygenases, which cleave adjacent to the enediol. The two reactions are presented in Fig. 10.9.[15,19]

The active site structure of protocatechuate 3,4-dioxygenase (intradiol-cleaving catechol dioxygenase) and 2,3-dihydroxybiphenyl 1,2-dioxygenase (extradiol-cleaving catechol dioxygenase) are presented in Fig. 10.10.

According to the proposed mechanism for intradiol cleavage (Fig. 10.11), the first step is the binding of the catecholate substrate to the

---

[19](a) L. Que, Jr. and R. Y. N. Ho, *Chem. Rev.* **1996**, *96*, 2607. (b) M. Costas, M. P. Mehn, M. P. Jensen, and L. Que, Jr., *Chem. Rev.* **2004**, *104*, 939.

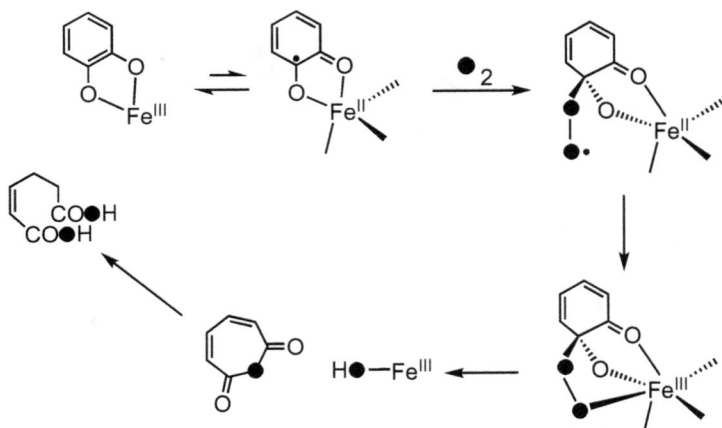

Fig. 10.11   Proposed mechanism for intradiol catechol dioxygenases.

Fe(III) center, displacing the hydroxide and the axial Tyr residue to generate a square pyramidal [Fe(his)$_2$(tyr)(catecholate)] complex. The covalency of the Fe(III)-catecholate interaction introduces semiquinonate radical character to bound substrate and makes it susceptible to O$_2$ attack, generating a transient alkylperoxoiron(III) intermediate. This intermediate then undergoes a Criegee-type rearrangement to form muconic anhydride, and the Fe^{III}–OH species thus formed acts as the nucleophile to convert the anhydride into the ring-opened product.

According to the proposed mechanism for extradiol cleavage (Fig. 10.12, the first step involves the binding of the substrate, followed by the binding of O$_2$ to the Fe(II) center. Some electron transfer from Fe(II) to O$_2$ results in a superoxide-like species, which gives the bound O$_2$ nucleophilic character. The bound O$_2$ attacks the carbon adjacent to the enediol unit in a Michael-type addition, to form a 'peroxy intermediate' that decomposes by a Criegee-type rearrangement to the observed product. This attack of a nucleophilic superoxide (vs an electrophilic dioxygen in the intradiol cleavage mechanism (Fig. 10.11) on the bound substrate serves as the cornerstone for the current proposed mechanism for extradiol cleavage. It provides a very attractive rationale for the regiospecificity of dioxygen attack, as the bound substrate will undoubtedly have differing loci for nucleophilic and electrophilic attack.

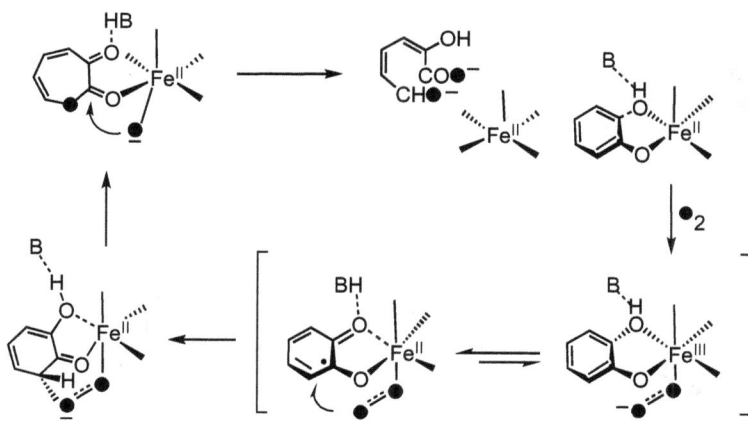

Fig. 10.12 Proposed mechanism for the extradiol cleavage of catechol. B stands for base.

Hemerythrin, ribonucleotide reductase, and methane monooxygenase

Hemerythrin is an oxygen-transport protein (Fig. 10.13).[20] Oxygen-activating enzymes with carboxylate-bridged nonheme diiron sites have attracted intense interest (Fig. 10.13). Prominent among these are methane monooxygenase (MMO), which converts methane to methanol, and ribonucleotide reductase (RNR), which is a key enzyme in the biosynthesis of DNA.[20] MMO consists of three components, a hydroxylase, a reductase, and a coupling protein, and the diiron site responsible for dioxygen activation and alkane hydroxylation resides in the hydroxylase component (MMOH). RNR, on the other hand, consists of R1 and R2 components. Ribonucleotide reduction on R1 is initiated by a stable Tyr radical found in R2; the tyrosyl radical is formed by dioxygen activation on the diiron site in R2.[7a,20]

10.6 Nickel enzyme, urease

Urease (urea amidohydrolase) is a nickel-dependent enzyme found in a large variety of organisms, including plants, algae, fungi, and several prokaryotes. It is critically involved in the mineralization step of the global

---

[20](a) A. L. Feig and S. J. Lippard, *Chem. Rev.* **1994**, *94*, 759. (b) L. Que, Jr., *J. Chem. Soc., Dalton Trans.* **1997**, 3933.

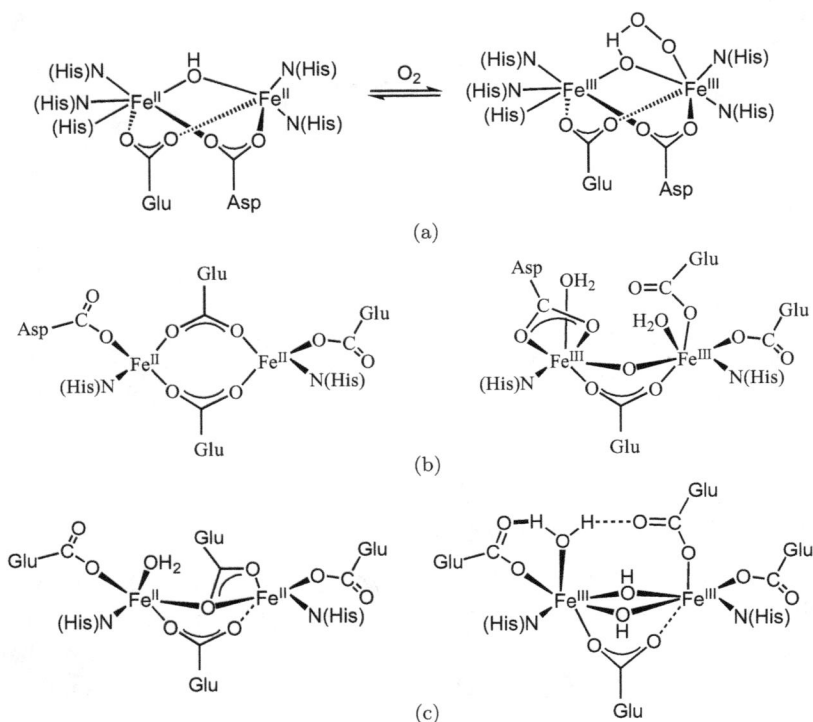

Fig. 10.13 Active site structures of (a) hemerythrin, (b) ribonucleotide reductase, and (c) methane monooxygenase.

nitrogen cycle, being able to catalyze the rapid hydrolytic decomposition of urea to produce ammonia and carbamate, the latter eventually decomposes spontaneously into a second molecule of ammonia and bicarbonate.[21] The active site structure of urease is displayed in Fig. 10.14.

[21](a) L. Mazzei, F. Musiani, and S. Ciurli, *J. Biol. Inorg. Chem.* **2020**, *25*, 829. (b) K. Kappaun, A. R. Piovesan, C. R. Carlini, and R. Ligabue-Braun, *J. Adv. Res.* **2018**, *13*, 3.
[21](a) N. Kitajima and Y. Moro-oka, *Chem. Rev.* **1994**, *94*, 737. (b) E. I. Solomon, P. Chen, M. Metz, S.-K. Lee, and A. E. Palmer, *Angew. Chem. Int. Ed.* **2001**, *40*, 4570. (c) L. Q. Hatcher and K. D. Karlin, *J. Biol. Inorg. Chem.* **2004**, *9*, 669. (d) L. M. Mirica, X. Ottenwaelder, and T. D. P. Stack, *Chem. Rev.* **2004**, *104*, 1013. (e) E. A. Lewis and W. B. Tolman, *Chem. Rev.* **2004**, *104*, 1047. (f) S. Itoh and S. Fukuzumi, *Acc. Chem. Res.* **2007**, *40*, 592. (f) A. De, S. Mandal, and R. Mukherjee, *J. Inorg. Biochem.* **2008**, *102*, 1170

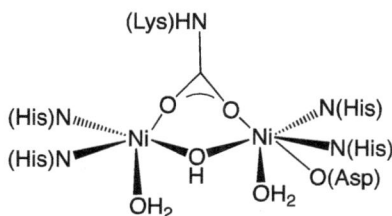

Fig. 10.14 The active site structure (ChemDraw) of urease.

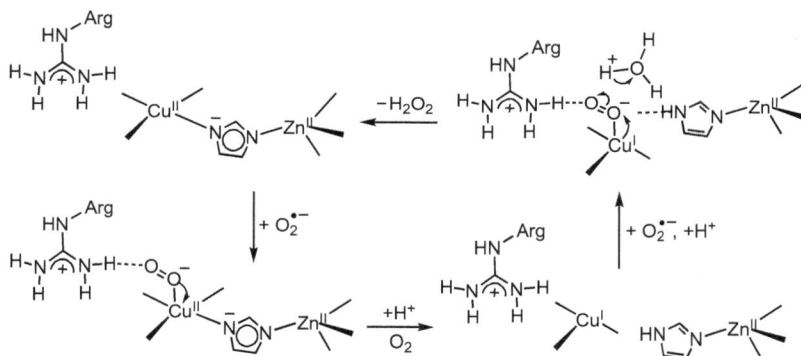

Fig. 10.15 The catalytic cycle of Cu,Zn-containing SOD.

10.7 Copper proteins

Copper proteins are involved in numerous processes that impact life and the environment. Well-known blue copper proteins function as electron transfer agent (plastocyanin, azurin; see above) and dismutation of superoxide dismutase (Cu,Zn active site, see below). Copper proteins are involved in reversible dioxygen binding (hemocyanin, see below), two-electron reduction to peroxide coupled to oxidation of substrates (galactose and catechol oxidase, see below).[22]

We discuss first superoxide dismutase with Cu,Zn-containing active site with one copper center and then a mononuclear copper protein galactose oxidase, with a protein radical at the active site.

---

[22](a) J. W. Whittaker, *Chem. Rev.* **2003**, *103*, 2347. (b) F. Thomas, *Eur. J. Inorg. Chem.* **2007**, 2379.

Superoxide dismutase

The toxicity of dioxygen has received widespread attention. Bio-friendly reduced form of O_2 is H_2O. However, partial reduction of O_2 causes problem in biology. One of the products O_2^{2-} is processed by the heme-containing enzyme *catalase* and the other product is processed by the nonheme enzyme *superoxide dismutase*. Superoxide dismutase (SOD) catalyzes the disproportionation of superoxide into dioxygen and peroxide $(2O^{\cdot -} + 2H^+ \rightarrow O_2 + H_2O_2)$. The functioning of well-known Cu,Zn-containing SOD is presented in Fig. 10.15.[8]

Deprotonated histidine imidazolate ion bridges the two metal ions Cu(II) and Zn(II). The metal centers are terminally coordinated by imidazole nitrogens of histidines. The reaction between Cu(II) and $O_2^{\cdot -}$ gives rise to the formation of Cu(I) and liberation of O_2 (Cu(II) + $O_2^{\cdot -} \rightarrow$ Cu(I) + O_2). Subsequent reaction between Cu(I) and $O_2^{\cdot -}$ brings back Cu(I) to Cu(II) state with liberation of H_2O_2 (Cu(I) + $O_2^{\cdot -}$ + $2H^+ \rightarrow$ Cu(II) + H_2O_2). Notably, proximal arginine subtly controls protonation/deprotonation phenomena and H-bonding interactions.

Galactose oxidase

The use of metal complexes, with radicals on the supporting ligands, as catalysts draws inspiration from enzymatic reactions of certain metalloproteins. One of the best understood examples is galactose oxidase (GOase), which catalyzes the two-electron oxidation of alcohols to aldehydes. A Cu(II) ion is coordinated to a modified tyrosyl radical, and this intricate bonding situation gives rise to the function of the enzyme (Fig. 10.16). This phenomenon may be pervasive in metal-containing redox proteins.[22]

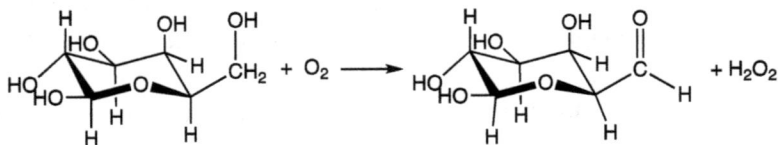

Hemocyanin, tyrosinase, and catechol oxidase

Hemocyanin, tyrosinase, and catechol oxidase are the members of the type 3 copper protein family[21] with antiferromagnetically coupled dinuclear

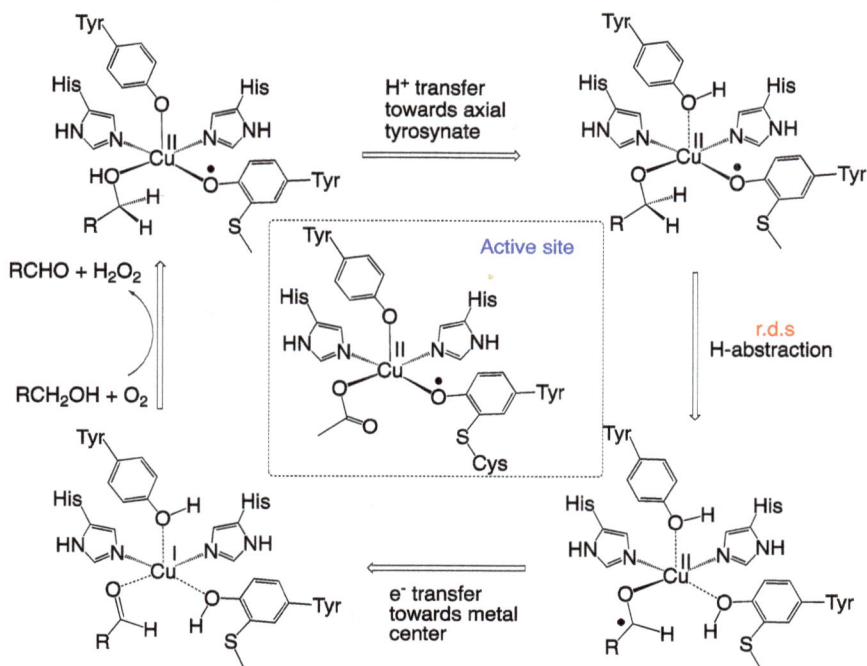

Fig. 10.16 Mechanism of the function of GOase.

copper reaction center. Hemocyanin has only the reversible dioxygen binding ability, thus acting as the oxygen storage and carrier protein of mollusks and arthropods (Fig. 10.14). Tyrosinase and catechol oxidase exhibit different reactivity toward exogenous substrates. Tyrosinase catalyzes *ortho*-hydroxylation of phenols to the corresponding catechols (phenolase activity), as well as dehydrogenation of catechols to the corresponding *o*-quinones (catecholase activity) (Fig. 10.18). *o*-Quinone formation from tyrosine is the initial step for the synthesis of melanin pigments in nature. Catechol oxidase exhibits only catecholase activity (Fig. 10.19), without acting on phenols, showing the monooxygenase activity. These differences in the chemical reactivity among the type 3 copper proteins can be attributed to the different architectures of the enzyme active sites as in the case of heme proteins (see above) The chemical reactivity of the metalloproteins is thereby controlled by the arrangement of amino acid side chains in the active sites.

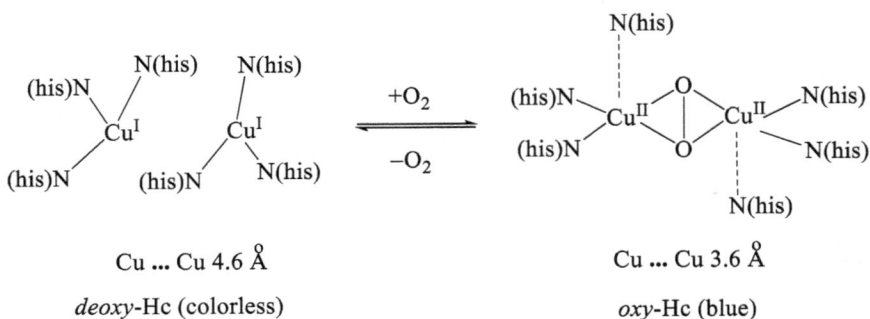

Cu ... Cu 4.6 Å

deoxy-Hc (colorless)

Cu ... Cu 3.6 Å

oxy-Hc (blue)

Fig. 10.17 Schematic drawing of the structures of *deoxy*- and *oxy*-state of Hc.

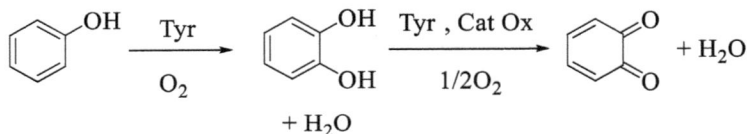

Fig. 10.18 Tyrosinase and catechol oxidase activity.

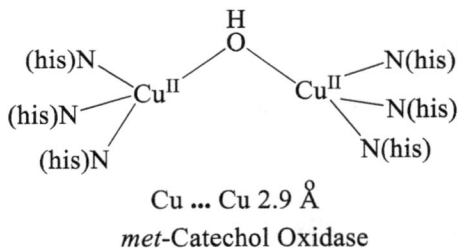

Cu ... Cu 2.9 Å

met-Catechol Oxidase

Fig. 10.19 Structure of the *met*-form of catechol oxidase.

Figure 10.20 depicts the structure of oxygenated hemocyanin, tyrosinase, and catechol oxidase.

Figure 10.21 shows a proposed catalytic mechanism of tyrosinase and catechol oxidase. The native enzymatic reaction involves many fundamental catalytic steps such as binding of dioxygen to deoxy-tyrosinase [dicopper(I) form] to generate peroxo species (oxy-tyrosinase), association of the substrate with oxy-tyrosinase, oxygen atom transfer from the peroxo species to the substrate to give catecholate product, and dehydrogenation of

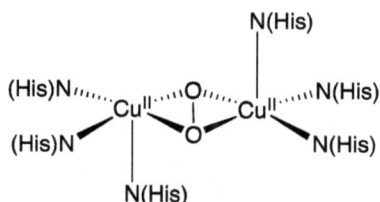

Fig. 10.20 Active site of oxy-tyrosinase, depicting a μ-η^2:η^2-peroxo-dicopper(II) core.

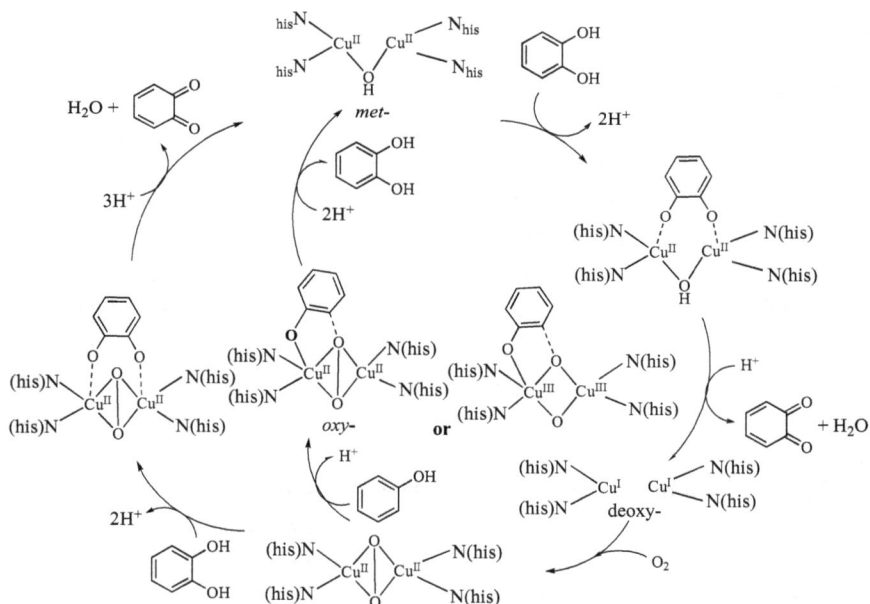

Fig. 10.21 Mechanism of the functioning of hemocyanin, tyrosinase, and catechol oxidase.

catechol to *o*-quinone. Among these processes, the oxygenation of phenols by the peroxo species is most attractive from the viewpoints of synthetic inorganic/organic chemistry and catalytic oxidation chemistry. However, the mechanism of the oxygen atom-transfer process has yet to be fully clarified due to the complicated side reactions such as nonenzymatic transformation of the *o*-quinone products to melanin pigments.

10.8   Zinc enzymes

Zinc is an essential element — one that is necessary for the occurrence of reactions that are required in the metabolic processes of living organisms.[10] The reason why zinc is used for many hydrolytic, condensation, or other atom and group transfer reactions, include its (i) flexible coordination geometry, (ii) fast ligand exchange, (iii) Lewis acidity, (iv) intermediate polarizability (hard-soft character), (v) availability, (vi) strong binding to suitable sites, and (vii) lack of redox activity (no generation of reactive radicals). It is the combination of these factors that makes zinc so attractive. Among many metal ions Co(II) is most suitable to replace Zn(II) efficiently.

The most numerous of the zinc enzymes are classified as hydrolases; they catalyze the hydrolysis of such condensed bonds as those in pyrophosphate, in esters (both phosphate and carboxylate), and in various types of peptides. Bond breaking itself is caused by the nucleophile water. The first question that arises is why nature has chosen Zn(II) for this Lewis acid role. The crucial step in these enzymic reactions is polarization of some relevant bond, since polarization would enhance the rate of subsequent attack by a nucleophile. The function of Zn(II) in these enzymes must then be that of a Lewis acid. A zinc(II) entity can also facilitate catalysis by a precise alignment of substrates through their coordination.

The functioning of two mononuclear zinc-containing enzymes carbonic anhydrase and carboxypeptidase are discussed here. The first enzyme known to require Zn as a cofactor is *carbonic anhydrase* (see above), which catalyzes the reversible formation of bicarbonate and a hydrogen ion from water and carbon dioxide.[10]

In the active site of carbonic anhydrase, a Zn(II) ion is coordinated to the protein by the imidazole groups of three histidine residues, with the remaining tetrahedral site being occupied by a water molecule (or hydroxide ion, depending upon pH). The coordinated water molecule is also involved in a H-bonding interaction with a Thr (threonine) residue that in turn is H-bonded to a Glu (glutamate) residue. The overall features of the mechanism of action of carbonic anhydrase are illustrated in Fig. 10.22 comprising the following steps: (i) deprotonation of the coordinated water with a pKa $\approx$ 7 to give the active Zn(II)-hydroxide derivative

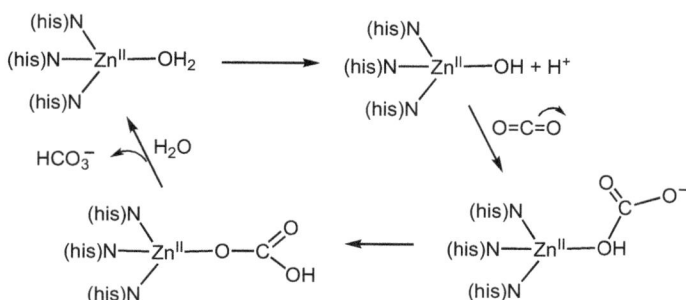

Fig. 10.22 Proposed mechanism for carbonic anhydrase.

$[(his)_3Zn^{II}-OH]^+$, (ii) nucleophilic attack of the Zn(II)-bound hydroxide ion at the carbon of noncoordinating CO_2 substrate. This yields bicarbonate which binds with its OH group coordinated to Zn(II) ion to give a hydrogen carbonate intermediate $[(his)_3Zn^{II}-OCO_2H]^+$, and (iii) displacement of the bicarbonate anion by H_2O to complete the catalytic cycle.

The enzyme carboxypeptidase catalyzes hydrolysis of peptide bonds.[10] The active-site of Zn(II) ion in *carboxypeptidase A* is bound to two histidine, a glutamate, and to a water molecule (Fig. 10.23). The Zn(II)-bound water molecule is H-bonded to proximal Glu. The mechanisms of action of carboxypeptidase is controversial, with two proposals which are principally differentiated according to whether hydrolysis of the peptide linkage occurs via attack of a Zn(II)-bound "hydroxide" or by attack of a noncoordinated water molecule. In one proposal, the "hydroxide mechanism" (Fig. 10.22), the Zn(II) ion serves two roles: (i) activating the water towards deprotonation by a glutamate residue and (ii) activating the carbonyl group of the peptide unit towards nucleophilic attack. Subsequent transfer of the incipient hydroxide to the carbonyl group, followed by a return transfer of the proton from the glutamate residue to the nitrogen, achieves the peptide cleavage.[10]

10.9 Photosynthesis

Photosynthesis is the system of reactions by which higher plants, some algae, and bacteria capture solar energy, convert it to chemical energy, and

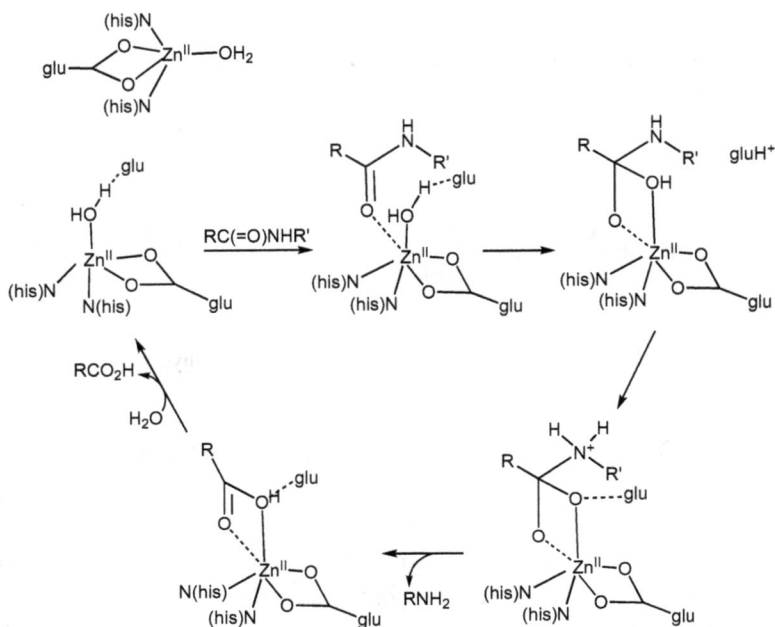

Fig. 10.23 Proposed mechanism for carboxypeptidase A.

use this energy in the reduction of carbon dioxide to sugars. The absorption of light energy by photosynthetic pigments, the transfer of the energy among pigment molecules, and the stabilization of the energy by charge separation are often referred to as the "light reactions" of photosynthesis. The products of the light reactions, stored as reducing power and as chemical energy in form of the compounds nicotinamide adenine dinucleotide phosphate (NADPH) and adenosine triphosphate (ATP), respectively, are used in the reduction or fixation of the carbon source, carbon dioxide. The biochemical processes involved in carbon dioxide fixation are not directly light dependent and are called "dark reactions".

The primary light-absorbing pigment in all photosynthesizing plants is chlorophyll a. Chlorophyll is a magnesium-containing porphyrin structure similar to the iron-containing porphyrin structure of heme in the hemoglobin of blood (see above). Chlorophyll contains a long, nonpolar hydrocarbon chain called a phytyl group. The phytyl group associates with nonpolar lipids in the cell membranes. Chlorophyll found in photosynthetic

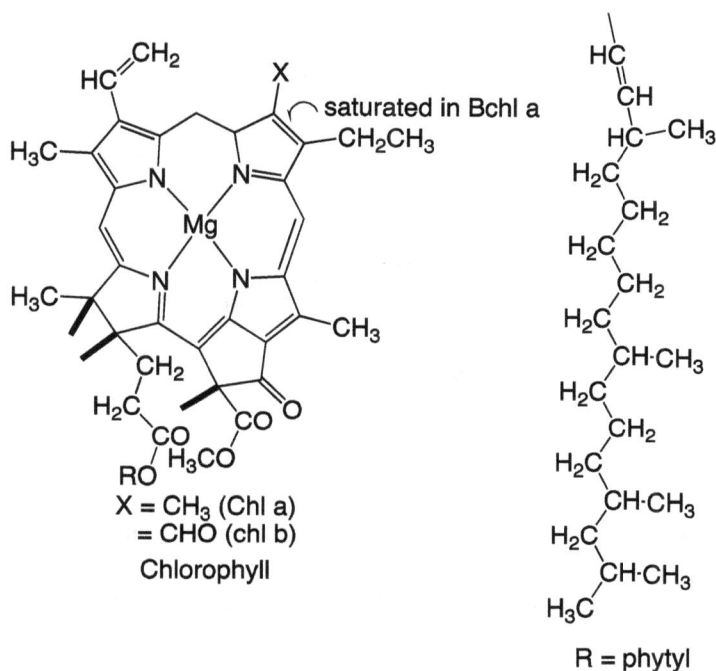

Fig. 10.24 The structure of chlorophyll and the phytyl group.

bacteria is slightly different in structure and is called bacteriochlorophyll a (Fig. 10.24).[23]

The term photosystem is used to identify two distinguishable apparatuses called Photosystem I (PSI) and Photosystem II (PSII).

The overall process in photosynthesis in green plants is the absorption of CO_2 and H_2O in the presence of light to give carbohydrates and dioxygen,

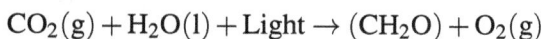

$$CO_2(g) + H_2O(l) + Light \rightarrow (CH_2O) + O_2(g)$$

(CH_2O) represents a "proto-carbohydrate." Photosynthesis can be divided in three steps:

---

[23]P. Borrell and D. T. Dixon, *J. Chem. Educ.* **1984**, *61*, 83. (b) M. B. Bishop and C. B. Bishop, *J. Chem. Educ.* **1987**, *64*, 302

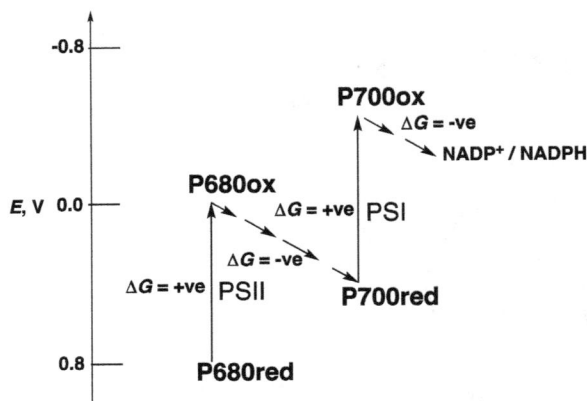

Fig. 10.25 Simplified Z-scheme.

1) Light collection via chlorophyll and other pigments and conveying the energy to a reaction center.
2) The oxidation of H_2O to O_2 and the reduction of $NADP^+$ to NADPH. The overall reaction is

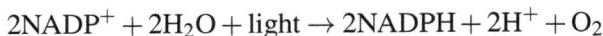

$$2NADP^+ + 2H_2O + \text{light} \rightarrow 2NADPH + 2H^+ + O_2$$

ATP is also formed from the adenosine diphosphate, ADP, in this step.
3) The absorption of CO_2, oxidation of NADPH, and formation of (CH_2O):

$$2NADPH + 2H^+ + CO_2 \rightarrow (CH_2O) + 2NADP^+ + H_2O$$

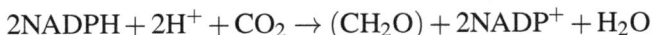

The process requires three molecules of ATP.

Z-scheme

The Z-scheme (Fig. 10.25) is a suggested mechanism for the overall reaction during photosynthesis. It is simplest to follow an electron through the sequence but it must be understood that in the dark all the various reactants and products are at concentrations determined by the equilibrium constants for the reactions (in the absence of side processes). Illumination shifts the equilibria to steady state values which will depend on the light intensity. The process can be looked at sequentially as follows.

Step 1. Light is entrapped by PSII in which it is thought a complex of the photo-receptor, P680, with a quinone electron acceptor, Q, is excited and undergoes charge transfer:

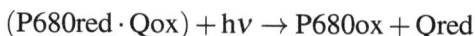

$$(P680red \cdot Qox) + h\nu \rightarrow P680ox + Qred$$

$$\Delta G = +ve \text{ (endothermic reaction)}$$

The oxidized form, $P680_{OX}$ undergoes, via intermediate reactions, a reaction with water,

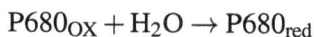

$$P680_{OX} + H_2O \rightarrow P680_{red}$$

So the net effect is the transfer of an electron as water is oxidized

$$1/2H_2O \rightarrow 1/4O_2 + H^+ + e^-$$

and Q_{ox} reduced,

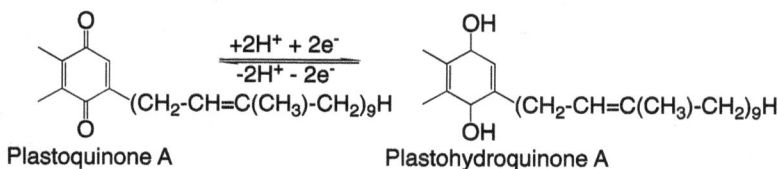

$$Qox + e^- \rightarrow Qred$$

Plastoquinone A   Plastohydroquinone A

Actual reaction sequence (not stoichiometrically balanced),[24]

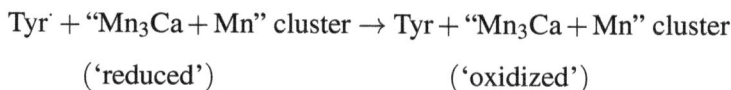

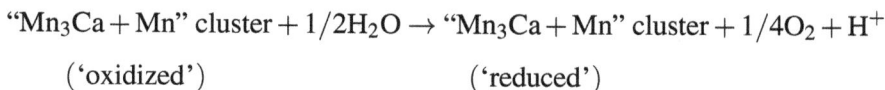

$$P680ox + Tyr \text{ (tyrosine)} \rightarrow P680red + Tyr^{\cdot} \text{ (tyrosyl radical)}$$

$$Tyr^{\cdot} + \text{“Mn}_3Ca + Mn\text{” cluster} \rightarrow Tyr + \text{“Mn}_3Ca + Mn\text{” cluster}$$

(‘reduced’)   (‘oxidized’)

$$\text{“Mn}_3Ca + Mn\text{” cluster} + 1/2H_2O \rightarrow \text{“Mn}_3Ca + Mn\text{” cluster} + 1/4O_2 + H^+$$

(‘oxidized’)   (‘reduced’)

---

[24] (a) J. P. McEvoy and G. W. Brudvig, *Chem. Rev.* **2006**, *106*, 4455. (b) C. Chen, C. Zhang, H. Dong, and J. Zhao, *Dalton Trans.* **2015**, *44*, 4431.

As the diagram shows, this requires work to be done on the system.

Step 2. The reduced quinone (Q_{red}) reacts with a plastocyanin via a series of reactions involving a quinone, a plastoquinone, and a cytochrome; again an electron is transferred.

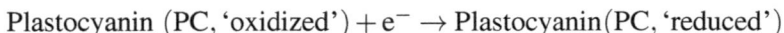

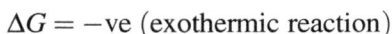

$$\text{Plastocyanin (PC, 'oxidized')} + e^- \rightarrow \text{Plastocyanin(PC, 'reduced')}$$

$$\Delta G = -\text{ve (exothermic reaction)}$$

Overall reaction:

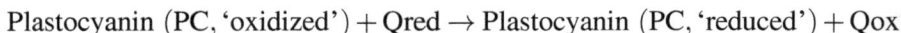

$$\text{Plastocyanin (PC, 'oxidized')} + \text{Qred} \rightarrow \text{Plastocyanin (PC, 'reduced')} + \text{Qox}$$

$$\text{Qox} + e^- \rightarrow \text{Qred}$$

The transfer is a downhill process so it is a spontaneous step and work is available.

Some of the available work is thought to be used for the conversion of ADP to ATP.

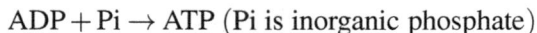

$$\text{ADP} + \text{Pi} \rightarrow \text{ATP (Pi is inorganic phosphate)}$$

It is known that half a mole of ATP is formed per mole of electrons transferred and so it appears that less than half of the available work is used in this conversion.

Step 3. In photosystem I, the photoreceptor P700 reacts in an electron-transfer reaction with the substance, X, (P700red · Xox) + hν → P700ox + Xred and the oxidized, P700ox reacts with the reduced plastocyanin,

$$\text{P700ox} + \text{PCred} \rightarrow \text{P700red} + \text{PCox}$$

so that the overall process is electron transfer from PCred to Xred.

The minimum work required is provided by the photon.

Step 4. The final steps involve the reduction of $NADP^+$ and oxidation of Xred again through a sequence involving a ferredoxin.

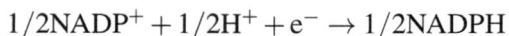

$$\text{Xox} + e^- \rightarrow \text{Xred}$$

$$1/2\text{NADP}^+ + 1/2\text{H}^+ + e^- \rightarrow 1/2\text{NADPH}$$

The process is spontaneous. The NADPH produced is used in CO_2 fixation.

The active site of PSII, responsible for oxidation of water is depicted in Fig. 10.26.[24]

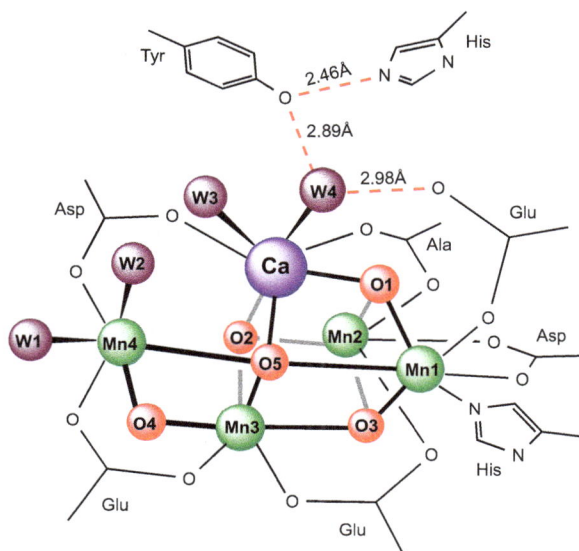

Fig. 10.26 Active site structure of PSII.

10.10 Nitrogenase

Nitrogen-fixing bacteria utilize the enzyme nitrogenase to catalyze the reduction of dinitrogen (N_2) to two NH_3 molecules, the major contribution of fixed nitrogen into the biogeochemical nitrogen cycle, which sustains life on Earth. The most widely studied nitrogenase is the Mo-dependent enzyme. The reduction of N_2 by this enzyme involves the transient interaction of two component proteins, designated the Fe protein and the MoFe protein, and minimally requires sixteen MgATP, eight protons, and eight electrons (Pi is inorganic phosphate)[25]:

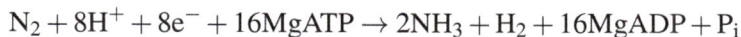

$$N_2 + 8H^+ + 8e^- + 16MgATP \rightarrow 2NH_3 + H_2 + 16MgADP + P_i$$

In this catalytic reaction H_2 is a by-product. This enzyme functions with involvement of three electron transfer proteins: Fe_4S_4 cluster (see above), P Cluster two Fe_4S_4 clusters clubbed together), and FeMo-cofactor cluster. The P-cluster in the MoFe protein functions in nitrogenase catalysis as an

[25] (a) O. Einsle and D. C. Rees, *Chem. Rev.* **2020**, *120*, 4969. (b) P. V. Rao and R. H. Holm, *Chem. Rev.* **2004**, *104*, 527; (c) C.-H. Wang and S. DeBeer, *Chem. Soc. Rev.* **2021**, *50*, 8743.

intermediate electron carrier between the external electron donor, the Fe protein (Fe_4S_4 cluster), and the FeMo-co sites of the MoFe protein. The FeMo-cofactor $Fe_7Mo(S)_9$C-homocitrate with two protein side chains, one S(cysteinate) coordinated to Fe and one N(histidine) bound to Mo. The role of carbide (C^{4-}) is not known. The cluster can be viewed as composed of one Fe_4S_3 cluster and one $MoFe_3S_3$ cluster. The site of N_2 reduction is Mo center. The mechanism is not known with certainty.

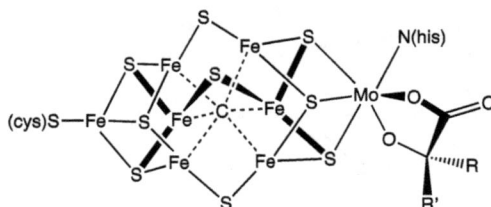

Further reading

W. Kaim, B. Schwederski, and A. Klein, *Bioinorganic Chemistry: Inorganic Elements in the Chemistry of Life — An Introduction and Guide*, Second Edition, Wiley (2013)

J. E. Huheey, E. A. Keiter, and R. L. Keiter, *Inorganic Chemistry: Principles of Structure and Reactivity*, 4th edition, Addison-Wesley Publishing Company (1993)

S. J. Lippard and J. M. Berg, *Principles of Bioinorganic Chemistry*, University Science Books (1994)

W. L. Jolly, Modern Inorganic Chemistry, 2nd edition, McGraw-Hill Inc., McGraw-Hill International Editions Chemistry Series (1991)

F. A. Cotton and G. Wilkinson, *Advanced Inorganic Chemistry*, 5th edition, John Wiley & Sons (1988)

K. F. Purcell and J. C. Kotz, Inorganic Chemistry, Saunders Golden Sunburst Series, W. B. Saunders Company, Holt-Saunders Japan (1985)

L. Stryer, *Biochemistry*, 2nd ed., CBS Publishers and Distributors (1986)

Exercises

10.1 Write the steps involved in the catalysis of HRP and CPO. Justify that the net reactions are

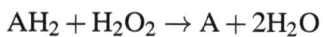

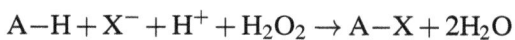

$$AH_2 + H_2O_2 \rightarrow A + 2H_2O$$

$$A{-}H + X^- + H^+ + H_2O_2 \rightarrow A{-}X + 2H_2O$$

10.2 Comment on the expected mechanism of the enzyme urease.

Answers to Exercises

Chapter 1

Hints/Answers:

1.1 $SOF_2 \rightarrow AB_2B'E_1$, pyramidal. Due to availability of lone pair, SOF_2 exhibits Lewis base behavior.

1.2 $S(=O)_3 \rightarrow AB_3$, triangular planar

1.3 $S_2O_3^{2-} \rightarrow (O^-)-S(=O)_2-S(^-)$

1.4 $^-O_3Cl-O-ClO_3{}^-$

1.5 $XeOF_2 \rightarrow AB_2B'E_2$, T-shape; $XeOF_4 \rightarrow AB_4B'E_1$, square pyramidal; $XeO_2F_2 \rightarrow AB_2B'_2E_1$, see-saw; $XeO_3 \rightarrow AB_3E_1$, pyramidal

1.6 $ClO_3^- \rightarrow AB_3E_1$, pyramidal; $I_3^- \rightarrow AB_2E_3$, linear; $NH_2^- \rightarrow AB_2E_2$, V-shape; $IO_6^{5-} \rightarrow$ octahedral

1.7 NO_2 (AB_2E; odd electron), NO_2^- (AB_2E_1), NO_2^+ (AB_2). Thus, the observed angles are justified.

1.8 XeF_6 (AB_6E_1, distorted octahedral as the lone pair is stereochemically active), $SbCl_6^{3-}$ (AB_6E_1, perfect octahedral as the lone pair is stereochemically inactive)

Chapter 2

2.1 Cl_2O (AB_2E_2, C_{2v}), NH_2^- (AB_2E_2, C_{2v}), $POCl_3$ (AB_3B', C_{3v}), XeO_2F_2 ($AB_2B'_2E_1$, C_{2v}), ICl_2^- (AB_2E_3, $D_{\infty h}$)

2.2 C_{2v}: $+1/-1/+1/-1(T_x)$, B_1; $+1/+1/-1/-1(R_z)$, A_2

Chapter 3

3.1 Same treatment as BF_3, as both belong to the point group D_{3h}.

3.2 The molecule PF_5 belongs to the D_{3h} point group. From character table for D_{3h}, the valence orbitals of P: s, dz^2 (a_1'); (px, py) (e'), pz (a_2''); (dx^2-y^2, dxy) (e'), (dxz, dyz) (e'').

The molecule contains both axial and equatorial bonds. They are calculated separately as they are different. The number of bonds that remain unshifted by the operations are represented by the following:

D_{3h}	E	$2C_3$	$3C_2$	σ_h	$2S_3$	$3\sigma_v$
$\Gamma\sigma$	5	2	1	3	0	3
$\Gamma\sigma$ (axial)	2	2	0	0	0	2
$\Gamma\sigma$ (equatorial)	3	0	1	3	0	1

Now the reduction will be performed on the representation of the five P–F bonds.

The total number of symmetry operation is 12. There is one E element, two C_3, three C_2, one σ_h, two S_3, and three σ_v planes.

The irreducible representations are calculated as before.

The number of times the A_1', A_2', E', A_1'', A_2'', E'' occur in the representation is 2, 0, 1, 0, 1, 0 (see the case for BH_3).

Therefore, $\Gamma\sigma = 2A_1' + E' + A_2''$.

Five a_1' orbitals involving s orbitals is expected to remain effectively nonbonding and lowest in energy scale. The a_1' orbital involving dz^2 is expected to have higher energy because of the high energy of the d orbitals. The energy of a_2'' (pz orbital) is expected to be lower than that of e' (px, py) because of the more favorable overlap. The remaining nonbonding d orbitals belonging to the representations e' (dx^2-y^2, dxy) and e'' (dxz, dyz). The e' pair in the xy plane should be higher in energy than the e'' pair.

3.3 The molecule SF_6 belongs to the point group O_h. Follow text.

Sulfur has 6 valence electrons. Consider the character table of O_h. The F $2px$, $2py$, $2pz$ orbitals from 6 F's give rise to 18 LGO's, which transform as $a_{1g} + e_g + t_{1g} + t_{2g} + t_{1u} + t_{1u} + t_{2u}$. Six F $2s$ orbitals give 6 LGO's that transform as $a_{1g} + e_g + t_{1u}$, but as they are very low in energy they remain effectively nonbonding. Remaining 14 LGO's remain nonbonding in the MO energy-level diagram. It is to be remembered that the five $3d$ AO's of S transform as $t_{2g} + e_g$.

Four LGO's (one a_{1g} and one of the t_{1u} sets) from six F's can interact with four S AO's [$3s$ (transforms as a_{1g}) and $3px$, $3py$, $3pz$ (transform as t_{1u})] giving four bonding and four antibonding MO's.

Total number of nonbonding MO's is 14 ($e_g + t_{1g} + t_{2g} + t_{1u} + t_{2u}$).

Sulfur has a $3s$ orbital (transforms as a_{1g}) and three $3p$ orbitals (transform as t_{1u}) with 6 valence electrons. Six F contributes 42 electrons. So, the total number of valence electron = 48.

From the concept of reducible representations (σ and π orbitals),

O	E	$6C_4^z$	$6C_2^z$	$8C_3$	$3C'_2$	i (O_h)
$\Gamma\sigma$	6	2	2	0	0	0

σ orbitals transform as $a_{1g} + e_g + t_{2g}$ (total is 6).

O	E	$6C_4^z$	$6C_2^z$	$8C_3$	$3C'_2$	i (O_h)
$\Gamma\pi$	12	0	0	0	-4	0

π orbitals transform as $t_{1g} + t_{2g} + t_{1g} + t_{2g}$ (total is 12).

Construct the MO energy-level diagram of SF_6 with only $3s$ and $3p$ orbitals of S.

Total bond-order equals 4, or an 'average' of only $4/6 = 2/3$ bond per S–F interaction.

SF_6 is an example of hypervalent molecule. Hypervalent molecules comprise main-group elements with more than 8 electrons in their valence shells.

Remember, the Lewis structure predicts 6 bond pairs and 18 lone pairs.

3.4 For HF_2^-: Follow MO treatment for C_{2v}

For BrF_5: Follow MO treatment for C_{4v}

For SF_4 (Chapter 1; p-12): Follow MO treatment for C_{2v}

Treat four F atoms as simple spherical F orbitals. Then,

	E	$C_2^{(z)}$	$\sigma_v(xz)$	$\sigma_v(yz)$
$\Gamma_\sigma =$	4	0	2	2

As discussed in Chapter 3 (p-77) the reducible representation Γ_σ reduces to $2A_1 + B_1 + B_2$ (Character Table for C_{2v}: Chapter 2; Table 2.2, p-38).

Overall, the bonding orbitals can be with the s and p_z or d_z^2 orbital with A_1 symmetry (dsp^2 or d^2sp),

p_x or d_{xy} with B_1 symmetry, and p_y or d_{yz} with B_2 symmetry.

Chapter 4

4.1

$$O_2(g) + 4H^+(aq) + 4e^- \rightarrow 2H_2O(l)$$

$$E^\circ = 1.23 \text{ V} ([H^+] = 1.0 \text{ M i.e. pH} = 0)$$

For the $O_2(g)/H_2O(l)$ redox process: $\Delta G^\circ = -4F \times 1.23$ kcal/mol

$$E = E^\circ - 0.059 \text{ pH} \text{ (see } Stability\ field\ of\ water)$$

At pH $= 7$, $E^\circ = 0.82$ V
For the $O_2(g)/H_2O(l)$ redox process: $\Delta G^\circ = -4F \times 0.82$ kcal/mol
For the $2Cu^{2+}(aq) + 4e^- \rightarrow 2Cu(s)$ redox process: $\Delta G^\circ = -4F \times 0.34$
For the reverse reaction, $2Cu(s) \rightarrow 2Cu^{2+}(aq) + 4e^-$, $\Delta G^\circ = 4F \times 0.34$ kcal/mol
The difference in ΔG° for the oxidation of Cu(s) by atmospheric dioxygen (pH $= 0$):

$$2Cu(s) + O_2(g) + 4H^+(aq) \rightarrow 2Cu^{2+}(aq) + 2H_2O(l)$$

$$\Delta G^\circ = -4F \times (1.23 - 0.34)$$

$$= -4F \times 0.89 \text{ kcal/mol}$$

The difference in ΔG° for the oxidation of Cu(s) by atmospheric dioxygen (pH $= 7$):

$$2Cu(s) + O_2(g) + 4H^+(aq) \rightarrow 2Cu^{2+}(aq) + 2H_2O(l)$$
$$\Delta G^\circ = -4F \times (0.82 - 0.34)$$
$$= -4F \times 0.48 \text{ kcal/mol}$$

So, atmospheric oxidation of copper metal by molecular oxygen is thermodynamically favorable when the pH is on the acidic side of neutrality.

The well-known green coating on copper is a passive layer of an almost impenetrable hydrated copper(II) carbonate and copper(II) sulfate formed, due to oxidation of copper in the presence of atmospheric CO_2 and SO_2, respectively.

4.2

$$Fe^{2+}(aq) + 2e^- \rightarrow Fe(s), \Delta G^\circ = -2F \times -0.44 = 2F \times 0.44$$
$$Fe^{3+}(aq) + e^- \rightarrow Fe^{2+}(aq), \Delta G^\circ = -1F \times 0.77$$

Therefore, for $Fe^{2+}(aq) \rightarrow Fe^{3+}(aq) + e^-$; $\Delta G^\circ = 1F \times 0.77$
Then, $2Fe^{2+}(aq) \rightarrow 2Fe^{3+}(aq) + 2e^-$; $\Delta G^\circ = 2F \times 0.77$
Adding two half-cell reactions, $3Fe^{2+}(aq) \rightarrow 2Fe^{3+}(aq) + Fe(s)$

$$\Delta G^\circ = 2F \times (0.44 + 0.77) = 2F \times 1.21 = -2F \times -1.21 \text{ kcal/mol}$$

Then, $E^\circ = -1.21$ V
From eq 2, $\log K = 2 \times -1.21/0.059 \simeq -41$; $K \simeq (10)^{-41}$
Hence, disproportionation reaction $3Fe^{2+}(aq) \rightarrow 2Fe^{3+}(aq) + Fe(s)$ will not take place.

4.3

$$Cu^{2+}(aq) + 2e^- \rightarrow Cu(s); \Delta G^\circ = -2F \times 0.34 \text{ kcal/mol}$$

Therefore, for $Cu(s) \rightarrow Cu^{2+}(aq) + 2e^-$; $\Delta G^\circ = 2F \times 0.34$

$$Cu(OH)_2(s) + 2e^- \rightarrow Cu(s) + 2OH^-(aq)$$
$$\Delta G^\circ = -2F \times -0.22 = 2F \times 0.22 \text{ kcal/mol}$$

Adding two half-cell reactions, $Cu(OH)_2(s) \rightarrow Cu^{2+}(aq) + 2OH^-(aq)$

$$\Delta G^o = 2F \times (0.34 + 22) = 2F \times 0.56 = -2F \times -0.56\ kcal/mol$$

Then, $E^o = -0.56$ V

Therefore, $K = [Cu^{2+}(aq)][OH^-(aq)]^2/[Cu(OH)_2(s)]$

$K_{sp} = [Cu^{2+}(aq)]$ (since in alkaline medium, $[OH^-(aq)] \simeq 1.0$ M)

From eq 2, $\log K = 2 \times -0.56/0.059 \simeq -19$; $K \simeq (10)^{-19}$

Therefore, the solubility product of $Cu(OH)_2(s) \simeq (10)^{-19}$

4.4 Considering thermodynamic justification for HAT (see text), we write,

$$BDE_{O-H}\ for\ [Mn^{III}(L)(OH)]^- = [1.37 \times 28.3 + (-1.51 \times 23.06)$$
$$+ 73.3) = 77.25\ kcal/mol$$

$$BDE_{O-H}\ for\ [Fe^{III}(L)(OH)]^- = [1.37 \times 25.0 + (-1.79 \times 23.06)$$
$$+ 73.3) = 66.27\ kcal/mol$$

$$BDE_{O-H}\ for\ [Mn^{III}(L)(O)]^{2-} = [1.37 \times 28.3 + (-0.076 \times 23.06)$$
$$+ 73.3] = 110.32\ kcal/mol$$

$$BDE_{O-H}\ for\ [Fe^{III}(L)(O)]^{2-} = [1.37 \times 25.0 + (0.34 \times 23.06)$$
$$+ 73.3] = 115.39\ kcal/mol$$

This analysis provides an estimate of O–H bond strength of the OH^- ligand in $[M^{II}(L)(OH)]^-$ of 77 (Mn) and 66 (Fe) and in $[Mn^{III}(L)(OH)]^{2-}$ of 110 (Mn) and 115 (Fe) kcal/mol.

Chapter 5

5.1 Follow worked out examples and proceed.

5.2 Follow worked out examples and proceed.

Chapter 6

6.1 In ls d^6 complex *trans*-$[Co^{III}(NH_3)_4Cl_2]^+$ two Cl^- ligands are in the axial position and four NH_3 ligands are in the equatorial plane. Two different ligand fields are exerted around Co(III) ion with even $(t_{2g})^6$

electron distribution. This leads to tetragonally-distorted octahedral structure.

In d^9 complex $[Cu^{II}(NH_3)_6]^{2+}$ all six ligands are same but the complex assumes tetragonally-distorted octahedral structure with $(t_{2g})^6\,(e_g)^3$ electron distribution due to Jahn-Teller distortion.

6.2 Considering Fig. 6.13 the electron distributions are:

d^8 tetrahedral case: $(e)^4\,(t_2)^4$
elongation is stabilized by $-2/3\delta'$ and compression (flattening) is stabilized by $-1/3\delta'$.

d^9 tetrahedral case: $(e)^4\,(t_2)^5$
elongation is stabilized by $-1/3\delta'$ and compression (flattening) is stabilized by $-2/3\delta'$.

6.3

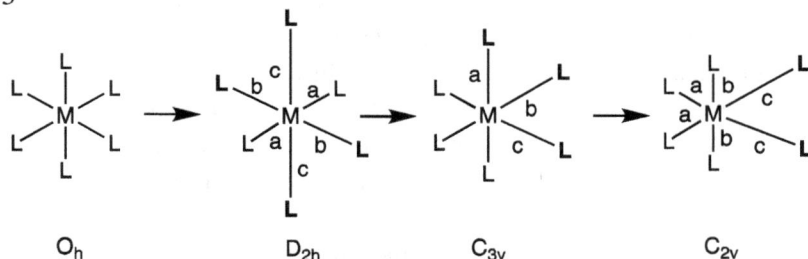

In O_h all M–L bond lengths are equal. In D_{2h}, three pairs of M–L distances are equal to each other (a/a, b/b, c/c; $c > b > a$, $a \neq b \neq c$; three M–L distances are different from that in O_h). In C_{3v} three M–L bond distances of a face are equal ($a = b = c$ and each one is different from original a in O_h; in other words, three M–L bond distances are longer than that in O_h). In C_{2v} four M–L distances are equal ($a = b$) and two M–L bond distances are equal ($c = c$; $c >$ a or b; in other words, two M–L distances are longer than that in O_h).

6.4 An octahedral Co(II) complex can have either $(t_{2g})^6\,(e_g)^1$ (ls; one unpaired electron) or $(t_{2g})^5\,(e_g)^2$ (hs; three unpaired electrons). Given the spin-only value of 3.87 μ_B it is obvious that the electronic configuration is $(t_{2g})^5\,(e_g)^2$. The CFSE for the complex is $-8\,Dq_o$.

6.5 Normal: $Mn^{II}[Fe^{III}_2O_4]$
Inverse: $Fe^{III}[Mn^{II}Fe^{III}O_4]$
Since O^{2-} is a weak field ligand, for both d^5 hs Mn(II) and d^5 hs Fe(III) the CFSE $= 0$.

As there is no driving force for the spinel to get inverted, it remains in a normal spinel structure. It means the metal ion with higher charge will stay in the octahedral environment.

6.6 As Cl^- ion is a weak field ligand, $[CoCl_4]^{2-}$ has a tetrahedral structure. $[Co(H_2O)_6]^{2+}$ is octahedral. The separation between the t_{2g} and e_g levels is much smaller in the former in comparison to the latter. Therefore, the absorption energy is more towards the red in the former (it appears blue) while in the latter the higher energy absorption corresponds to appearance of pink color.

6.7 Follow MO treatment for T_d (CH_4).

6.8 As H_2O is a weak field ligand, $[Mn^{II}(H_2O)_6]^{2+}$ is a d^5 hs complex. There is only one sextet $^6A_{1g}$ state and hence no spin-allowed transitions are expected. So, the case is unique; it is both Laporte- and spin-forbidden. Similar to d^5 hs $[Fe^{III}(H_2O)_6]^{3+}$, $[Mn^{II}(H_2O)_6]^{2+}$ is faintly colored. The observed weak spin-forbidden transitions are to quartet states since it is even less probable that the spins of two electrons would be reversed to give a state of even lower spin multiplicity. Nevertheless, there are many quartet states (see Tanabe-Sugano diagram for octahedral hs d^5 ion). It is noted that the unresolved $^4A_{1g}$, ^4E_g peak is very sharp and that the line representing the energies of these states (hs case) in the Tanabe-Sugano diagram is nearly horizontal, implying that the energy is independent of Dq.

It is to be remembered that the CF strength is very much dependent on the M–L distance. Usually the electronic energy levels are very sensitive to changes in the M–L distance and Dq. The ligands vibrate all the time back and forth about an equilibrium position. This results in the variation of the M–L distance and hence covers a range of Dq values. Those electronic transitions which involve Dq will therefore occur over a range of energies. This situation leads to broadening of *d-d* absorption bands. On the other hand, if there are some transitions which do not involve Dq [in this case ($^6A_{1g} \rightarrow {}^4A_{1g}$, ^4E_g), since the energy is independent of Dq], it is most likely that those transitions will not be affected by the variation in the M–L distance due to vibrations and sharp lines will result.

For $[Mn^{II}(H_2O)_6]^{2+}$ no electronic transition from the ground state $^6A_{1g}$ to any other excited state involves Dq. This is the reason why absorption spectra of hs d^5 $[Mn^{II}(H_2O)_6]^{2+}$ is very weak and sharp.

Chapter 7

7.1 Both the starting $Co^{3+}(d^6)$ and product $Cr^{3+}(d^3)$ chloride complexes are substitutionally inert so Cl^- transfer must have occurred via a bridged species.

a) Step 1:

It will have very little effect, since mostly affected by lability of leaving group on partner (assuming I_D mechanism is operative).

Step 2:

There will be some effect because the stronger the bridge, the better electronic effects are transmitted.

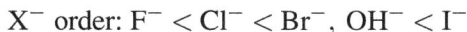

$$X^- \text{ order: } F^- < Cl^- < Br^-, OH^- < I^-$$

7.2 $[Fe(bipy)_3]^{2+/3+}$: adds only one electron to a ls d^5 configuration in going to ls d^6 (t_{2g}^5 to t_{2g}^6) and this results in rapid ($>10^6$) electron transfer rate.

$[Co(NH_3)_6]^{2+/3+}$: changes from ls d^6 Co(III) to hs d^7 Co(II). The change in spin state (t_{2g}^6 to $t_{2g}^5 e_g^2$) requires expenditure of energy. So, the rate is very slow (10^{-6}).

Chapter 8

8.1

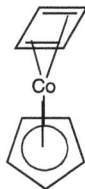

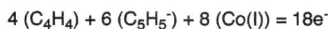

$$4 (C_4H_4) + 6 (C_5H_5^-) + 8 (Co(I)) = 18e^-$$

8.2

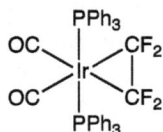

Oxidative addition

8.3 Ferrocene has 18 electrons in the valence shell. Nickelocene has 20 valence electrons. The two extra electrons occupy the antibonding orbitals (Fig. 8.5) thereby weakening (lengthening) the Ni–C bond.

Chapter 9

9.1

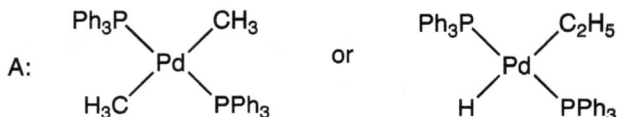

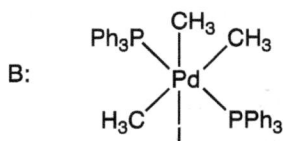

9.2

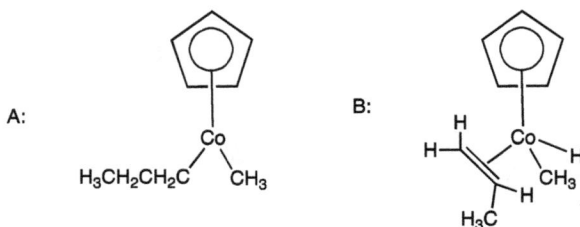

9.3

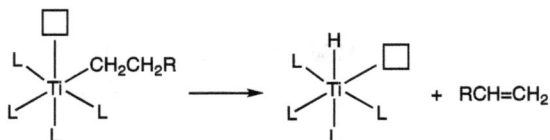

Chapter 10

10.1 Follow text p-333

Adding three HRP reaction steps:

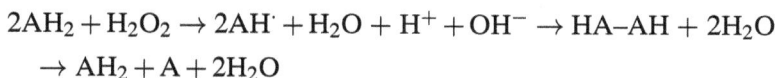

$$2AH_2 + H_2O_2 \rightarrow 2AH^{\cdot} + H_2O + H^+ + OH^- \rightarrow HA\text{–}AH + 2H_2O$$
$$\rightarrow AH_2 + A + 2H_2O$$

Therefore, $AH_2 + H_2O_2 \rightarrow A + 2H_2O$

Adding three CPO reaction steps:

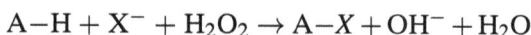

$$A\text{–}H + X^- + H_2O_2 \rightarrow A\text{–}X + OH^- + H_2O$$

Therefore, $A\text{–}H + X^- + H^+ + H_2O_2 \rightarrow A\text{–}X + 2H_2O$

10.2 Each Ni(II) center has a coordinated water. At the appropriate pH the active form of the enzyme is $\{(H_2O)Ni(II)\text{-bridge-}Ni(II)\text{-OH}\}$. Now the substrate $(NH_2C(=O)NH_2)$ binds replacing coordinated water. The binding is with carbonyl O. Nucleophilic attack by OH^- coordinated to other Ni(II) site takes place forming O-bridged species $\{Ni(II)\text{-O-}C(NH_2)_2\text{-O-}Ni(II)\}$. A base present near the active site removes H^+ from Ni(II)-coordinated OH^-. Now, elimination of NH_4^+ takes place. A carbamate $(\text{-OC}(=O)NH_2)$ remains coordinated by O at one of the Ni(II) centers and the other Ni(II) remains vacant at the site where OH^- was present. Finally, two water molecules get coordinated to two Ni(II) centers and the carbamate ion NH_2COO^- is released. The active site $\{(H_2O)Ni(II)\text{-bridge-}Ni(II)\text{-}OH\}$ is regenerated.

Index